AF615572

LINEAR ALGEBRA

LINEAR ALGEBRA

Roger C. Baker

Department of Mathematics
Brigham Young University

RP

Rinton Press, Inc.

12 Castleton Road
Princeton, New Jersey 08540
USA

Published by Rinton Press, Inc.

Printed in the United State of America

ISBN 1-58949-012-6

Preface

This book is designed for a one-semester first course in linear algebra. My main idea was to write a book that an average hard-working student, not necessarily a math major, could read from cover to cover during a semester. There is nothing like reading an entire book on a topic to give a sense of accomplishment.

Reading the book includes solving most of the exercises. These are gathered at the end of each chapter and labeled to show when to attempt them. Thus Exercises 2.1 through 2.5 of Chapter 1 should be worked after reading Section 2. Do try to work in a team of two or three students outside class. Your instructor can probably supply a list of e-mail addresses of other students. Joint learning is fun and effective. If you figure out a solution jointly, make sure you write it up on your own, either for your revision notes or for credit. Writing up is a key part of learning. Always write in sentences, and include explanation of what you are doing throughout. If a question is asked with a yes or no answer, you must explain why you chose your particular answer to it.

Many of the processes taught in this book can be done quickly by computer. Indeed, there is no other way for the linear algebra problems in many variables that are needed in the 'real world'–oil exploration, air freight and so on. However, to learn the first principles of the subject, you are better off with hand calculation, just as fourth graders are better off doing arithmetic without a calculator. Feel free to check your matrix inverse (for instance) with the program available in your computer lab; a popular one is MATLAB. These programs are user friendly, and you should not find it difficult to input your matrix and get out the answer. You will, of course, not learn any linear algebra this way!

I would like to thank my students in linear algebra classes since 1971 for their shared sense of enjoyment of the subject and for helping me to develop, by trial, good ways of explaining the hard parts. I am particularly grateful to my Fall 2000 class for constructive criticism of the final draft of this book, and to Lonette Stoddard for tireless help in preparing the typescript and figures.

Your comments on this book are most welcome, and can be sent to me at

baker@math.byu.edu

Contents

Chapter 1

Euclidean space

1 The Euclidean plane

We begin with the Euclidean plane $\mathbb{R}^2$, the setting for high school coordinate geometry, because it is familiar and intuitive. We then flex our muscles by working in $\mathbb{R}^3$, three dimensional Euclidean space. When we have got some results, we will conclude the chapter with an introduction to Euclidean n-dimensional space $\mathbb{R}^n$. This consists of points with n coordinates $(x_1, x_2, \dots, x_n)$. Each x_i is a real number. We cannot draw subsets of $\mathbb{R}^n$ on a blackboard, but there is no doubt that $\mathbb{R}^n$ is useful in modeling a chemical reaction with n ingredients in masses $x_1, \dots, x_n$. To take another example, the first attempt to model the U.S. economy involved a system of 42 equations for 42 unknowns, set up by W. Leontief in 1949 and solved by an early computer. Modern applications may involve hundreds of equations. If there are n unknowns, the natural description of the problem takes place in $\mathbb{R}^n$. You may be surprised how easy it is to think and calculate in n dimensions.

Let us briefly review the notion of coordinate geometry. We locate points in a plane extending to infinity in all directions by specifying an origin $\mathbf{0}$ and two perpendicular **directed** (arrowed) segments $\boldsymbol{e}_1, \boldsymbol{e}_2$, of length 1. See Figure 1.

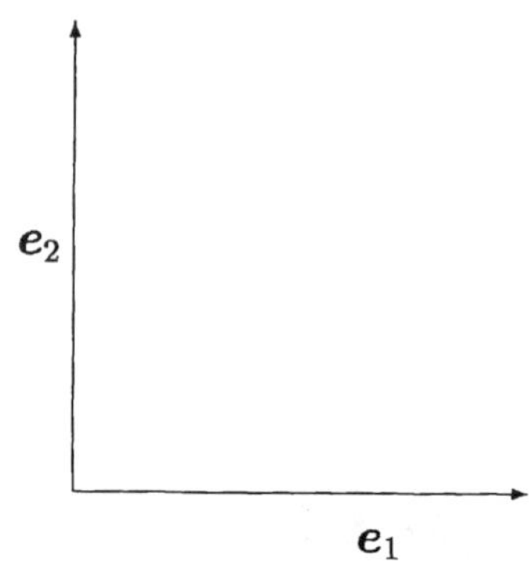

Figure 1. Perpendicular directed segments $\boldsymbol{e}_1$ and $\boldsymbol{e}_2$.

The infinite straight line containing $\boldsymbol{e}_1$ is called the x axis and the one containing $\boldsymbol{e}_2$ is called the y axis. We then specify points in the plane by giving their positive or negative projections onto these axes, as an ordered pair (x, y). Thus $(2, 1)$ represents a point obtained by starting at $\mathbf{0}$, moving distance 2 in the direction of $\boldsymbol{e}_1$, and then moving distance 1 in the direction of $\boldsymbol{e}_2$, while $\left(-\frac{1}{2}, -3\right)$ is reached by starting at $\mathbf{0}$ and moving $\frac{1}{2}$ in the direction opposite to $\boldsymbol{e}_1$, then 3 in direction opposite to $\boldsymbol{e}_2$. See Figure 2. We now think of the plane as **equal** to the set of real ordered pairs. We write $\mathbb{R}$ for the real numbers and $\mathbb{R}^2$ for the set of real ordered pairs.

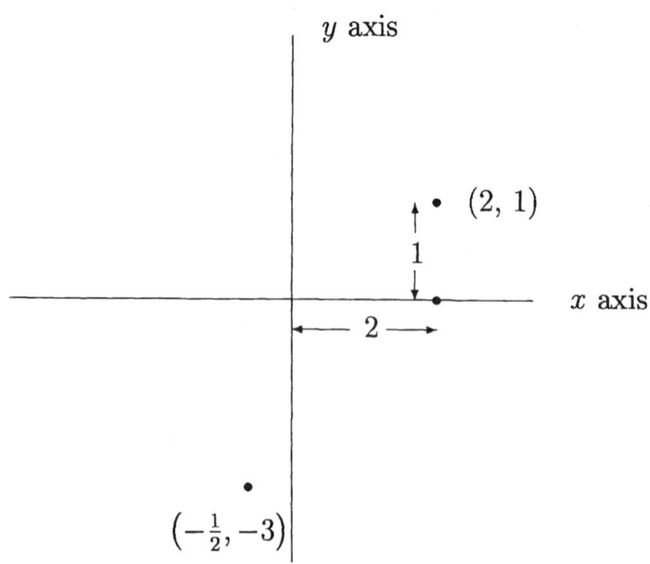

Figure 2. Points in the plane as ordered pairs.

We can now write down equations of curves. For instance

$$y = x^2$$

signifies the set of all points (x, x^2) (x real). The straight line through (a, b) and (c, d) $(a \neq c)$ is

$$y - b = \frac{d - b}{c - a}(x - a)$$

since this expresses the chord slope $(y - b)/(x - a)$ being a constant. See Figure 3. A vertical line is

$$x = a$$

(see Figure 4); so an arbitrary straight line is

$$ux + vy = w \qquad (u, v \text{ not both zero}).$$

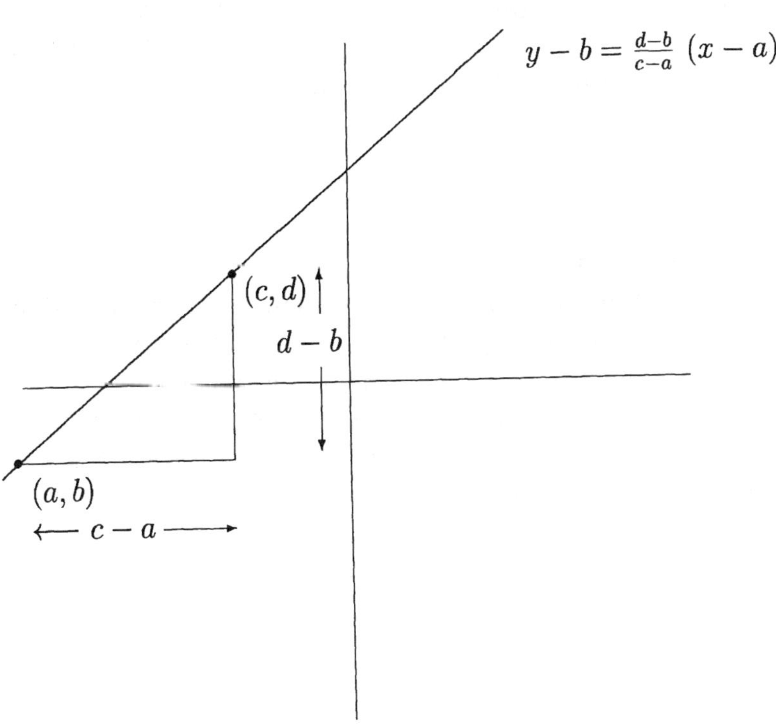

Figure 3. Chord slope gives equation of straight line.

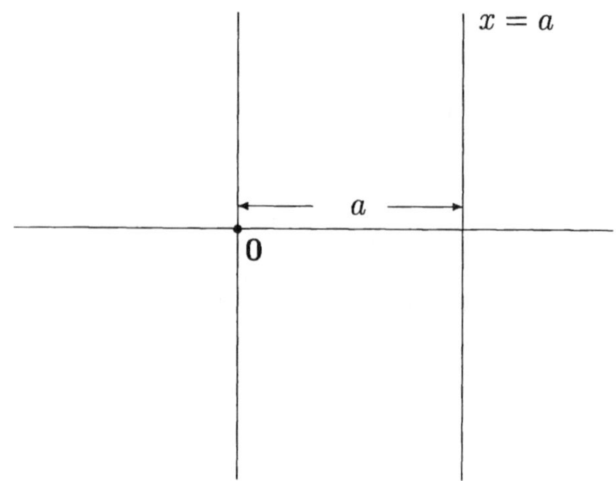

Figure 4. Equation of vertical straight line.

Now let us switch notation. Since, later in the chapter, we have ordered sets $(x_1, \dots, x_n)$, we write (x_1, x_2) instead of (x, y). We write $\boldsymbol{x} = (x_1, x_2)$, and similarly $\boldsymbol{a} = (a_1, a_2), \boldsymbol{b} = (b_1, b_2)$ and so on. Our arbitrary straight line is now

$$ux_1 + vx_2 = w. \tag{1}$$

It passes through $\mathbf{0}$ when $w = 0$.

Another mental image of $\boldsymbol{x}$ is as a **vector**, or directed segment from $\mathbf{0}$ to $\boldsymbol{x}$. Thus $\boldsymbol{x}$ may be called a point or a vector. A **translate** of $\boldsymbol{x}$ is a directed segment obtained by moving the initial point of $\boldsymbol{x}$ from $\mathbf{0}$ to (a_1, a_2) and the terminal point from $\boldsymbol{x}$ to $(x_1 + a_1, x_2 + a_2)$. The translate is parallel to $\boldsymbol{x}$ and has the same length. When it does not cause confusion, we give the translate the same name, $\boldsymbol{x}$. See Figure 5.

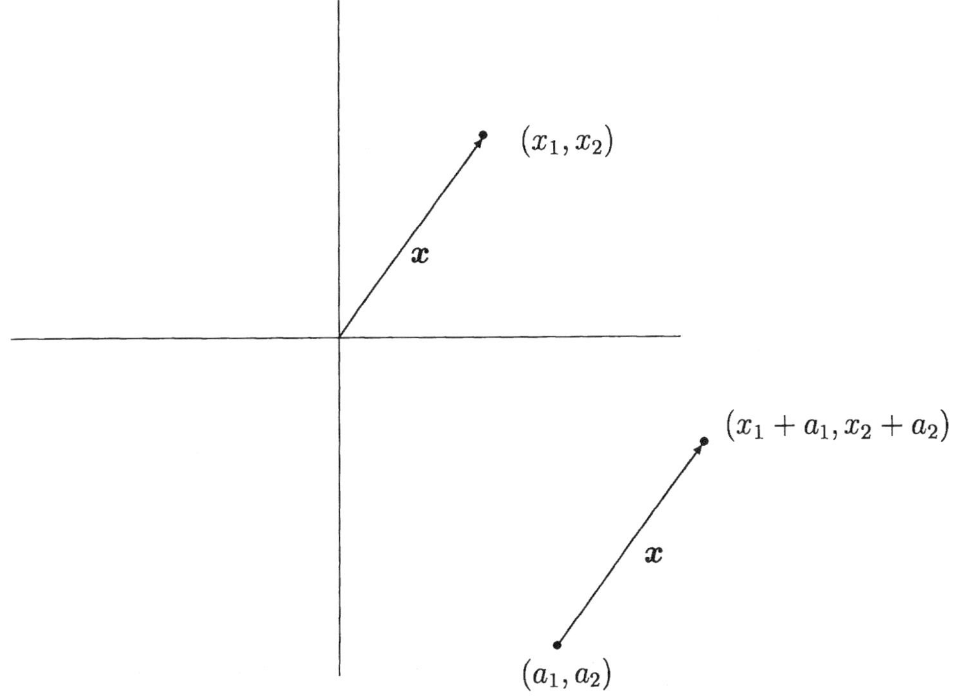

Figure 5. A vector is a point; **or**, is a directed segment.

We now come to a key definition.

Definition 1 The **sum** of the vectors $\boldsymbol{x} = (x_1, x_2)$ and $\boldsymbol{y} = (y_1, y_2)$ is

$$\boldsymbol{x} + \boldsymbol{y} = (x_1 + y_1, x_2 + y_2).$$

Given $\boldsymbol{x}$ and $\boldsymbol{y}$, place the initial point of a translate of $\boldsymbol{y}$ at $\boldsymbol{x}$. Then $\boldsymbol{x}+\boldsymbol{y}$ is the terminal point of the translate (Figure 6). Alternatively, $\boldsymbol{x} + \boldsymbol{y}$ is the fourth vertex of the parallelogram with vertices $\mathbf{0}, \boldsymbol{x}$ and $\boldsymbol{y}$ (Figure 7).

We often **scale** a vector by multiplying it by a real number c.

Definition 2 The **scalar product** $c\boldsymbol{x}$, where c is real and $\boldsymbol{x}$ is in $\mathbb{R}^2$, is

$$c\boldsymbol{x} = (cx_1, cx_2).$$

Note that $c\boldsymbol{x}$ points in the same direction as $\boldsymbol{x}$ if $c > 0$ and the opposite direction if $c < 0$ (Figure 8). We write $-\boldsymbol{x}$ instead of $(-1)\boldsymbol{x}$. Clearly

$$(-\boldsymbol{x}) + \boldsymbol{x} = \mathbf{0}.$$

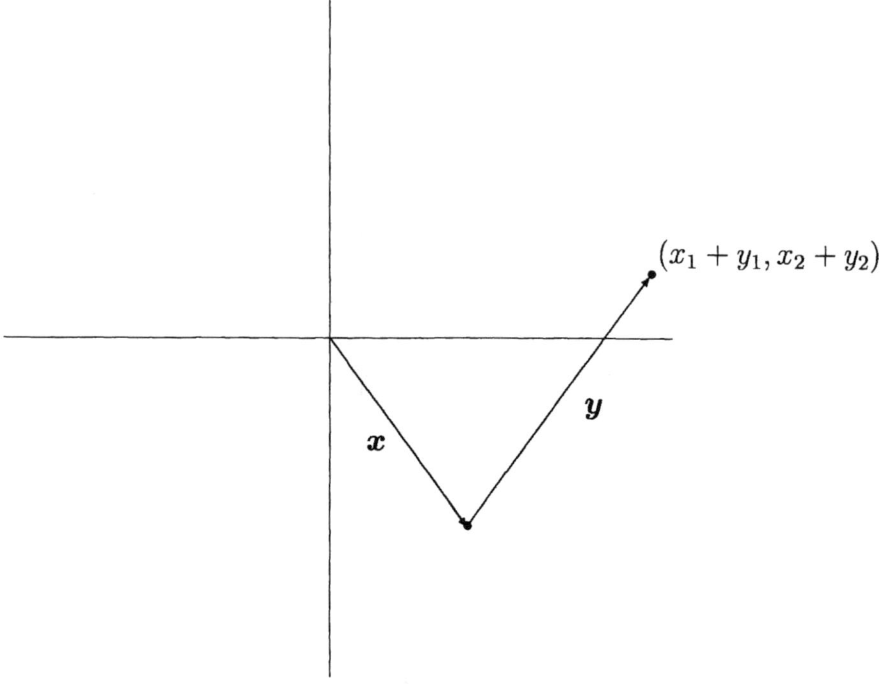

Figure 6. Addition by combining directed segments.

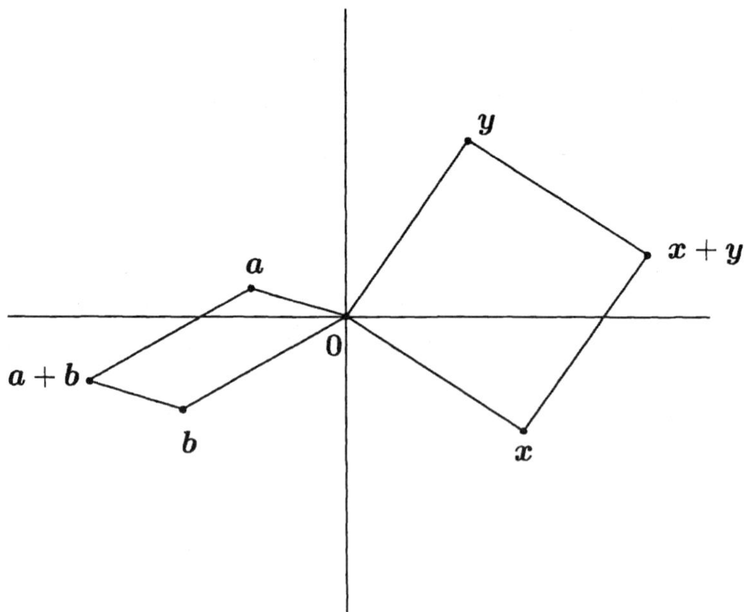

Figure 7. The fourth vertex of the parallellogram is the sum of the adjacent vertices.

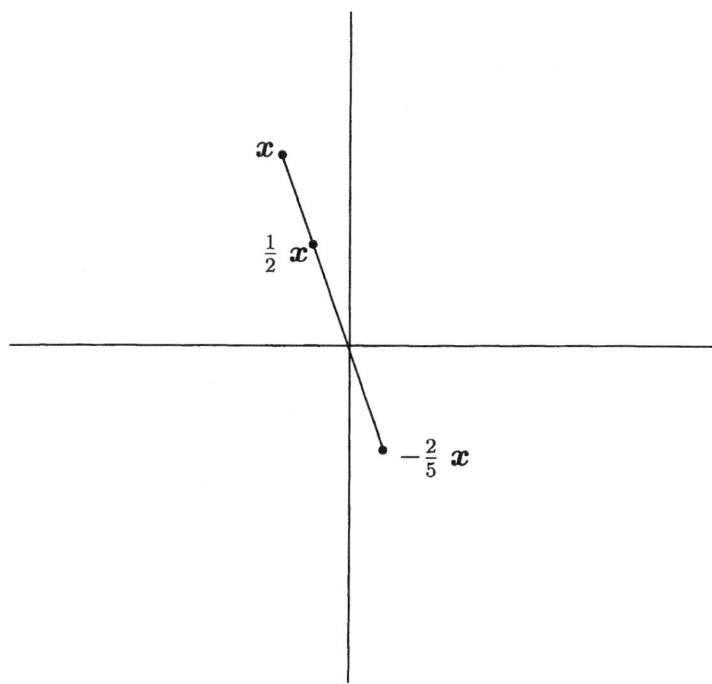

Figure 8. Scalar multiples of a vector.

We can **subtract** vectors. The difference $\boldsymbol{z} - \boldsymbol{y}$ is defined by the property that

$$(\boldsymbol{z} - \boldsymbol{y}) + \boldsymbol{y} = \boldsymbol{z}. \tag{2}$$

Obviously, then,

$$\boldsymbol{z} - \boldsymbol{y} = (z_1 - y_1, z_2 - y_2).$$

It is also true that

$$\boldsymbol{z} - \boldsymbol{y} = \boldsymbol{z} + (-\boldsymbol{y}).$$

Another helpful notion is that $\boldsymbol{z} - \boldsymbol{y}$ is the vector with initial point at $\boldsymbol{y}$ and terminal point at $\boldsymbol{z}$. This follows from (2).

Example 1 Find the point $\boldsymbol{v}$ that is one-third of the way along the segment joining $(-1, 2)$ to $(7, 11)$.

Solution. The directed segment in question is $(7, 11) - (-1, 2) = (8, 9)$. Clearly $\frac{1}{3}(8, 9)$ is one-third as long as this segment. Hence

$$\boldsymbol{v} = (-1, 2) + \frac{1}{3}(8, 9) = \left(\frac{5}{3}, 5\right).$$

The issue of length deserves a more formal treatment. The length $|\boldsymbol{a}|$ of the vector $\boldsymbol{a} = (a_1, a_2)$, or alternatively the distance from $\mathbf{0}$ to (a_1, a_2), is given by Pythagoras's theorem :

$$|\boldsymbol{a}| = (a_1^2 + a_2^2)^{1/2};$$

while the distance $d(\boldsymbol{u}, \boldsymbol{v})$ from $\boldsymbol{u}$ to $\boldsymbol{v}$ is $|\boldsymbol{v} - \boldsymbol{u}|$. See Figure 9.

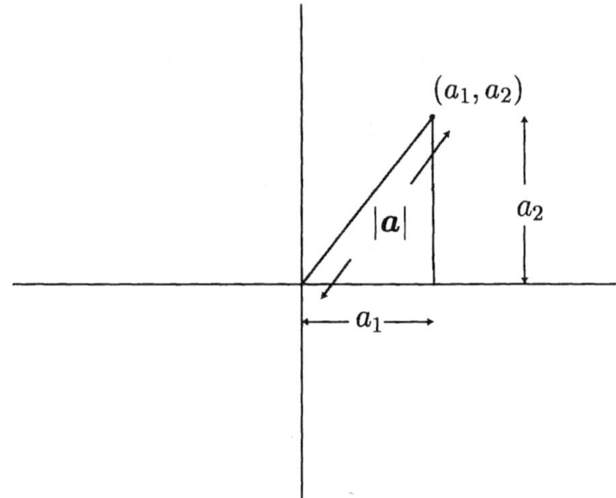

Figure 9. Length of a vector.

Proposition 1 *We have*

$$|c\boldsymbol{x}| = |c|\ |\boldsymbol{x}|.$$

Proof. $|c\boldsymbol{x}| = |(cx_1, cx_2)| = ((cx_1)^2 + (cx_2)^2)^{1/2} = |c|(x_1^2 + x_2^2)^{1/2} = |c|\ |\boldsymbol{x}|$, since the positive square root of c^2 is $|c|$.

The proposition shows–for instance–that $\frac{1}{3}(8, 9)$ is one third as long as $(8, 9)$.

When you are asked to do proofs in the exercises–very easy ones at first–begin by convincing yourself that the result is true. Your task is then to write out a series of linked statements to convince a reader that the result is true. The essence of a proof is to cover all possibilities that might occur. For instance, the above proof is written in a way that accounts for the three possibilities $c > 0, c = 0, c < 0$.

A common student error I call this 'proof by example'. In the case of the above proposition, the student works out both $\left|-\frac{1}{6}(4, 3)\right|$ and $\frac{1}{6}|(4, 3)|$, obtaining the same answer, and feels that the proof is complete. It is not, because all cases have certainly not been covered!

Definition 3 A vector of length 1 is said to be a **unit vector**.

We shall use the same definition in $\mathbb{R}^3$ and $\mathbb{R}^n$ after we define length. A unit vector $\boldsymbol{v}$ in $\mathbb{R}^2$ can be written in the form $\boldsymbol{v} = (\cos a, \sin a)$ where $0 \leq a < 2\pi$. (If $\boldsymbol{v}$ is obtained by rotating $(1,0)$ anticlockwise through a, a simple figure shows that $v_1 = \cos a, v_2 = \sin a$.)

The following result gives a useful alternative way of writing the equation of a straight line.

Proposition 2 *For $\boldsymbol{b} \neq \boldsymbol{0}$, the equation*

$$\boldsymbol{x} = \boldsymbol{a} + t\boldsymbol{b} \quad (t \text{ real}) \tag{3}$$

describes a straight line through $\boldsymbol{a}$ in the direction of $\boldsymbol{b}$.

We call this a **parametric representation** of a straight line, with t as **parameter** (a real number giving all the points $\boldsymbol{x}$ of the straight line via equation (3)).

Proof. The directed segment with initial point $\boldsymbol{a}$ and terminal point $\boldsymbol{a} + t\boldsymbol{b}$ is $t\boldsymbol{b}$, which points in the direction of $\boldsymbol{b}$ if $t > 0$ and the opposite direction if $t < 0$. By varying t we produce such a segment of arbitrary length ($t = 0$ yields $\boldsymbol{x} = \boldsymbol{a}$). Thus (3) is the desired straight line; see Figure 10.

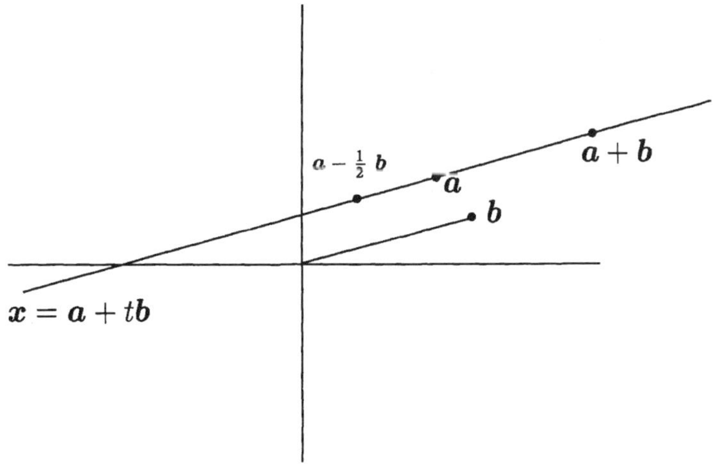

Figure 10. The straight line $\boldsymbol{x} = \boldsymbol{a} + t\boldsymbol{b}$.

Example 2 Write the straight line

$$7x_1 - x_2 + 5 = 0$$

in the form (3).

Solution. The line passes through (0, 5) and (1, 12). Take $\boldsymbol{a} = (0,5)$ and

$$\boldsymbol{b} = (1,12) - (0,5) = (1,7).$$

The line is then

$$\boldsymbol{x} = (0,5) + t(1,7)$$

or $\boldsymbol{x} = (t, 5+7t)$.

2 Inner product in $\mathbb{R}^2$

The following definition turns out to be very useful.

Definition 4 The **inner product** $\boldsymbol{u} \cdot \boldsymbol{v}$ of $\boldsymbol{u} = (u_1, u_2)$ and $\boldsymbol{v} = (v_1, v_2)$ is defined to be

$$\boldsymbol{u} \cdot \boldsymbol{v} = u_1 v_1 + u_2 v_2.$$

Inner product is sometimes called **dot product**. Note that the length of $\boldsymbol{u}$ can be expressed as

$$|\boldsymbol{u}| = (\boldsymbol{u} \cdot \boldsymbol{u})^{1/2}. \tag{4}$$

To connect inner product with familiar notions of plane geometry, we let a be the angle between two nonzero vectors $\boldsymbol{u}$ and $\boldsymbol{v}$; we take a to satisfy $0 \le a \le \pi$. See Figure 11.

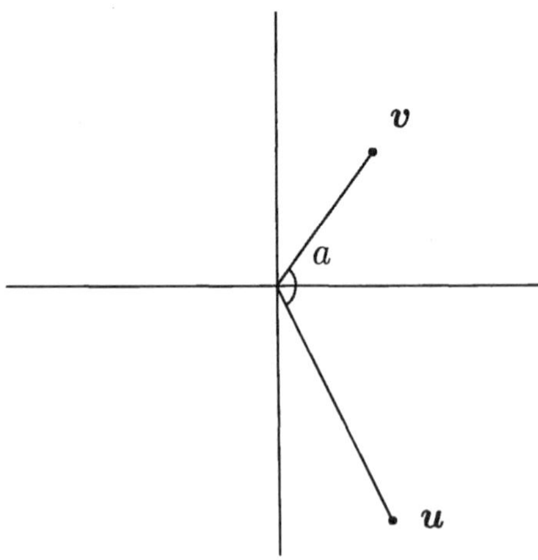

Figure 11. The angle a between $\boldsymbol{u}$ and $\boldsymbol{v}$.

Proposition 3 *We have*

$$\boldsymbol{u} \cdot \boldsymbol{v} = |\boldsymbol{u}|\ |\boldsymbol{v}| \cos a. \tag{5}$$

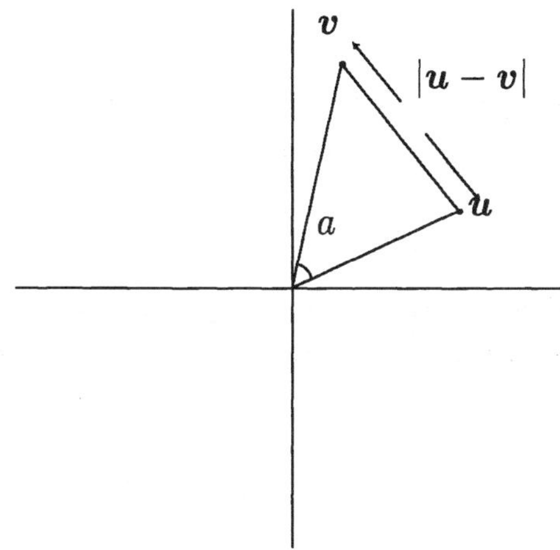

Figure 12. The triangle with vertices $\mathbf{0}, \boldsymbol{u}$ and $\boldsymbol{v}$.

Proof. Consider the triangle with vertices $\mathbf{0}, \boldsymbol{u}$ and $\boldsymbol{v}$ (Figure 12). The sides have length $|\boldsymbol{u}|$, $|\boldsymbol{v}|$ and $|\boldsymbol{u} - \boldsymbol{v}|$. By the law of cosines,

$$|\boldsymbol{u} - \boldsymbol{v}|^2 = |\boldsymbol{u}|^2 + |\boldsymbol{v}|^2 - 2|\boldsymbol{u}|\ |\boldsymbol{v}| \cos a.$$

However, using (3),

$$|\boldsymbol{u} - \boldsymbol{v}|^2 = (\boldsymbol{u} - \boldsymbol{v}) \cdot (\boldsymbol{u} - \boldsymbol{v}).$$

It is easy to see that

$$\begin{aligned}(\boldsymbol{u} - \boldsymbol{v}) \cdot (\boldsymbol{u} - \boldsymbol{v}) &= (\boldsymbol{u} - \boldsymbol{v}) \cdot \boldsymbol{u} - (\boldsymbol{u} - \boldsymbol{v}) \cdot \boldsymbol{v} \\ &= \boldsymbol{u} \cdot \boldsymbol{u} - 2\boldsymbol{u} \cdot \boldsymbol{v} + \boldsymbol{v} \cdot \boldsymbol{v} \\ &= |\boldsymbol{u}|^2 - 2\boldsymbol{u} \cdot \boldsymbol{v} + |\boldsymbol{v}|^2.\end{aligned}$$

Note that $\boldsymbol{u} \cdot \boldsymbol{v} = \boldsymbol{v} \cdot \boldsymbol{u}$ from the definition.

If we compare our two expressions for $|\boldsymbol{u} - \boldsymbol{v}|^2$, we get (5) after a little rearrangement.

Example 3 Find the angle between (1, 7) and $(-12, 16)$.

Solution. The angle is a, where

$$|(1,7)|\,|(-12,16)|\cos a = 1 \times -12 + 7 \times 16 = 100;$$

$$\cos a = \frac{100}{(50)^{1/2}(400)^{1/2}} = \frac{1}{\sqrt{2}}.$$

Hence $a = \pi/4$.

We use the term **perpendicular** for two vectors $\boldsymbol{u}, \boldsymbol{v}$ such that the angle between $\boldsymbol{u}$ and $\boldsymbol{v}$ is $\frac{\pi}{2}$. The expression **orthogonal** for a pair of vectors $\boldsymbol{u}, \boldsymbol{v}$ means that

$$\boldsymbol{u} \cdot \boldsymbol{v} = 0.$$

Of course, if one of $\boldsymbol{u}$ and $\boldsymbol{v}$ is zero, they are orthogonal; but if $\boldsymbol{u} \neq \boldsymbol{0}, \boldsymbol{v} \neq \boldsymbol{0}$ are orthogonal vectors, then they are perpendicular, since $\cos a = 0, 0 \leq a \leq \pi$, implies $a = \pi/2$.

Proposition 4 *Let* $\boldsymbol{u} \neq \boldsymbol{0}$. *The straight line through* $\boldsymbol{a}$ *perpendicular to* $\boldsymbol{u}$ *has equation*

$$\boldsymbol{u} \cdot \boldsymbol{x} = \boldsymbol{u} \cdot \boldsymbol{a}. \tag{6}$$

Proof. The equation (6) asserts that

$$\boldsymbol{u} \cdot (\boldsymbol{x} - \boldsymbol{a}) = 0.$$

Other than $\boldsymbol{x} = \boldsymbol{a}$, the points $\boldsymbol{x}$ that satisfy this are such that the segment with initial point $\boldsymbol{a}$ and terminal point $\boldsymbol{x}$ is perpendicular to $\boldsymbol{u}$ (Figure 13). As $\boldsymbol{x}$ varies, we get the straight line through $\boldsymbol{a}$ perpendicular to $\boldsymbol{u}$.

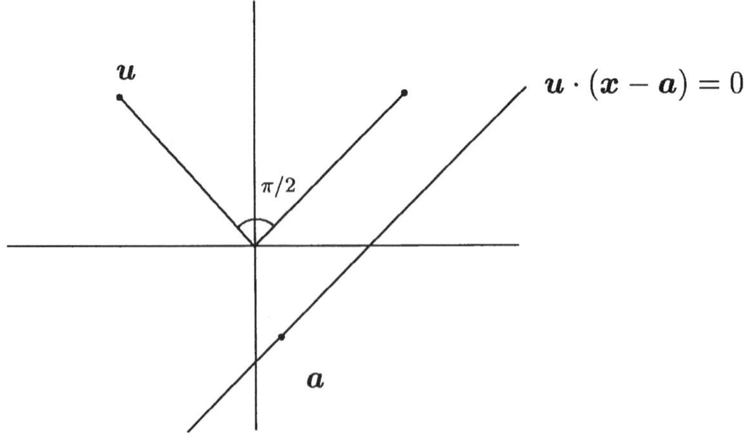

Figure 13. The straight line $\boldsymbol{u} \cdot (\boldsymbol{x} - \boldsymbol{a}) = 0$.

Example 4 Find the straight line through $(-1, 3)$ perpendicular to (5, 2).

Solution. The line is

$$(5,2) \cdot (\boldsymbol{x} - (-1,3)) = 0$$

or $5(x_1 + 1) + 2(x_2 - 3) = 0$, that is,

$$5x_1 + 2x_2 = 1.$$

We now begin a theme which will recur with variations later: what is the minimum distance from a point $\boldsymbol{a}$ to a straight line L? We write the distance as $d(\boldsymbol{a}, L)$. Let L have equation

$$v_1 x_1 + v_2 x_2 = c. \tag{7}$$

Then $\boldsymbol{v} = (v_1, v_2)$ is perpendicular to L. Choose t so that the vector with initial point $\boldsymbol{a}$ and terminal point $\boldsymbol{a} + t\boldsymbol{v}$ lies on L (Figure 14). This amounts to

$$\boldsymbol{v} \cdot (\boldsymbol{a} + t\boldsymbol{v}) = c,$$
$$t = (\boldsymbol{v} \cdot \boldsymbol{v})^{-1}(c - \boldsymbol{v} \cdot \boldsymbol{a}) = |\boldsymbol{v}|^{-2}(c - \boldsymbol{v} \cdot \boldsymbol{a}).$$

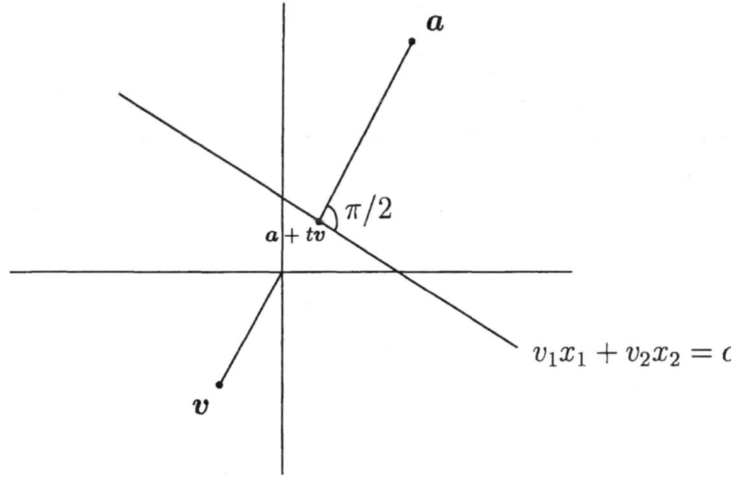

Figure 14. The distance from a point to a straight line.

Now the distance $d(\boldsymbol{a}, L)$ is equal to $|t\boldsymbol{v}|$. In fact,

$$d(\boldsymbol{a}, L) = |\boldsymbol{v}|^{-1}|\boldsymbol{v} \cdot \boldsymbol{a} - c|.$$

Example 5 Find the distance from the point $(-6, 1)$ to the straight line

$$x_2 = \frac{3}{4}x_1 - 7.$$

Solution. In the notation of (7), $\boldsymbol{v} = \left(\frac{3}{4}, -1\right), c = 7$, and the distance is

$$\frac{\left|\left(\frac{3}{4}, -1\right) \cdot (-6, 1) - 7\right|}{\left(\left(\frac{3}{4}\right)^2 + 1\right)^{1/2}} = \frac{25 \times 2}{(3^2 + 4^2)^{1/2}} = 10.$$

The distance $d(L_1, L_2)$ between two parallel lines L_1, L_2 can be found as follows. Clearly the lines will be of the form

$$u_1x_1 + u_2x_2 = c_1,$$
$$u_1x_1 + u_2x_2 = c_2.$$

Take any point $\boldsymbol{p}$ on the first line. The distance $d(L_1, L_2)$ is the same as $d(\boldsymbol{p}, L_2)$. So

$$d(L_1, L_2) = \frac{|\boldsymbol{u} \cdot \boldsymbol{p} - c_2|}{|\boldsymbol{u}|} = \frac{|c_1 - c_2|}{|\boldsymbol{u}|}.$$

A practical application of vectors in $\mathbb{R}^2$ is the topic of **velocity** . We say that an object has velocity $\boldsymbol{v}$ meters per second if it is moving in the direction of $\boldsymbol{v}$ at a speed of $|\boldsymbol{v}|$ meters per second. Other units of time and distance could be chosen. In our examples we assume that 'North' is the direction of vector $(0, 1)$.

Example 6 The velocity of a boat on a lake is $(7, 13)$. Find the speed at which the boat is travelling.

Solution. The speed is $|(7, 13)| = (7^2 + 13^2)^{1/2} = \sqrt{218}$.

Example 7 Two boats A and B start out with A 10 km. east of B. B sails northeast at 8 kph (km. per hour) and A sails northwest at 1 kph. What is the closest that A and B come to one another? See Figure 15.

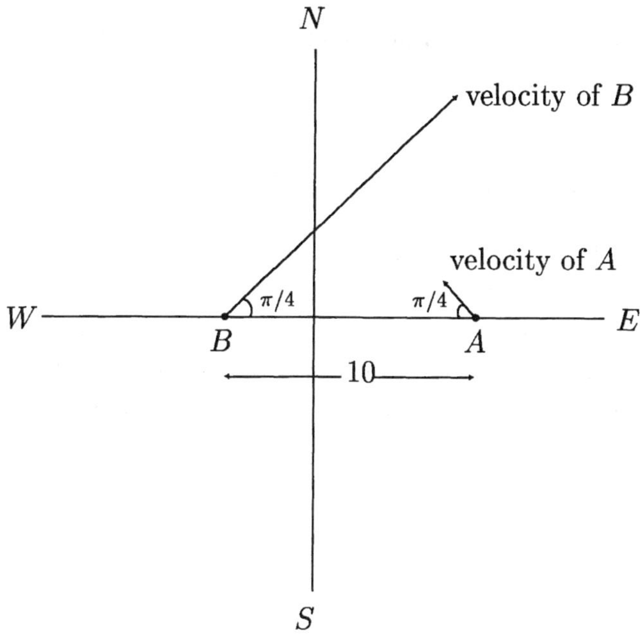

Figure 15. Velocities of A and B.

Solution. The velocity of A is $\boldsymbol{v} = \left(-\frac{1}{\sqrt{2}}, \frac{1}{\sqrt{2}}\right)$ and that of B is $\boldsymbol{w} = 8\left(\frac{1}{\sqrt{2}}, \frac{1}{\sqrt{2}}\right)$. We use a moving coordinate system with A at $\mathbf{0}$. If we add $-\boldsymbol{v}$ to the velocities of A and B, this produces velocities measured in the moving coordinate system. Now B has velocity $\boldsymbol{w} - \boldsymbol{v}$ (**relative velocity** of B with respect to A). With t as lapsed hours, the position $\boldsymbol{x}$ of B is

$$\begin{aligned}\boldsymbol{x} &= (\quad 10, 0) + (\boldsymbol{w} - \boldsymbol{v})t \\ &= (-10, 0) + \left(\frac{9}{\sqrt{2}}, \frac{7}{\sqrt{2}}\right)t.\end{aligned}$$

The nearest B comes to A is the distance of $\mathbf{0}$ from this line. Write the line as

$$\frac{x_2 - 0}{x_1 + 10} = \frac{7}{9}$$

or

$$9x_2 - 7x_1 - 70 = 0,$$

to get

$$d = \frac{|9.0 - 7.0 - 70|}{(7^2 + 9^2)^{1/2}} = \frac{70}{(130)^{1/2}},$$

or about 6.1 kilometers. The distance is easily seen to be initially decreasing, so that the minimum distance occurs for a positive value of t.

3 Three dimensional Euclidean space $\mathbb{R}^3$

Consider a third directed segment $\boldsymbol{e}_3$ of length 1 along with the pair $\boldsymbol{e}_1, \boldsymbol{e}_2$ discussed in Section 1, which is orthogonal to both $\boldsymbol{e}_1, \boldsymbol{e}_2$. We can now label any point in space as (x_1, x_2, x_3): this is the point reached by starting at the origin, moving distance x_1 (with appropriate interpretation according to sign) in direction $\boldsymbol{e}_1$, then x_2 in direction $\boldsymbol{e}_2$ and then x_3 in direction $\boldsymbol{e}_3$. See Figure 16.

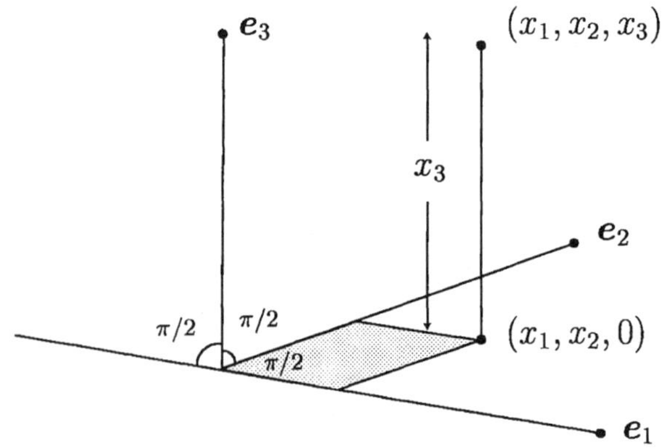

Figure 16. Points of three-dimensional space as ordered triples.

Think of $\boldsymbol{e}_1, \boldsymbol{e}_2$ as pointing east and north on the (flat) earth and $\boldsymbol{e}_3$ as pointing skywards. Now identify points in space with triples of real numbers to get $\mathbb{R}^3$. Formally, $\mathbb{R}^3$ is the set of ordered triples $\boldsymbol{x} = (x_1, x_2, x_3)$ with each x_i real. Again, we do not distinguish points from directed segments from $\mathbf{0}$ to $\boldsymbol{x}$, so $\mathbb{R}^3$ consists of vectors; and again, we may not always distinguish $\boldsymbol{x}$ from the translate whose initial point is $\boldsymbol{a} = (a_1, a_2, a_3)$ and terminal point $(x_1 + a_1, x_2 + a_2, x_3 + a_3)$ (Figure 17). A double use of Pythagoras's theorem, for the triangles with vertices $\mathbf{0}, (0, x_2, 0), (x_1, x_2, 0)$, and vertices $\mathbf{0}, (x_1, x_2, 0), (x_1, x_2, x_3)$ leads to the formula for the length $|\boldsymbol{x}|$ of vector $\boldsymbol{x}$ (Figure 18):

$$|\boldsymbol{x}| = (x_1^2 + x_2^2 + x_3^2)^{1/2}. \tag{8}$$

For, in the second triangle, $|\boldsymbol{x}|^2 = ((x_1^2 + x_2^2)^{1/2})^2 + x_3^2 = x_1^2 + x_2^2 + x_3^2$.

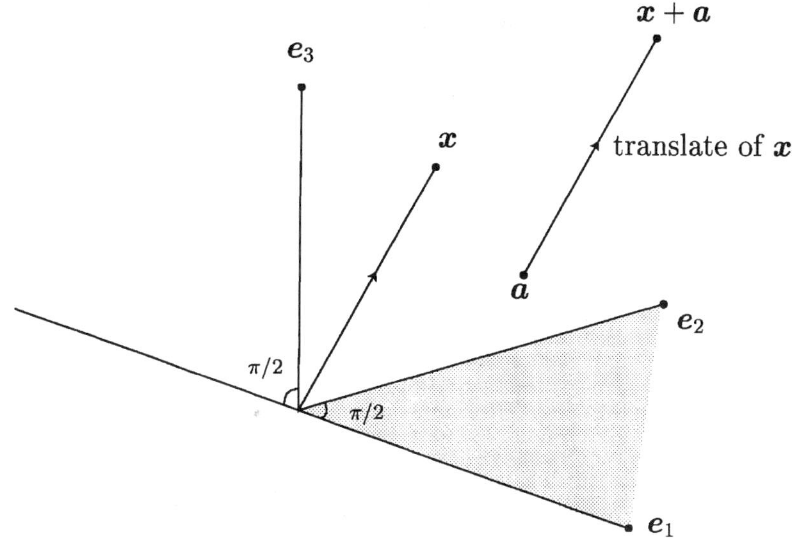

Figure 17. Translate of a vector in $\mathbb{R}^3$.

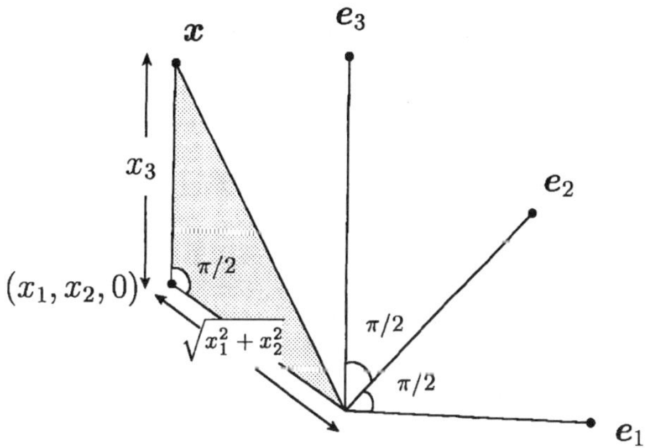

Figure 18. Double application of Pythagoras's theorem.

Definition 5 The **sum** of $\boldsymbol{a} = (a_1, a_2, a_3)$ and $\boldsymbol{b} = (b_1, b_2, b_3)$ is

$$\boldsymbol{a} + \boldsymbol{b} = (a_1 + b_1, a_2 + b_2, a_3 + b_3).$$

Geometrically, $\boldsymbol{a} + \boldsymbol{b}$ is at the terminal point of the vector $\boldsymbol{b}$ if $\boldsymbol{b}$ is placed with initial point at $\boldsymbol{a}$. Alternatively, the description of $\boldsymbol{a} + \boldsymbol{b}$, as the fourth vertex of the parallelogram with vertices $\mathbf{0}, \boldsymbol{a}, \boldsymbol{b}$, that we used in $\mathbb{R}^2$, is valid. This is often referred to as the **parallelogram law of addition**.

Subtraction of $\boldsymbol{z}$ from $\boldsymbol{y}$ is defined by

$$(\boldsymbol{y}-\boldsymbol{z})+\boldsymbol{z}=\boldsymbol{y}.$$

Thus

$$\boldsymbol{y}-\boldsymbol{z}=(y_1-z_1, y_2-z_2, y_3-z_3).$$

It is clear that the distance from $\boldsymbol{z}$ to $\boldsymbol{y}$ is $|\boldsymbol{y}-\boldsymbol{z}|$.

Definition 6 The **scalar product** $c\boldsymbol{x}$, where c is real and $\boldsymbol{x}$ is in $\mathbb{R}^3$, is

$$c\boldsymbol{x}=(cx_1, cx_2, cx_3).$$

We can, as before, describe $c\boldsymbol{x}$ as the vector pointing in the same direction as $\boldsymbol{x}$, but with length c times that of $\boldsymbol{x}$, if $c>0$; or the opposite direction, with length $-c$ times that of $\boldsymbol{x}$, if $c<0$. This depends on the formula

$$|c\boldsymbol{x}|=|c|\ |\boldsymbol{x}|$$

which we can get in a very similar way to the proof in $\mathbb{R}^2$:

$$|c\boldsymbol{x}|=((cx_1)^2+(cx_2)^2+(cx_3)^2)^{1/2}=|c|(x_1^2+x_2^2+x_3^2)^{1/2}=|c|\ |\boldsymbol{x}|.$$

You can probably guess the formula that we use to define inner product. It is

Definition 7 The **inner product** of $\boldsymbol{u}$ and $\boldsymbol{v}$ is

$$\boldsymbol{u}\cdot\boldsymbol{v}=u_1v_1+u_2v_2+u_3v_3.$$

This is consistent with Definition 4 if we think of $\mathbb{R}^2$ as being the same as the set of all $(x_1, x_2, 0)$ (or, the set of all $x_1\mathbf{e}_1+x_2\mathbf{e}_2$) where x_1 and x_2 are real.

Note that the equation (4) from $\mathbb{R}^2$,

$$|\boldsymbol{u}|=(\boldsymbol{u}\cdot\boldsymbol{u})^{1/2},$$

holds good in $\mathbb{R}^3$.

The angle a between nonzero vectors $\boldsymbol{u}, \boldsymbol{v}$ can be defined by drawing a plane that contains $\mathbf{0}, \boldsymbol{u}, \boldsymbol{v}$ and measuring a (between 0 and π) in that plane.

Proposition 5 *We have*

$$\boldsymbol{u} \cdot \boldsymbol{v} = |\boldsymbol{u}|\ |\boldsymbol{v}| \cos a.$$

Proof. We may repeat the proof of Proposition 3 verbatim.

'Verbatim' means 'word for word.' When you want to use for yourself a proof along these lines, bear in mind that students regularly arrive at wrong conclusions this way. If you try it with Proposition 4, you arrive at the false conclusion that

$$\boldsymbol{u} \cdot \boldsymbol{x} = \boldsymbol{u} \cdot \boldsymbol{a}$$

is the equation of a line in $\mathbb{R}^3$ (it is a plane; see below!). Thus you have to be very careful that arguments still work in a new context.

As in $\mathbb{R}^2$, for *nonzero* vectors $\boldsymbol{u}$ and $\boldsymbol{v}$ the relation

$$\boldsymbol{u} \cdot \boldsymbol{v} = 0 \tag{9}$$

is equivalent to $\boldsymbol{u}, \boldsymbol{v}$ being perpendicular. We say $\boldsymbol{u}$ and $\boldsymbol{v}$ are **orthogonal** if (9) holds.

Example 8 The angle between the vectors $(1, 7, b)$ and $(-2, 2, 1)$ is $a = \cos^{-1}(1/3)$. Find b.

Solution. We know that

$$\begin{aligned}(1,7,b) \cdot (-2,2,1) &= |(1,7,b)|\ |(-2,2,1)| \cos a \\ &= |(1,7,b)|\ |(-2,2,1)| 1/3.\end{aligned}$$

Thus

$$\begin{aligned}-2 + 14 + b &= (1^2 + 7^2 + b^2)^{1/2}(2^2 + 2^2 + 1)^{1/2} 1/3; \\ b + 12 &= (b^2 + 50)^{1/2}.\end{aligned}$$

If we square both sides and cancel b^2 we find that

$$24b + 144 = 50; \ b = -47/12.$$

4 Lines and planes in $\mathbb{R}^3$

Proposition 6 *Given* $\boldsymbol{a}$ *and* $\boldsymbol{b}$ *in* $\mathbb{R}^3$, *the equation*

$$\boldsymbol{x} = \boldsymbol{a} + t\boldsymbol{b} \quad (t \text{ real})$$

is a straight line through $\boldsymbol{a}$ *in the direction of* $\boldsymbol{b}$.

Proof. We may repeat the proof of Proposition 2 verbatim. (Don't forget to convince yourself of the soundness of this!)

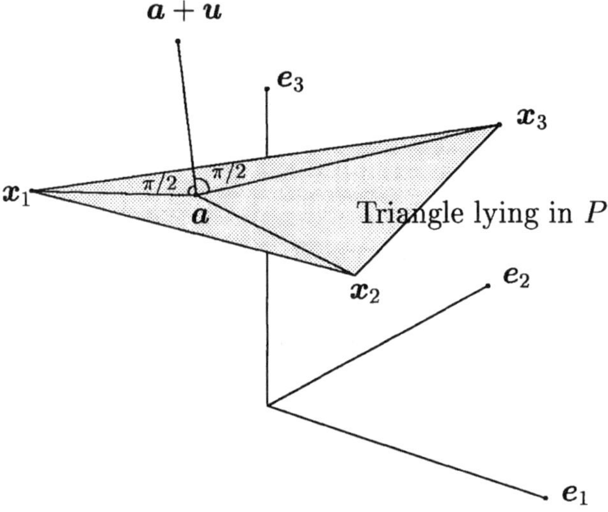

Figure 19. Any of the $\boldsymbol{x}_i$ satisfies $\boldsymbol{u} \cdot (\boldsymbol{x}_i - \boldsymbol{a}) = 0$.

Example 9 Find the equation of the straight line through $(-11, -1, 12)$ and $(2, 4, 5)$.

Solution. We may take

$$\boldsymbol{a} = (-11, -1, 12) \text{ and}$$
$$\boldsymbol{b} = (2, 4, 5) - (-11, -1, 12) = (13, 5, -7).$$

Thus the line is

$$\begin{aligned}\boldsymbol{x} &= (-11, -1, 12) + t(13, 5, -7)\\ &= (-11 + 13t, -1 + 5t, 12 - 7t).\end{aligned}$$

We now turn to an arbitrary plane P in $\mathbb{R}^3$. A **normal** to P is a nonzero vector $\boldsymbol{u}$ perpendicular to P. Take any $\boldsymbol{a}$ in P. The equation (Figure 19)

$$\boldsymbol{u} \cdot (\boldsymbol{x} - \boldsymbol{a}) = 0$$

gives precisely the set of $\boldsymbol{x}$ that lie in P. So a general plane is

$$u_1x_1 + u_2x_2 + u_3x_3 = c$$

where $\boldsymbol{u} \neq \mathbf{0}$, and c is real.

Example 10 Find the equation of the plane P passing through $\boldsymbol{a} = (1, -1, 0)$, $\boldsymbol{b} = (0, 3, 1), \boldsymbol{c} = (2, 2, -4)$.

Solution. If $\boldsymbol{a}, \boldsymbol{b}, \boldsymbol{c}$ are **collinear** (on a straight line), there will be no unique solution. However, $\boldsymbol{b} - \boldsymbol{a} = (-1, 4, 1)$ clearly cannot be a scalar multiple of $\boldsymbol{c} - \boldsymbol{a} = (1, 3, -4)$, so the points are not collinear.

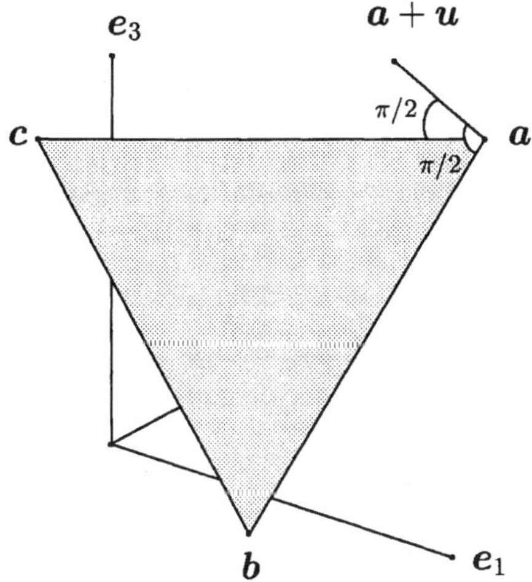

Figure 20. The normal $\boldsymbol{u}$ to the plane given by $\boldsymbol{a}, \boldsymbol{b}, \boldsymbol{c}$.

To find a normal $\boldsymbol{u}$ we solve the equations (Figure 20)

$$\boldsymbol{u} \cdot (\boldsymbol{b} - \boldsymbol{a}) = 0 \quad \text{and} \quad \boldsymbol{u} \cdot (\boldsymbol{c} - \boldsymbol{a}) = 0.$$

That is,

$$-u_1 + 4u_2 + u_3 = 0 \tag{10}$$

$$u_1 + 3u_2 - 4u_3 = 0. \tag{11}$$

Adding the equations,

$$7u_2 - 3u_3 = 0.$$

Take $u_2 = 3, u_3 = 7$. The equation (10) yields $u_1 = 19$, so

$$\boldsymbol{u} = (19, 3, 7).$$

A scalar multiple of $\boldsymbol{u}$ would also satisfy (10), (11), but this is not important. Since $\boldsymbol{u} \cdot \boldsymbol{a} = 16$, the plane P is

$$19x_1 + 3x_2 + 7x_3 = 16. \tag{12}$$

Checking. You can check the working here by substituting $\boldsymbol{a}, \boldsymbol{b}, \boldsymbol{c}$ into (12). Make this kind of checking–especially of your own work–a habit.

A plane could also be presented in **parametric** form . If $\boldsymbol{a}, \boldsymbol{b}, \boldsymbol{c}$ are three noncollinear points of P, then write $\boldsymbol{p} = \boldsymbol{b} - \boldsymbol{a}, \boldsymbol{q} = \boldsymbol{c} - \boldsymbol{a}$. The points

$$\boldsymbol{x} = \boldsymbol{a} + t\boldsymbol{p} + u\boldsymbol{q} \quad (t, u \text{ real}) \tag{13}$$

are all in P. Conversely, since t and u can be used to alter the size and shape of the parallelogram with vertices $\boldsymbol{a}, \boldsymbol{a}+t\boldsymbol{p}, a+u\boldsymbol{q}, \boldsymbol{x}$ (Figure 21), we see that all points of P can be written in the form (13). So (13) is the parametric form of P; t, u are the parameters.

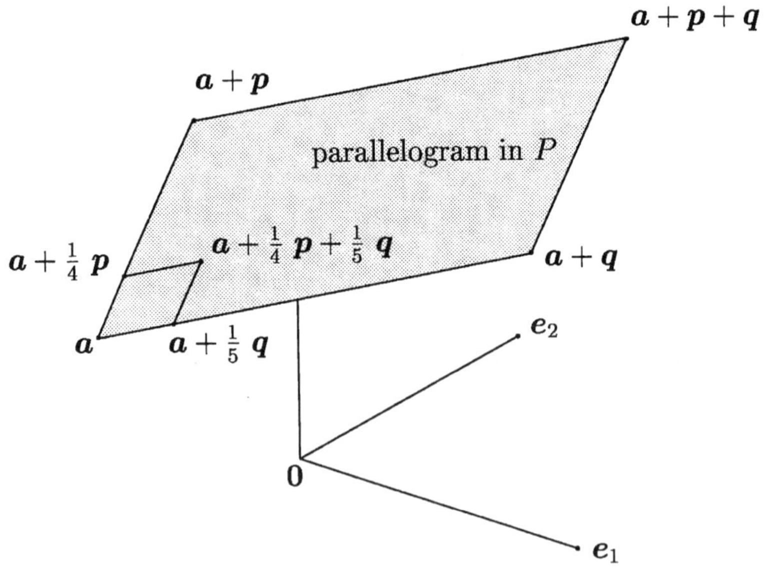

Figure 21. A plane as the set of $\boldsymbol{a} + t\boldsymbol{p} + u\boldsymbol{q}$ $(t, u \text{ real})$.

Example 11 Write the plane P

$$x_1 - 5x_2 + 6x_3 = 4$$

in parametric form.

Solution. Take $\boldsymbol{a} = (4,0,0)$ and $\boldsymbol{b} = (0,0,2/3)$ in the plane. So $\boldsymbol{b} - \boldsymbol{a} = (-4,0,2/3)$. It is not hard to pick $\boldsymbol{c}$ in the plane so that $\boldsymbol{c} - \boldsymbol{a}$ is not a scalar multiple of $\boldsymbol{b} - \boldsymbol{a}$; for instance $\boldsymbol{c} = (0,-4/5,0), \boldsymbol{c} - \boldsymbol{a} = (-4,-4/5,0)$. Now

$$\begin{aligned}\boldsymbol{x} &= (4,0,0) + t(-4,0,2/3) + u(-4,-4/5,0) \\ &= (4 - 4t - 4u, -4u/5, 2t/3)\end{aligned}$$

is a parametric form of P.

The intersection of two non-parallel planes is a straight line. It follows that the set of $\boldsymbol{x}$ satisfying *two* linear equations

$$\begin{aligned}u_1x_1 + u_2x_2 + u_3x_3 &= c_1 \\ v_1x_1 + v_2x_2 + v_3x_3 &= c_2\end{aligned}$$

is a straight line, provided the planes are not parallel. (Parallel planes would have $\boldsymbol{v}$ a scalar multiple of $\boldsymbol{u}$, so this is easy enough to check.) The parametric form of a straight line is usually more convenient.

The distance $d(\boldsymbol{a}, P)$ from a point $\boldsymbol{a}$ to a plane P with equation

$$\boldsymbol{v} \cdot \boldsymbol{x} = c$$

may be found by placing a vector with initial point $\boldsymbol{a}$ and terminal point $\boldsymbol{a} + t\boldsymbol{v}$ in such a way that $\boldsymbol{a} + t\boldsymbol{v}$ is on P. (The vector $t\boldsymbol{v}$ is, of course, perpendicular to P.) So

$$\boldsymbol{v} \cdot (\boldsymbol{a} + t\boldsymbol{v}) = c.$$

A calculation just like the one for $d(\boldsymbol{a}, L)$ in Section 2 gives

$$d(\boldsymbol{a}, P) = |\boldsymbol{v}|^{-1}|c - \boldsymbol{v} \cdot \boldsymbol{a}|.$$

As for the distance between two parallel planes

$$P_1 : \boldsymbol{u} \cdot \boldsymbol{x} = c_1 \text{ and } P_2 : \boldsymbol{u} \cdot \boldsymbol{x} = c_2,$$

pick any point $\boldsymbol{a}$ on P_1. Then

$$\begin{aligned}d(P_1, P_2) &= d(\boldsymbol{a}, P_2) \\ &= \frac{|\boldsymbol{u} \cdot \boldsymbol{a} - c_2|}{|\boldsymbol{u}|} = \frac{|c_1 - c_2|}{|\boldsymbol{u}|}.\end{aligned}$$

Example 12 A tetrahedron (pyramid with four faces) has vertices at $\mathbf{0}, \boldsymbol{a} = (1,5,6), \boldsymbol{b} = (2,1,2)$, $\boldsymbol{c} = (3,6,14)$. Find the length d of the perpendicular from $\mathbf{0}$ to the opposite face.

Solution. A normal $\boldsymbol{u}$ to the opposite face satisfies

$$\boldsymbol{u} \cdot (\boldsymbol{b} - \boldsymbol{a}) = 0, \boldsymbol{u} \cdot (\boldsymbol{c} - \boldsymbol{a}) = 0.$$

That is,

$$u_1 - 4u_2 - 4u_3 = 0, \tag{14}$$

$$2u_1 + u_2 + 8u_3 = 0. \tag{15}$$

Add 2 × (14) to (15):

$$4u_1 - 7u_2 = 0.$$

Put $u_1 = 7, u_2 = 4$. Then $u_3 = \frac{1}{4}(7 - 16) = \frac{-9}{4}$ from (14). Our normal is $(7, 4, -9/4)$ or, more conveniently, $(28, 16, -9)$. Calculating $28a_1 + 16a_2 - 9a_3$, the opposite face is part of the plane

$$28x_1 + 16x_2 - 9x_3 = 54.$$

Consequently

$$d = \frac{|(28,16,-9) \cdot \mathbf{0} - 54|}{|(28,16,-9)|} = \frac{54}{\sqrt{1121}}.$$

Example 13 Show that the line segment joining $\boldsymbol{a}$ to $\boldsymbol{b}$ is the set of $t\boldsymbol{a} + (1-t)\boldsymbol{b}$ with $0 \le t \le 1$.

Solution. We have $t\boldsymbol{a} + (1-t)\boldsymbol{b} = \boldsymbol{a} + t(\boldsymbol{b} - \boldsymbol{a})$. This gives the portion of the line through $\boldsymbol{a}$ and $\boldsymbol{b}$ for which a vector in the direction of $\boldsymbol{b} - \boldsymbol{a}$, but of any length $\le |\boldsymbol{b} - \boldsymbol{a}|$, is added to $\boldsymbol{a}$. This gives the desired line segment (Figure 22).

Example 14 Show that the lines joining the vertices of a triangle to the midpoints of the opposite sides are concurrent (Figure 23).

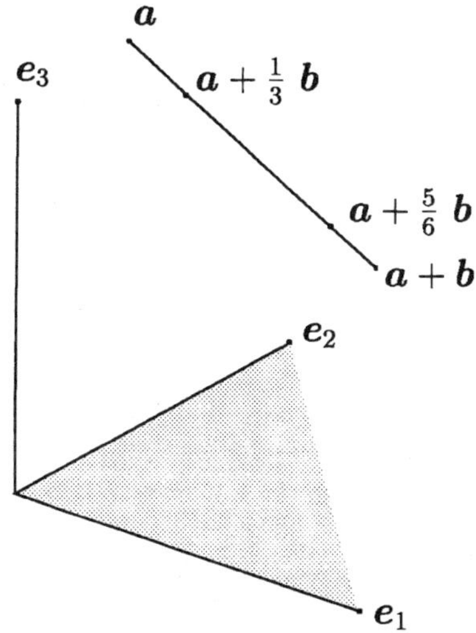

Figure 22. A segment as the set of $\boldsymbol{a} + t\boldsymbol{b}$, $0 \le t \le 1$.

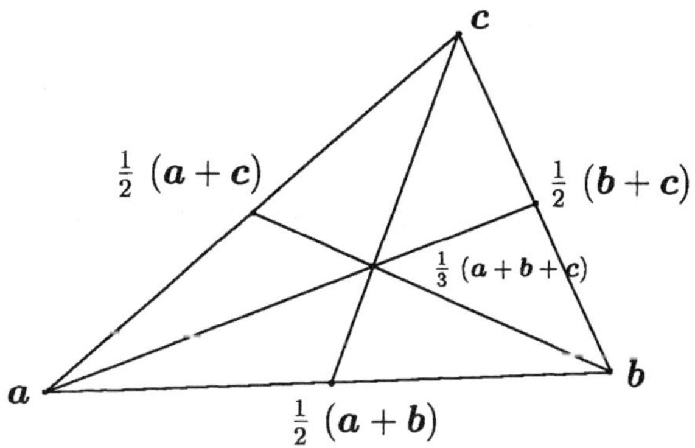

Figure 23. The lines joining the midpoints of the sides of a triangle to the opposite vertices.

'**Concurrent**' means 'passing through a single point.'

Solution. Let $\boldsymbol{a}, \boldsymbol{b}, \boldsymbol{c}$ be the vertices. The midpoint of the vector with initial point $\boldsymbol{b}$ and terminal point $\boldsymbol{c}$ is clearly

$$\boldsymbol{b} + \frac{1}{2}(\boldsymbol{c} - \boldsymbol{b}) = \frac{1}{2}(\boldsymbol{b} + \boldsymbol{c}).$$

The points on the line joining $\boldsymbol{a}$ to $\frac{1}{2}(\boldsymbol{b} + \boldsymbol{c})$ are

$$t_1\boldsymbol{a} + \frac{1}{2}(1 - t_1)(\boldsymbol{b} + \boldsymbol{c}) \quad (0 \le t_1 \le 1).$$

Similarly the other segments joining vertices to midpoints are

$$t_2\boldsymbol{b}+\frac{1}{2}\,(1-t_2)(\boldsymbol{a}+\boldsymbol{c})\ (0\le t_2\le 1)\text{ and }t_3\boldsymbol{c}+\frac{1}{2}\,(1-t_3)(\boldsymbol{a}+\boldsymbol{b})(0\le t_3\le 1).$$

These three segments share the point $\frac{1}{3}\,(\boldsymbol{a}+\boldsymbol{b}+\boldsymbol{c})$: just take $t_1=t_2=t_3=\frac{1}{3}$.

Example 15 Find the distance $d(\boldsymbol{c},L)$ from $\boldsymbol{c}$ to the straight line L with equation

$$\boldsymbol{x}=\boldsymbol{a}+t\boldsymbol{b}\quad(t\text{ real}).$$

Solution. Suppose first $\boldsymbol{a}=\boldsymbol{0}$. Then

$$\boldsymbol{v}=\boldsymbol{c}-\frac{\boldsymbol{c}\cdot\boldsymbol{b}}{|\boldsymbol{b}|^2}\,\boldsymbol{b}$$

is perpendicular to $\boldsymbol{b}$, since

$$\left(\boldsymbol{c}-\frac{\boldsymbol{c}\cdot\boldsymbol{b}}{|\boldsymbol{b}|^2}\,\boldsymbol{b}\right)\cdot\boldsymbol{b}=\boldsymbol{c}\cdot\boldsymbol{b}-\boldsymbol{c}\cdot\boldsymbol{b}=0.$$

Place $\boldsymbol{v}$ with initial point at $\frac{\boldsymbol{c}\cdot\boldsymbol{b}}{|\boldsymbol{b}|^2}\,\boldsymbol{b}$; the terminal point is $\boldsymbol{c}$ (Figure 24). Clearly

$$d(\boldsymbol{c},L)=|\boldsymbol{v}|=\left|\boldsymbol{c}-\frac{\boldsymbol{c}\cdot\boldsymbol{b}}{|\boldsymbol{b}|^2}\,\boldsymbol{b}\right|.$$

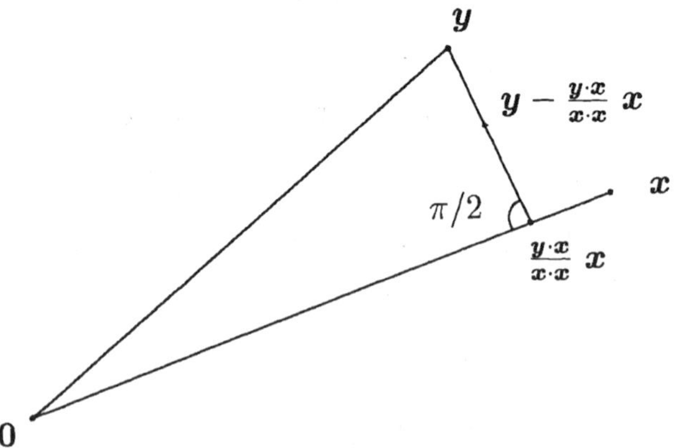

Figure 24. The projection of $\boldsymbol{y}$ on the vector $\boldsymbol{x}$.

Now consider a general $\boldsymbol{a}$. The distance from $\boldsymbol{c}$ to the line $\boldsymbol{x} = \boldsymbol{a} + t\boldsymbol{b}$ is the same as that from $\boldsymbol{c} - \boldsymbol{a}$ to the line $\boldsymbol{x} = t\boldsymbol{b}$. So

$$d(\boldsymbol{c}, L) = \left| \boldsymbol{c} - \boldsymbol{a} - \frac{(\boldsymbol{c} - \boldsymbol{a}) \cdot \boldsymbol{b}}{|\boldsymbol{b}|^2} \boldsymbol{b} \right|.$$

The vector $\frac{(\boldsymbol{c} \cdot \boldsymbol{b})}{|\boldsymbol{b}|^2} \boldsymbol{b}$ is said to be the **projection** of $\boldsymbol{c}$ on $\boldsymbol{b}$.

Example 16 Find the distance $d(L_1, L_2)$ between the straight lines

$$L_1 : \boldsymbol{x} = (2,1,4)t + (6,10,1) \quad (t \text{ real}), \text{and}$$
$$L_2 : \boldsymbol{z} = (1,1,3)u + (1,5,0) \quad (u \text{ real}).$$

Solution. We have

$$\boldsymbol{x} - \boldsymbol{z} = (2,1,4)t - (1,1,3)u + (5,5,1);$$

as t, u vary this covers a plane P. The smallest $\boldsymbol{x} - \boldsymbol{z}$ is the nearest point of P to $\boldsymbol{0}$. A normal $\boldsymbol{u}$ to P has $(2,1,4) \cdot \boldsymbol{u} = 0 = (1,1,3) \cdot \boldsymbol{u}$, since $(2,1,4), (1,1,3)$ are directions of vectors with initial points and terminal points in P. Thus

$$2u_1 + u_2 + 4u_3 = 0$$
$$u_1 + u_2 + 3u_3 = 0.$$

Subtract to get $u_1 + u_3 = 0$. Take $u_3 = 1, u_1 = -1$; now $u_2 = -2$. The plane is

$$-x_1 - 2x_2 + x_3 = -14;$$

and, using our formula for $d(\boldsymbol{0}, P)$,

$$d(L_1, L_2) = \frac{14}{(1^2 + 2^2 + 1^2)^{1/2}} = \frac{14}{\sqrt{6}}.$$

There is no difficulty in discussing velocity in $\mathbb{R}^3$. An object with **velocity** $\boldsymbol{v}$ is travelling in the direction of $\boldsymbol{v}$ at $|\boldsymbol{v}|$ meters per second (or mph, or ...).

Example 17 An airplane is heading west with groundspeed 400 kph and is gaining altitude at 30 meters per second. Find the velocity vector $\boldsymbol{v}$ of the airplane in the usual coordinate system.

Solution. The usual coordinate system has $(0,0,1)$ pointing vertically up, $(0,1,0)$ pointing north, $(1,0,0)$ pointing east. So

$$\boldsymbol{v} = (-400, 0, 0) + \left(0, 0, \frac{30 \times 3600}{1000}\right) = (-400, 0, 108).$$

5 Euclidean space $\mathbb{R}^n$

We have no spatial frame of reference for $n > 3$. We simply extend what we did for $\mathbb{R}^2$ and $\mathbb{R}^3$ and find that it leads to fruitful consequences over the next chapters.

Definition 8 Let $\mathbb{R}^n$ be the set of ordered n-tuples (or just 'tuples') $\boldsymbol{x} = (x_1, x_2, x_3, \ldots, x_n)$, where the x_i are real. We agree that '$\boldsymbol{x} = \boldsymbol{y}$' means $x_1 = y_1, x_2 = y_2, \ldots, x_n = y_n$. For $\boldsymbol{x} = (x_1, \ldots, x_n), \boldsymbol{y} = (y_1, \ldots, y_n)$, we define $\boldsymbol{x} + \boldsymbol{y}$ by

$$\boldsymbol{x} + \boldsymbol{y} = (x_1 + y_1, x_2 + y_2, \ldots, x_n + y_n).$$

For c in $\mathbb{R}$, we define the **scalar product** $c\boldsymbol{x}$ by

$$c\boldsymbol{x} = (cx_1, \ldots, cx_n).$$

We also refer to the tuples as **vectors**. It is very easy to verify that the following rules hold (we used these rules in $\mathbb{R}^2$ and $\mathbb{R}^3$ without drawing attention to them).

Algebraic rules of $\mathbb{R}^n$.

1° $\boldsymbol{x} + \boldsymbol{y} = \boldsymbol{y} + \boldsymbol{x}$ (commutative law of addition).

2° $(\boldsymbol{x} + \boldsymbol{y}) + \boldsymbol{z} = \boldsymbol{x} + (\boldsymbol{y} + \boldsymbol{z})$ (associative law of addition).

3° There is a point $\mathbf{0}$ such that $\boldsymbol{x} + \mathbf{0} = \boldsymbol{x}$ for all $\boldsymbol{x}$.

4° For each $\boldsymbol{x}$ there is a point $\boldsymbol{y}$ such that $\boldsymbol{x} + \boldsymbol{y} = \mathbf{0}$.

5° $1\boldsymbol{x} = \boldsymbol{x}$.

6° $(ab)\boldsymbol{x} = a(b\boldsymbol{x})$.

7° $a(\boldsymbol{x} + \boldsymbol{y}) = a\boldsymbol{x} + a\boldsymbol{y}$.

8° $(a + b)\boldsymbol{x} = a\boldsymbol{x} + b\boldsymbol{x}$.

Here $\boldsymbol{x}, \boldsymbol{y}, \boldsymbol{z}$ (in $\mathbb{R}^n$) and a, b (in $\mathbb{R}$) are arbitrary.

We need not go through proofs of all of these. The zero vector $\mathbf{0} = (0, 0, \ldots, 0)$ has the property needed for 3° and the point $\boldsymbol{y} = (-x_1, \ldots, -x_n)$, usually written $-\boldsymbol{x}$, has the property needed in 4°. The proof of 7° (for example):

$$\begin{aligned}a(\boldsymbol{x}+\boldsymbol{y}) &= a(x_1+y_1, x_2+y_2, \ldots, x_n+y_n)\\ &= (a(x_1+y_1), a(x_2+y_2), \ldots, a(x_n+y_n))\\ &= (ax_1+ay_1, ax_2+ay_2, \ldots, ax_n+ay_n)\\ &= a\boldsymbol{x}+a\boldsymbol{y}.\end{aligned}$$

Subtraction is defined by: $\boldsymbol{y}-\boldsymbol{z} = (y_1-z_1, y_2-z_2, \ldots, y_n-z_n)$, and we have

$$(\boldsymbol{y}-\boldsymbol{z})+\boldsymbol{z} = \boldsymbol{y}.$$

Example 18 Solve the equation

$$(1,1,3,2)+5(\boldsymbol{x}-(2,4,0,6)) = (11,1,-1,0)$$

in $\mathbb{R}^4$.

Solution.

$$\begin{aligned}(1,1,3,2)+5\boldsymbol{x}-(10,20,0,30) &= (11,1,-1,0);\\ 5\boldsymbol{x} &= -(1,1,3,2)+(10,20,0,30)+(11,1,-1,0)\\ &= (20,20,-4,28);\\ \boldsymbol{x} &= (4,4,-4/5,28/5).\end{aligned}$$

We can extend the previous definitions of inner product, length and distance in a quite natural way.

Definition 9 For $\boldsymbol{x} = (x_1, \ldots, x_n)$ and $\boldsymbol{y} = (y_1, \ldots, y_n)$ in $\mathbb{R}^n$, we define the **inner product** of $\boldsymbol{x}$ and $\boldsymbol{y}$ by

$$\boldsymbol{x}\cdot\boldsymbol{y} = x_1y_1 + x_2y_2 + \ldots + x_ny_n.$$

We note the following rules, in which the vectors $\boldsymbol{x}, \boldsymbol{y}, \boldsymbol{z}$ and the scalar c are arbitrary.

Algebraic rules of inner product

1° $(\boldsymbol{x}+\boldsymbol{z})\cdot\boldsymbol{y} = \boldsymbol{x}\cdot\boldsymbol{y}+\boldsymbol{z}\cdot\boldsymbol{y}$.

2° $(c\boldsymbol{x})\cdot\boldsymbol{y} = c\boldsymbol{x}\cdot\boldsymbol{y}$

3° $\boldsymbol{x} \cdot \boldsymbol{y} = \boldsymbol{y} \cdot \boldsymbol{x}$

4° $\boldsymbol{x} \cdot \boldsymbol{x} > 0$ if $\boldsymbol{x} \neq \mathbf{0}$.

These are very easy to check. For 4°, we note that $\boldsymbol{x} \cdot \boldsymbol{x} = x_1^2 + \cdots + x_n^2$; each summand is non-negative, and if $x_i \neq 0$, one summand is positive, so $\boldsymbol{x} \cdot \boldsymbol{x}$ is positive. (A **summand** is one of a group of terms that are added.)

Definition 10 We define the **length** $|\boldsymbol{x}|$ of $\boldsymbol{x}$ to be

$$|\boldsymbol{x}| = (\boldsymbol{x} \cdot \boldsymbol{x})^{1/2}.$$

We define the **distance** $d(\boldsymbol{x}, \boldsymbol{y})$ between $\boldsymbol{x}$ and $\boldsymbol{y}$ to be

$$d(\boldsymbol{x}, \boldsymbol{y}) = |\boldsymbol{x} - \boldsymbol{y}|.$$

We say that $\boldsymbol{x}$ and $\boldsymbol{y}$ are **orthogonal** if

$$\boldsymbol{x} \cdot \boldsymbol{y} = 0.$$

In $\mathbb{R}^2$ and $\mathbb{R}^3$ we have, for nonzero $\boldsymbol{x}$ and $\boldsymbol{y}$,

$$-1 \leq \frac{\boldsymbol{x} \cdot \boldsymbol{y}}{|\boldsymbol{x}|\ |\boldsymbol{y}|} \leq 1$$

because cosines lie between -1 and 1. It would be nice to extend this by proving that, for $\boldsymbol{x}$ and $\boldsymbol{y}$ in $\mathbb{R}^n$,

$$|\boldsymbol{x} \cdot \boldsymbol{y}| \leq |\boldsymbol{x}|\ |\boldsymbol{y}|.$$

The advantage of writing the inequality in this way is that it is true when $\boldsymbol{x}$ or $\boldsymbol{y}$ is $\mathbf{0}$, because both sides are 0.

Proposition 7 (Cauchy-Schwarz inequality). *Let $\boldsymbol{x}$ and $\boldsymbol{y}$ be in $\mathbb{R}^n$. Then*

$$|\boldsymbol{x} \cdot \boldsymbol{y}| \leq |\boldsymbol{x}|\ |\boldsymbol{y}|. \tag{16}$$

Proof. We can suppose $\boldsymbol{x} \neq \mathbf{0}$. We subtract from $\boldsymbol{y}$ the vector $\frac{\boldsymbol{y} \cdot \boldsymbol{x}}{\boldsymbol{x} \cdot \boldsymbol{x}} \boldsymbol{x}$, which we shall call the **projection** of $\boldsymbol{y}$ on $\boldsymbol{x}$ (Figure 24). Then we work out the square of the length of the result. It must be at least 0, so

$$\begin{aligned}\left|\boldsymbol{y} - \left(\frac{\boldsymbol{y} \cdot \boldsymbol{x}}{\boldsymbol{x} \cdot \boldsymbol{x}}\right) \boldsymbol{x}\right|^2 &= \boldsymbol{y} \cdot \boldsymbol{y} - 2\left(\frac{\boldsymbol{y} \cdot \boldsymbol{x}}{\boldsymbol{x} \cdot \boldsymbol{x}}\right) \boldsymbol{y} \cdot \boldsymbol{x} + \left(\frac{\boldsymbol{y} \cdot \boldsymbol{x}}{\boldsymbol{x} \cdot \boldsymbol{x}}\right)^2 \boldsymbol{x} \cdot \boldsymbol{x} \\ &= \boldsymbol{y} \cdot \boldsymbol{y} - \frac{(\boldsymbol{y} \cdot \boldsymbol{x})^2}{\boldsymbol{x} \cdot \boldsymbol{x}} \geq 0.\end{aligned}$$

The Cauchy-Schwarz inequality now follows on rearranging. Note that we used the definition of length together with the algebraic rules of inner product.

The following simple result is an 'n-dimensional Pythagoras's theorem' .

Proposition 8 *If $\boldsymbol{x}$ and $\boldsymbol{y}$ are orthogonal, then*

$$|\boldsymbol{x}+\boldsymbol{y}|^2 = |\boldsymbol{x}|^2 + |\boldsymbol{y}|^2.$$

Proof. We have

$$\begin{aligned}|\boldsymbol{x}+\boldsymbol{y}|^2 = (\boldsymbol{x}+\boldsymbol{y}).(\boldsymbol{x}+\boldsymbol{y}) &= \boldsymbol{x}\cdot\boldsymbol{x} + 2\boldsymbol{x}\cdot\boldsymbol{y} + \boldsymbol{y}\cdot\boldsymbol{y}\\ &= \boldsymbol{x}\cdot\boldsymbol{x} + \boldsymbol{y}\cdot\boldsymbol{y} = |\boldsymbol{x}|^2 + |\boldsymbol{y}|^2,\end{aligned}$$

since $\boldsymbol{x}\cdot\boldsymbol{y} = 0$.

Here is another easy extension of a result from $\mathbb{R}^2$ or $\mathbb{R}^3$.

Proposition 9 *We have*

$$|c\boldsymbol{x}| = |c|\ |\boldsymbol{x}|.$$

Proof. The left side is

$$\begin{aligned}|(cx_1, \dots, cx_n)| &= (c^2x_1^2 + \cdots + c^2x_n^2)^{1/2}\\ &= |c|(x_1^2 + \cdots + x_n^2)^{1/2} = |c|\ |\boldsymbol{x}|.\end{aligned}$$

A consequence of Proposition 7 is the following result, which would be clear from a simple drawing in $\mathbb{R}^2$ or $\mathbb{R}^3$.

Proposition 10 *We have*

$$d(\boldsymbol{x}, \boldsymbol{z}) \le d(\boldsymbol{x}, \boldsymbol{y}) + d(\boldsymbol{y}, \boldsymbol{z}). \tag{17}$$

Proof. First of all, we observed earlier that

$$|\boldsymbol{u}+\boldsymbol{v}|^2 = |\boldsymbol{u}|^2 + 2\boldsymbol{u}\cdot\boldsymbol{v} + |\boldsymbol{v}|^2.$$

The right side is

$$\le |\boldsymbol{u}|^2 + 2|\boldsymbol{u}|\ |\boldsymbol{v}| + |\boldsymbol{v}|^2$$

by the Cauchy-Schwarz inequality. So

$$|\boldsymbol{u}+\boldsymbol{v}|^2 \le (|\boldsymbol{u}| + |\boldsymbol{v}|)^2,$$

that is,

$$|\boldsymbol{u}+\boldsymbol{v}| \leq |\boldsymbol{u}| + |\boldsymbol{v}|. \tag{18}$$

Now put $\boldsymbol{u} = \boldsymbol{x} - \boldsymbol{y}, \boldsymbol{v} = \boldsymbol{y} - \boldsymbol{z}$. Since $\boldsymbol{u} + \boldsymbol{v} = \boldsymbol{x} - \boldsymbol{z}$, we deduce (17) from (18).

Both (17) and (18) are referred to as the **triangle inequality**.

We can define a **line** in $\mathbb{R}^n$ as the set of $\boldsymbol{x}$ satisfying

$$\boldsymbol{x} = \boldsymbol{a} + t\boldsymbol{b} \quad (t \text{ real})$$

for given $\boldsymbol{a}$ and (nonzero) $\boldsymbol{b}$. A **plane** in $\mathbb{R}^n$ is the set of $\boldsymbol{x}$ satisfying

$$\boldsymbol{x} = \boldsymbol{a} + t\boldsymbol{p} + u\boldsymbol{q} \quad (t, u \text{ real})$$

for given $\boldsymbol{a}$ and a pair of nonzero vectors $\boldsymbol{q}$ and $\boldsymbol{p}$, with $\boldsymbol{p}$ not a multiple of $\boldsymbol{q}$.

In Chapter 3 we will develop the ideas of subspace, basis and dimension. This will permit us to see lines and planes of $\mathbb{R}^n$ as 'special cases'.

For the moment, let us define a **hyperplane** to be the set of $\boldsymbol{x}$ satisfying an equation

$$\boldsymbol{u} \cdot \boldsymbol{x} = c, \tag{19}$$

with $\boldsymbol{u}$ nonzero. A hyperplane in $\mathbb{R}^2$ is a straight line; a hyperplane in $\mathbb{R}^3$ is a plane.

Proposition 11 *The distance from the point* $\boldsymbol{a}$ *to the hyperplane* P *with equation*

$$\boldsymbol{u} \cdot \boldsymbol{x} = c \tag{20}$$

is

$$d(\boldsymbol{a}, P) = \frac{|\boldsymbol{u} \cdot \boldsymbol{a} - c|}{|\boldsymbol{u}|}.$$

We need to formalize the notion of the distance $d(\boldsymbol{a}, S)$ from a point $\boldsymbol{a}$ to a set of points S. If there is a point $\boldsymbol{w}$ in S such that

$$|\boldsymbol{a} - \boldsymbol{w}| \leq |\boldsymbol{a} - \boldsymbol{s}|$$

for all $\boldsymbol{s}$ in S, then $d(\boldsymbol{a}, S)$ is defined to be $|\boldsymbol{a} - \boldsymbol{w}|$.

Proof of Proposition 11. Take the vector $\boldsymbol{u}$, which we may call a **normal** to our hyperplane, and choose a real number t such that

$$\boldsymbol{w} = \boldsymbol{a} + t\boldsymbol{u}$$

lies in the hyperplane. That is,

$$(\boldsymbol{a} + t\boldsymbol{u}) \cdot \boldsymbol{u} = c,$$

leading to

$$t = \frac{c - \boldsymbol{u} \cdot \boldsymbol{a}}{\boldsymbol{u} \cdot \boldsymbol{u}}.$$

The distance from $\boldsymbol{a}$ to $\boldsymbol{w}$ is

$$|\boldsymbol{a} - \boldsymbol{w}| = |t\boldsymbol{u}| = |\boldsymbol{u} \cdot \boldsymbol{a} - c|/|\boldsymbol{u}|,$$

since $|\boldsymbol{u}|^2 = \boldsymbol{u} \cdot \boldsymbol{u}$. To complete the proof, we have to show that

$$|\boldsymbol{a} - \boldsymbol{w}| \leq |\boldsymbol{a} - \boldsymbol{s}|$$

for any $\boldsymbol{s}$ in P. (This was left to geometric intuition in $\mathbb{R}^2, \mathbb{R}^3$.)

We find that $\boldsymbol{s} - \boldsymbol{w}$ is orthogonal to $\boldsymbol{a} - \boldsymbol{w}$. For, using equation (20) for both $\boldsymbol{w}$ and $\boldsymbol{s}$,

$$\boldsymbol{u} \cdot (\boldsymbol{s} - \boldsymbol{w}) = \boldsymbol{u} \cdot \boldsymbol{s} - \boldsymbol{u} \cdot \boldsymbol{w} = c - c = 0,$$

and so

$$(\boldsymbol{a} - \boldsymbol{w}) \cdot (\boldsymbol{s} - \boldsymbol{w}) = (-t\boldsymbol{u}) \cdot (\boldsymbol{s} - \boldsymbol{w}) = 0.$$

Now we appeal to Proposition 8:

$$\begin{aligned} |\boldsymbol{a} - \boldsymbol{s}|^2 = |(\boldsymbol{a} - \boldsymbol{w}) + (\boldsymbol{w} - \boldsymbol{s})|^2 &= |\boldsymbol{a} - \boldsymbol{w}|^2 + |\boldsymbol{w} - \boldsymbol{s}|^2 \\ &\geq |\boldsymbol{a} - \boldsymbol{w}|^2. \end{aligned}$$

This completes the proof.

We will get more general results later when we study orthogonal sets. To me, a lot of the appeal of linear algebra is being able to work **as if** I can picture $\mathbb{R}^n$.

Exercises for Chapter 1

The first digit of the exercise number shows which section to read before attempting it. For instance, read section 3 before attempting exercises 3.1, 3.2 and so on.

1.1 Find two unit vectors whose distance from $\left(\frac{2}{5}, -\frac{1}{5}\right)$ is $\sqrt{2}$.

1.2 The curve

$$\frac{x_1^2}{a^2} + \frac{x_2^2}{b^2} = 1$$

where $0 < b < a$, is called an **ellipse**.

Show that the points of the ellipse are $\boldsymbol{x} = (a\cos t, b\sin t)$ $(0 \le t \le 2\pi)$.

1.3 The **eccentricity** of the above ellipse is defined to be $e = (1 - b^2/a^2)^{1/2}$. Let $\mathbf{c}_1 = (ae, 0), \mathbf{c}_2 = (-ae, 0)$. Show that the distance from $\boldsymbol{x} = (a\cos t, b\sin t)$ to $\mathbf{c}_1$ is $a - ae\cos t$ and the distance from $\boldsymbol{x}$ to $\mathbf{c}_2$ is $a + ae\cos t$. Deduce that

$$|\boldsymbol{x} - \mathbf{c}_1| + |\boldsymbol{x} - \mathbf{c}_2| = 2a$$

on the ellipse.

Hint: $|\boldsymbol{x} - \mathbf{c}_1|^2 = (a\cos t - ae)^2 + a^2(1 - e^2)(1 - \cos^2 t)$.

2.1 Find the distance from (1, 2) to the line $x_2 = 3x_1 - 5$.

2.2 Find the angle between the lines $x_2 = x_1 + 1$ and $x_2 = 4x_1 - 5$ (to the nearest degree).

2.3 A swimmer crosses a straight river of width 30 meters. His stroke is perpendicular to the bank and in still water would carry him at 1 meter per second. If there is a current of 4 meters per second, how far downstream does he land? How long does the crossing take?

2.4 Two toy cars A, B are released simultaneously. A is heading southeast at 10 mph, B is heading northeast at 5 mph. If B is initially 50 yards south of A, find the closest that the two cars get to each other.

2.5 Find the base point of the perpendicular from (1, 1) to the opposite side of the triangle with vertices $(1,1), (5,4), (3,8)$.

3.1 (Parametric form of sphere.) Show that for any real t_1 and t_2, $(\cos t_1 \cos t_2, \cos t_1 \sin t_2, \sin t_1)$ is a unit vector. Conversely show that any unit vector in $\mathbb{R}^3$ can be written in this way.

3.2 Find a parametric form of the ellipsoid

$$\frac{x_1^2}{a_1^2} + \frac{x_2^2}{a_2^2} + \frac{x_3^2}{a_3^2} = 1$$

similar to the one in Exercise 3.1.

3.3 Show that the midpoints of the sides of a quadrilateral in $\mathbb{R}^2$ are the vertices of a parallelogram.

3.4 Three consecutive vertices of a parallelogram in $\mathbb{R}^3$ are $\boldsymbol{a}, \boldsymbol{b}, \boldsymbol{c}$. Is the fourth vertex $\boldsymbol{a}+\boldsymbol{b}+\boldsymbol{c}$, $\boldsymbol{a}-\boldsymbol{b}+\boldsymbol{c}$ or $-\boldsymbol{a}+\boldsymbol{b}-\boldsymbol{c}$?

4.1 A submarine is 3 km. west of a destroyer and is moving southeast at 2 km. per hour while diving at the rate of $\frac{1}{\sqrt{2}}$ km. per hour. The destroyer is travelling south at $\frac{3}{\sqrt{2}}$ km. per hour. Show that the shortest distance between them will be $\sqrt{3}$ km.

4.2 Find the distance between the straight lines

$$\boldsymbol{x} = (1,0,1)t \ (t \text{ real}) \quad \text{and } \boldsymbol{z} = (-1,1,0)u + (0,1,0) \ (u \text{ real}).$$

4.3 A tetrahedron has vertices $\mathbf{0}, (1,2,0), (2,0,3), (0,1,b)$ where $b > 0$. If the distance from $\mathbf{0}$ to the opposite face of the tetrahedron is $\frac{4}{\sqrt{5}}$, find b.

4.4 Find the equation of the plane that includes the line $\boldsymbol{x} = (1,5,-2)t + (0,1,6)$ and the point $(1,1,4)$.

4.5 Find the equation of the plane that passes through $(1,0,1), (2,0,3)$ and $(1,7,-1)$.

4.6 Let $\boldsymbol{a}, \boldsymbol{b}$ be vectors in $\mathbb{R}^3$. The angle between them is t and the area of the parallelogram with vertices $\mathbf{0}, \boldsymbol{a}, \boldsymbol{b}, \boldsymbol{a}+\boldsymbol{b}$ is A. Show that

$$A = |\boldsymbol{a}|\ |\boldsymbol{b}| \sin t = \{(\boldsymbol{a}\cdot\boldsymbol{a})(\boldsymbol{b}\cdot\boldsymbol{b}) - (\boldsymbol{a}\cdot\boldsymbol{b})^2\}^{1/2}.$$

4.7 Show that the distance between the parallel lines

$$\boldsymbol{x} = (1,1,2)t + (3,1,0) \text{ and } \boldsymbol{z} = (1,1,2)u + (6,6,1)$$

is $\sqrt{165}/3$.

4.8 Show that the distance between $(-1,2,7)$ and the line $\boldsymbol{x} = (3,1,4)t + (1,0,1)$ is approximately 5.35.

5.1 Find the distance from the point $(1,-2,5,0,1)$ to the hyperplane in $\mathbb{R}^5$ with equation

$$x_1 + 3x_2 + x_4 + 3x_5 = 0.$$

5.2 Find the equation of the hyperplane in $\mathbb{R}^4$ that is perpendicular to $(2,0,-3,3)$ and contains the point $(1,4,4,7)$.

5.3 Solve the following equation in $\mathbb{R}^5$:

$$(2,4,-1,-2,1) - 3((1,0,2,0,1) - \boldsymbol{x}) = (4,5,7,0,1).$$

5.4 Show that $\boldsymbol{y} - \frac{\boldsymbol{y}\cdot\boldsymbol{x}}{\boldsymbol{x}\cdot\boldsymbol{x}}\,\boldsymbol{x}$ is orthogonal to $\boldsymbol{x}$ for any $\boldsymbol{x} \neq \mathbf{0}$ and $\boldsymbol{y}$ in $\mathbb{R}^n$ (Figure 24).

Chapter 2

Linear systems and matrices

1 Linear systems

A **linear equation** in the variables $x_1, \ldots, x_n$ is an equation that can be written in the form

$$a_1x_1 + \ldots + a_nx_n = b.$$

The expression $a_1x_1 + \ldots + a_nx_n$ is called a **linear form** in $x_1, \ldots, x_n$. The numbers $a_1, \ldots, a_n$ are called **coefficients** .

We may seek numbers $x_1, \ldots, x_n$ ('unknowns') that satisfy several linear equations **simultaneously**. Suppose there are m equations, say

$$(1) \qquad \begin{aligned} a_{11}x_1 + \cdots + a_{1n}x_n &= b_1 \\ a_{21}x_1 + \cdots + a_{2n}x_n &= b_2 \\ &\cdots \\ a_{m1}x_1 + \cdots + a_{mn}x_n &= b_m. \end{aligned}$$

We call (1) a **linear system** . A **solution** is a vector $\boldsymbol{x} = (x_1, \ldots, x_n)$ for which all m equations are valid.

In applications the numbers $x_1, \ldots, x_n$ might be currents in different parts of an electric circuit; or populations of n given cities, in which case $\boldsymbol{x}$ would change from year to year; or temperatures at assigned points of a steel bar. The possibilities are myriad.

Example 1 A solution of the linear system

$$\begin{aligned} 2x_1 - x_2 + 3x_3 + x_4 &= 9 \\ x_1 \qquad + 4x_3 + x_4 &= 10 \end{aligned}$$

is $\boldsymbol{x} = (1, 0, 2, 1)$. You should check that $\boldsymbol{x}$ satisfies both equations.

A convenient algorithm called **Gaussian elimination** generates the set of all vectors $\boldsymbol{x}$ that are solutions of (1), which we call the **general solution** or **solution set** . An **algorithm** is a systematic process where the next step is determined by what has happened so far, there is a finite number of steps, and we can give a definite upper bound for the number of steps needed. We can illustrate Gaussian elimination with a very simple linear system:

$$x_1 + x_2 + x_3 = 1 \tag{2}$$

$$x_2 + 3x_3 = 4. \tag{3}$$

Subtracting the second equation from the first,

$$x_1 - 2x_3 = -3. \tag{4}$$

This gives a new linear system (3), (4) with the same solutions as the original one (2), (3). We can choose x_3 freely, so we call it a **free variable** . Now $x_1 = -3 + 2x_3, x_2 = 4 - 3x_3$. The general solution is

$$\boldsymbol{x} = (-3 + 2x_3, 4 - 3x_3, x_3) \ (x_3 \text{ real}). \tag{5}$$

Geometrically, the intersection of the planes (2), (3) is the line (5). Simple as this example is, it contains all the elements of our algorithm.

In order to handle large systems we set up some notation. We write a general member of $\mathbb{R}^n$ vertically, as a **column vector**,

$$\boldsymbol{x} = \begin{bmatrix} x_1 \\ x_2 \\ \vdots \\ x_n \end{bmatrix}$$

in computations involving linear systems. Thus $\begin{bmatrix} x_1 \\ x_2 \\ x_3 \end{bmatrix}$ is exactly the same vector as (x_1, x_2, x_3). If you find a solution to a linear system, such as $\begin{bmatrix} 1 \\ 4 \\ -1 \\ 2 \end{bmatrix}$, you have the option of going back to horizontal mode and saying that $(1, 4, -1, 2)$ is a solution. You may find this strange at first, but it would consume a lot of space to always use column vectors. The linear system (1) can be written very succinctly as

$$x_1 \begin{bmatrix} a_{11} \\ a_{21} \\ \vdots \\ a_{m1} \end{bmatrix} + x_2 \begin{bmatrix} a_{12} \\ a_{22} \\ \vdots \\ a_{m2} \end{bmatrix} + \cdots + x_n \begin{bmatrix} a_{1n} \\ a_{2n} \\ \vdots \\ a_{mn} \end{bmatrix} = \begin{bmatrix} b_1 \\ b_2 \\ \vdots \\ b_m \end{bmatrix} \tag{6}$$

or, if we write $\boldsymbol{a}_1, \ldots, \boldsymbol{a}_n, \boldsymbol{b}$ for the vectors appearing from left to right,

$$x_1\boldsymbol{a}_1 + \cdots + x_n\boldsymbol{a}_n = \boldsymbol{b}. \tag{7}$$

In other words, a linear system of m equations in n unknowns is equivalent to a **vector equation** in $\mathbb{R}^m$.

Example 2 Write the system

$$\begin{aligned} x_1 + 2x_2 + 3x_3 &= 9 \\ x_1 \qquad - \ x_3 &= 2 \\ x_2 + 5x_3 &= 7 \end{aligned}$$

as a vector equation.

Solution. $x_1 \begin{bmatrix} 1 \\ 1 \\ 0 \end{bmatrix} + x_2 \begin{bmatrix} 2 \\ 0 \\ 1 \end{bmatrix} + x_3 \begin{bmatrix} 3 \\ -1 \\ 5 \end{bmatrix} = \begin{bmatrix} 9 \\ 2 \\ 7 \end{bmatrix}.$

2 Matrices

Even more convenient than the vector equation for our algorithm is the use of matrix notation. A **matrix** is a rectangular array

$$A = \begin{bmatrix} a_{11} & \cdots & a_{1n} \\ \vdots & & \vdots \\ a_{m1} & \cdots & a_{mn} \end{bmatrix} \tag{8}$$

or, succinctly,

$$A = [a_{ij}]_{1 \le i \le m, 1 \le j \le n}.$$

The a_{ij} are real numbers called the **entries** of A. The **rows** or **row vectors** of A are $\boldsymbol{r}_1 = [a_{11} \dots a_{1n}], \dots, \boldsymbol{r}_m = [a_{m1} \dots a_{mn}]$ and the **columns**, or **column vectors** , are the $\boldsymbol{a}_i$ that appear in (6) and (7). We say that A is $m \times n$ if there are m rows and n columns, so that, for example,

$$\begin{bmatrix} 1 & 0 & -1 & 2 \\ 3 & 7 & 0 & 8 \end{bmatrix}$$

is 2×4. A matrix that is $m \times m$ for some m is said to be **square**, examples being

$$\begin{bmatrix} 2 & 0 \\ 1 & 2 \end{bmatrix}, \begin{bmatrix} 3 & 7 & 3 \\ 0 & 6 & 2 \\ 1 & 5 & -1 \end{bmatrix}.$$

The plural of 'matrix' is 'matrices'.

Matrices are interesting algebraic objects in their own right and have many uses besides linear systems. However, in this chapter we are concerned with two kinds of matrix that are attached to a linear system (1). The **coefficient matrix** of the linear system (1) is the matrix of coefficients, which appears in (8). For example, the coefficient matrix for the linear system in Example 1 is

$$A = \begin{bmatrix} 2 & -1 & 3 & 4 \\ 1 & 0 & 4 & 1 \end{bmatrix}$$

and, for Example 2,

$$A = \begin{bmatrix} 1 & 2 & 3 \\ 1 & 0 & -1 \\ 0 & 1 & 5 \end{bmatrix}.$$

Now we add an extra column, consisting of right hand side numbers from (1),

$$\boldsymbol{b} = \begin{bmatrix} b_1 \\ \vdots \\ b_m \end{bmatrix}$$

to A, getting $[A|\boldsymbol{b}]$; this is called the **augmented matrix** of the system. The augmented matrix of the system in Example 2 is

$$\begin{bmatrix} 1 & 2 & 3 & 9 \\ 1 & 0 & -1 & 2 \\ 0 & 1 & 5 & 7 \end{bmatrix}.$$

We now define the **product** of an $m \times n$ matrix A and a vector $\boldsymbol{x}$ in $\mathbb{R}^n$. Let

$$A\boldsymbol{x} = \begin{bmatrix} a_{11} & \cdots & a_{1n} \\ \vdots & & \vdots \\ a_{m1} & \cdots & a_{mn} \end{bmatrix} \begin{bmatrix} x_1 \\ \vdots \\ x_n \end{bmatrix} = \begin{bmatrix} a_{11}x_1 + \cdots + a_{1n}x_n \\ \vdots \\ a_{m1}x_1 + \cdots + a_{mn}x_n \end{bmatrix}.$$

The product is in $\mathbb{R}^m$; its j-th coordinate is the inner product of the j-th row of A and $\boldsymbol{x}$. Whenever a product such as $A\boldsymbol{x}$ or $B\boldsymbol{y}$ appears, the capital letter is a matrix and the bold face letter is a column vector.

Example 3 The 3×3 coefficient matrix of the system in Example 2 is

$$\begin{bmatrix} 1 & 2 & 3 \\ 1 & 0 & -1 \\ 0 & 1 & 5 \end{bmatrix}.$$

The system can be written as

$$\begin{bmatrix} 1 & 2 & 3 \\ 1 & 0 & -1 \\ 0 & 1 & 5 \end{bmatrix} \begin{bmatrix} x_1 \\ x_2 \\ x_3 \end{bmatrix} = \begin{bmatrix} 9 \\ 2 \\ 7 \end{bmatrix}.$$

More generally, any linear system can be written–with extraordinary brevity–as

$$A\boldsymbol{x} = \boldsymbol{b}.$$

Notice that this is simply a rewriting of (7), because $x_1\boldsymbol{a}_1 + \cdots + x_n\boldsymbol{a}_n = A\boldsymbol{x}$. We should make a formal record of this.

Proposition 1 *Let A be an $m \times n$ matrix with columns $\boldsymbol{a}_1, \ldots, \boldsymbol{a}_n$. Briefly, we write*

$$A = [\boldsymbol{a}_1\ \boldsymbol{a}_2 \ldots \boldsymbol{a}_n]. \tag{9}$$

Then for $\boldsymbol{x}$ in $\mathbb{R}^n$, we have

$$A\boldsymbol{x} = x_1\boldsymbol{a}_1 + \cdots + x_n\boldsymbol{a}_n.$$

Proof. Both sides of the last equation are vectors in $\mathbb{R}^m$. The i-th entry on the left is

$$a_{i1}x_1 + \cdots + a_{in}x_n.$$

This is also the i-th entry on the right, which proves the proposition.

We will consistently stick with the 'letter notation' in (9) for columns, so that if B is $m \times p$ with columns $\boldsymbol{b}_1, \ldots, \boldsymbol{b}_p$ we write

$$B = [\boldsymbol{b}_1 \ldots \boldsymbol{b}_p].$$

3 Row reduction

We single out two operations on an augmented matrix that 'do not change the solutions': that is, they give a new augmented matrix for a linear system with the same solutions as the original one.

Operation 1 Add c times row j to row k (where $j \neq k$ and c is real).

Operation 2 Multiply row j by c (where $c \neq 0$).

For example, suppose that $n = 1, k = 2, j = 1$. Operation 1 applied to A gives

$$\begin{bmatrix} a_{11} & a_{12} & \cdots & a_{1n} & b_1 \\ a_{21} + ca_{11} & a_{22} + ca_{12} & \cdots & a_{2n} + ca_{1n} & b_2 + cb_1 \end{bmatrix}$$

which is the augmented matrix of the system

$$\begin{aligned} a_{11}x_1 + \cdots + a_{1n}x_n &= b_1 \\ (a_{21} + ca_{11})x_1 + \cdots + (a_{2n} + ca_{1n})x_n &= b_2 + cb_1. \end{aligned}$$

Any solution of the original system is a solution of the new one, and vice versa. Similarly, the equation

$$ca_{11}x_1 + \cdots + ca_{1n}x_n = cb_1$$

is equivalent to the equation

$$a_{11}x_1 + \cdots + a_{1n}x_n = b_1$$

provided $c \neq 0$. So Operation 2 also gives the augmented matrix of a new system with the same solutions as the original one.

Operations 1 and 2 are called **row operations**.

Interchange of two rows k and j can be achieved by doing several row operations. Suppose the rows of A are $\boldsymbol{r}_1, \ldots, \boldsymbol{r}_m$. In an obvious shorthand, we get in succession the pair of rows $(\boldsymbol{r}_k, \boldsymbol{r}_j), (\boldsymbol{r}_k, \boldsymbol{r}_j + \boldsymbol{r}_k), (-\boldsymbol{r}_j, \boldsymbol{r}_j + \boldsymbol{r}_k)$, $(-\boldsymbol{r}_j, \boldsymbol{r}_k), (\boldsymbol{r}_j, \boldsymbol{r}_k)$ by applying Operations 1, 1, 1, and 2.

Example 4 Simplify the linear system

$$\begin{aligned} x_1 + 2x_2 - \ x_3 &= 0 \\ 8x_1 - 5x_2 \qquad &= 13 \\ 4x_1 + 7x_2 - 8x_3 &= 5. \end{aligned}$$

Find the general solution.

Solution. We show application of row operations to the augmented matrix B in shorthand form:

$$B = \begin{bmatrix} 1 & 2 & -1 & 0 \\ 8 & -5 & 0 & 13 \\ 4 & 7 & -8 & 5 \end{bmatrix} \sim \begin{bmatrix} 1 & 2 & -1 & 0 \\ 0 & -21 & 8 & 13 \\ 0 & -1 & -4 & 5 \end{bmatrix} \begin{matrix} \scriptstyle \text{II} - 8\text{I} \\ \scriptstyle \text{III} - 4\text{I} \\ \ \end{matrix}$$

The symbol $\sim$ ('tilde') in $B \sim C$ means 'C is obtained from B by doing one or more row operations'. The Roman numerals I, II, III indicate rows; II $-$ 8I means 'add $-8 \times$ row 1 to row 2'; and one reads from top to bottom to get the order in which the operations are done; here, II $-$ 8I first, then III $-$4I. Simplifying further, and using II $\leftrightarrow$ III to show interchange of rows 2 and 3, and III $\times c$ to show multiplication of row 3 by c,

$$\begin{bmatrix} 1 & 2 & -1 & 0 \\ 0 & -21 & 8 & 13 \\ 0 & -1 & -4 & 5 \end{bmatrix} \sim \begin{bmatrix} 1 & 2 & -1 & 0 \\ 0 & 1 & 4 & -5 \\ 0 & -21 & 8 & 13 \end{bmatrix} \begin{matrix} \scriptstyle \text{II} \leftrightarrow \text{III} \\ \scriptstyle \text{II} \times -1 \\ \ \end{matrix}$$

$$\sim \begin{bmatrix} 1 & 2 & -1 & 0 \\ 0 & 1 & 4 & -5 \\ 0 & 0 & 1 & -1 \end{bmatrix} \begin{matrix} \scriptstyle \text{III} + 21\text{II} \\ \scriptstyle \text{III} \times 1/92 \\ \ \end{matrix}$$

$$\sim \begin{bmatrix} 1 & 0 & 0 & 1 \\ 0 & 1 & 0 & -1 \\ 0 & 0 & 1 & -1 \end{bmatrix} \begin{matrix} \scriptstyle \text{II} - 4\,\text{III} \\ \scriptstyle \text{I} + \text{III} \\ \scriptstyle \text{I} - 2\text{II} \\ \ \end{matrix} = C.$$

The final linear system is $x_1 = 1, x_2 = -1, x_3 = -1$. Accordingly $\boldsymbol{x} = (1, -1, -1)$ is the unique solution to the original system.

The algorithm we are using to simplify our matrix is Gaussian elimination or **row reduction** ; we have **row reduced** B to C. The last but one matrix we got is in **echelon form** and the very last matrix, C, is in **reduced echelon form**. Here are the definitions of these terms. A **leading entry** in a nonzero row, say row i, is the entry $a_{ij} \neq 0$ with j least.

Definition 1 A matrix is in **echelon** form if

1° The nonzero rows are above any zero rows.

2° The leading entry in row i is to the right of the leading entry in row $i-1$ (for all possible i).

Notice that 2° implies that the entries in a column *below* a leading entry are 0.

Definition 2 A matrix is in **reduced echelon form** if

1° The nonzero rows are above any zero rows.

2° The leading entries are 1.

3° The leading entry in row i is to the right of the leading entry in row $i-1$ (for all possible i).

4° The other entries in a column with a leading entry are 0.

Example 5 The following matrices are in echelon form. The stars indicate arbitrary real numbers: this shorthand will occur quite often.

$$\begin{bmatrix} 6 & * & * & * & * \\ 0 & 2 & * & * & * \\ 0 & 0 & 0 & 0 & 7 \end{bmatrix}, \begin{bmatrix} 1 & * \\ 0 & -1 \end{bmatrix}, \begin{bmatrix} 0 & 2 & * & * \\ 0 & 0 & 5 & * \\ 0 & 0 & 0 & 0 \end{bmatrix}, \begin{bmatrix} 0 & -1 \\ 0 & 0 \\ 0 & 0 \\ 0 & 0 \\ 0 & 0 \end{bmatrix}.$$

Example 6 The following matrices are in reduced echelon form.

$$\begin{bmatrix} 1 & 0 & * & * & 0 \\ 0 & 1 & * & * & 0 \\ 0 & 0 & 0 & 0 & 1 \end{bmatrix}, \begin{bmatrix} 1 & 0 \\ 0 & 1 \end{bmatrix}, \begin{bmatrix} 0 & 1 & 0 & 0 \\ 0 & 0 & 1 & 0 \\ 0 & 0 & 0 & 0 \end{bmatrix}, \begin{bmatrix} 0 & 1 \\ 0 & 0 \\ 0 & 0 \\ 0 & 0 \\ 0 & 0 \end{bmatrix}.$$

Let us show that the process 'always works,' in the following sense.

Proposition 2 *Any matrix can be row reduced to reduced echelon form.*

Proof. First we show that a matrix A can be row reduced to **echelon** form. Reading from the left, take the first nonzero column, say $\boldsymbol{a}_i$. (If there is no such i, A is 0 and we have finished already.) Bring a nonzero entry of column i to position a_{1i} (using Operation 1, if necessary). Bring all entries in column i below a_{1i} to zero by using Operation 1 (in the form II $-a_{i2}/a_{i1}$I, and so on).

We have finished with column i. We now choose our next row operation by looking at the matrix B obtained by deleting the first i columns and the first row of A. We bring a nonzero entry to the top of the first nonzero column of B and introduce zeros below it, just as we did for A. Of course the row operations will not change the columns already finished with.

Continue the process of eliminating 'finished columns,' and one row at a time, until we either run out of columns or reach a zero 'submatrix.' At this point it is easy to see that our matrix, C say, is in echelon form.

We now row reduce C to reduced echelon form. Turn the leading entries into 1 by using Operation 2. Introduce zeros in the columns above the leading entries by using Operation 1, starting on the right and working towards the left. The result is a matrix in reduced echelon form.

Example 7 We apply row reduction to the matrix A below.

$$A = \begin{bmatrix} 0 & 0 & 3 & 6 \\ 0 & 5 & 10 & \frac{45}{2} \\ 0 & 2 & 0 & 6 \\ 0 & 4 & 0 & 12 \end{bmatrix} \sim \begin{bmatrix} 0 & 2 & 3 & 12 \\ 0 & 5 & 10 & \frac{45}{2} \\ 0 & 2 & 0 & 6 \\ 0 & 0 & 0 & 0 \end{bmatrix} \begin{matrix} \text{I} + \text{III} \\ \text{IV} - 2\text{III} \\ \\ \\ \end{matrix} \sim \begin{bmatrix} 0 & 2 & 3 & 12 \\ 0 & 0 & \frac{5}{2} & -\frac{15}{2} \\ 0 & 0 & -3 & -6 \\ 0 & 0 & 0 & 0 \end{bmatrix} \begin{matrix} \text{II} - \frac{5}{2}\text{I} \\ \text{III} - \text{I} \\ \\ \\ \end{matrix}$$

$$\sim \begin{bmatrix} 0 & 2 & 3 & 12 \\ 0 & 0 & 1 & 2 \\ 0 & 0 & 0 & -\frac{25}{2} \\ 0 & 0 & 0 & 0 \end{bmatrix} \begin{matrix} \text{II} \leftrightarrow \text{III} \\ \text{II} \times -\frac{1}{3} \\ \text{III} - \frac{5}{2}\text{II} \\ \\ \end{matrix} .$$

At this point we can see that the reduced echelon form is

$$\begin{bmatrix} 0 & 1 & 0 & 0 \\ 0 & 0 & 1 & 0 \\ 0 & 0 & 0 & 1 \\ 0 & 0 & 0 & 0 \end{bmatrix}$$

(this is what results from the last paragraph of the above proof).

Note that we departed from the strict use of the procedure in that proof when we did IV 2 III, because it was an obvious simplifying step.

Definition 3 Let A be a matrix in echelon form. A column of A containing a leading entry is said to be a **pivot column** .

The pivot columns of the reduced echelon form of A are known when we reach *echelon* form. For some problems that we will discuss in Chapter 3, it is enough to know which columns are pivot columns.

Since there are many ways of carrying out a row reduction it is not obvious that **there is only one possible reduced echelon form** for a given matrix A. We will prove this in the next chapter, where it fits conveniently into our discussion of linear independence.

A common student error

It is not the case that

$$\begin{bmatrix} 1 & 1 & 1 \\ 3 & 3 & 3 \end{bmatrix} \sim \begin{bmatrix} 0 & 0 & 0 \\ 0 & 0 & 0 \end{bmatrix} \begin{matrix} \text{I} - \text{III} \times \frac{1}{3}\text{II} \\ \text{I} - 3\text{I} \end{matrix} .$$

The only way to reach this conclusion would be to do the two row operations simultaneously, which is not allowed. I usually receive at least one error of this kind each time I teach linear algebra.

4 General solution of a linear system

Example 8 Find the general solution of the linear system whose augmented matrix is

$$[A|\boldsymbol{b}] = \begin{bmatrix} 1 & -1 & 2 & 3 \\ 5 & -5 & 0 & 1 \\ -3 & 3 & 4 & 2 \end{bmatrix} .$$

Solution.

$$[A|\boldsymbol{b}] \sim \begin{bmatrix} 1 & -1 & 2 & 3 \\ 0 & 0 & -10 & -14 \\ 0 & 0 & 10 & 11 \end{bmatrix} \begin{matrix} \text{II - 5I} \\ \text{III + 3I} \\ \end{matrix}$$

$$\sim \begin{bmatrix} 1 & -1 & 2 & 3 \\ 0 & 0 & -10 & -14 \\ 0 & 0 & 0 & -3 \end{bmatrix} \begin{matrix} \text{III + II} \\ \\ \end{matrix} .$$

The final system includes the impossible equation $0 = -3$. Thus there is **no solution**. A linear system with no solution is said to be **inconsistent**. In general, if the last column of the echelon form of $[A|\boldsymbol{b}]$ is a pivot column, then the system is inconsistent. The opposite of 'inconsistent' is of course '**consistent**.'

Example 9 Find the general solution of the linear system whose augmented matrix is

$$[A|\boldsymbol{b}] = \begin{bmatrix} 1 & 2 & 1 & 5 & 29 \\ 1 & -1 & -3 & -1 & -13 \\ 2 & 1 & 4 & 4 & 34 \\ 1 & 0 & -1 & 1 & 3 \end{bmatrix}.$$

Solution.

$$[A|\boldsymbol{b}] \sim \begin{bmatrix} 1 & 2 & 1 & 5 & 29 \\ 0 & -3 & -4 & -6 & -42 \\ 0 & -3 & 2 & -6 & -24 \\ 0 & -2 & -2 & -4 & -26 \end{bmatrix} \begin{matrix} \text{II} - \text{I} \\ \text{III} - 2\text{I} \\ \text{IV} - \text{I} \\ \end{matrix} \sim \begin{bmatrix} 1 & 2 & 1 & 5 & 29 \\ 0 & 1 & 1 & 2 & 13 \\ 0 & 0 & 5 & 0 & 15 \\ 0 & 0 & -1 & 0 & -3 \end{bmatrix} \begin{matrix} \text{II} \leftrightarrow \text{IV} \\ \text{II} \times -\frac{1}{2} \\ \text{III} + 3\text{II} \\ \text{IV} + 3\text{II} \end{matrix}$$

$$\sim \begin{bmatrix} 1 & 2 & 1 & 5 & 29 \\ 0 & 1 & 1 & 2 & 13 \\ 0 & 0 & 1 & 0 & 3 \\ 0 & 0 & 0 & 0 & 0 \end{bmatrix} \begin{matrix} \text{III} \times \frac{1}{5} \\ \text{IV} + \text{III} \\ \\ \end{matrix} \sim \begin{bmatrix} 1 & 2 & 0 & 5 & 26 \\ 0 & 1 & 0 & 2 & 10 \\ 0 & 0 & 1 & 0 & 3 \\ 0 & 0 & 0 & 0 & 0 \end{bmatrix} \begin{matrix} \text{II} - \text{III} \\ \text{I} - \text{III} \\ \\ \end{matrix}$$

$$\sim \begin{bmatrix} 1 & 0 & 0 & 1 & 6 \\ 0 & 1 & 0 & 2 & 10 \\ 0 & 0 & 1 & 0 & 3 \\ 0 & 0 & 0 & 0 & 0 \end{bmatrix} \begin{matrix} \text{I - 2II} \\ \\ \\ \end{matrix}.$$

We can choose x_4 arbitrarily to satisfy the final system, which then reads $x_1 = 6 - x_4, x_2 = 10 - 2x_4, x_3 = 3$, or

$$\boldsymbol{x} = (6 - x_4, 10 - 2x_4, 3, x_4) = (6, 10, 3, 0) + x_4(-1, -2, 0, 1).$$

In the terminology of the following definition, x_4 is a 'free variable'.

Definition 4 Let A be the coefficient matrix of a linear system and suppose column j of A is not a pivot column. We say that the variable x_j is a **free variable** . The variables that are not free are said to be **basic variables** .

In Example 9 the basic variables are x_1, x_2, x_3.

We can now state the rule for the general solution of a linear system $\boldsymbol{Ax} = \boldsymbol{b}$.

1. Row reduce $[A|\boldsymbol{b}]$ to reduced echelon form $[B|\boldsymbol{c}]$.

2. If the last column $\boldsymbol{c}$ is a pivot column, the system is inconsistent.

3. Suppose $\boldsymbol{c}$ is not a pivot column. We rewrite the non-zero equations of the final system in the form

$$x_j = c_j + \ell_j \quad \text{(for every basic variable } x_j) \tag{10}$$

where c_j is the j-th entry of $\boldsymbol{c}$ and ℓ_j is a linear form in the free variables. The formula (10), with the free variables chosen arbitrarily, is the general solution. If there are *no* free variables, then $\boldsymbol{x} = \boldsymbol{c}$ is the unique solution.

This formula sounds rather complicated but is very easy to use in a concrete example like Example 9. We can summarize by saying that the basic variables are determined by the free variables.

Example 10 The augmented matrix $[A|\boldsymbol{b}]$ below is already in reduced echelon form. Write down the general solution of the linear system $\boldsymbol{Ax} = \boldsymbol{b}$.

$$[A|\boldsymbol{b}] = \begin{bmatrix} 1 & 0 & 5 & -2 & 0 & 3 \\ 0 & 1 & -1 & -3 & 0 & 4 \\ 0 & 0 & 0 & 0 & 1 & 8 \end{bmatrix}.$$

Solution. $\boldsymbol{x} = (3 - 5x_3 + 2x_4, 4 + x_3 + 3x_4, x_3, x_4, 8)$ with x_3, x_4 free. We could write this in the form

$$\boldsymbol{x} = (3, 4, 0, 0, 8) + x_3(-5, 1, 1, 0, 0) + x_4(2, 3, 0, 1, 0).$$

It is worth paying special attention to square coefficient matrices.

Proposition 3 *Let A be a square matrix. If A can be row reduced to*

$$I = \begin{bmatrix} 1 & 0 & 0 & \cdots & 0 \\ 0 & 1 & 0 & & 0 \\ 0 & 0 & 1 & & 0 \\ \vdots & & \vdots & & \vdots \\ 0 & & 0 & & 1 \end{bmatrix}$$

(the matrix with 1's on the main diagonal, and 0's elsewhere), then every system $\boldsymbol{Ax} = \boldsymbol{b}$ has a unique solution.

The **main diagonal** is the set of entries $a_{11}, a_{22}, \ldots, a_{nn}$.

This is not really in need of a proof. The row reduced form of $[A|\boldsymbol{b}]$ will be $[I|\boldsymbol{c}]$ and the final system is $x_1 = c_1, \ldots, x_n = c_n$. Notice that if all columns of a square matrix A are pivot columns, then $A \sim I$.

5 Homogeneous systems

A linear system in which the right-hand sides of the equations are all 0 is said to be **homogeneous** . The system can be written as

$$A\boldsymbol{x} = \boldsymbol{0}.$$

A homogeneous system has the zero solution, $\boldsymbol{x} = \boldsymbol{0}$, so it is of course consistent.

Example 11 Show that the unique solution of

$$\begin{aligned} x_1 + 5x_2 &= 0 \\ x_1 + 2x_2 + x_3 &= 0 \\ 2x_1 - x_2 + 2x_3 &= 0 \end{aligned}$$

is the zero solution.

Solution. The coefficient matrix A satisfies

$$A \sim \begin{bmatrix} 1 & 5 & 0 \\ 0 & -3 & 1 \\ 0 & -11 & 2 \end{bmatrix} \begin{matrix} \text{II} - \text{I} \\ \text{III} - 2\text{I} \end{matrix} \sim \begin{bmatrix} 1 & 5 & 0 \\ 0 & 1 & -\frac{1}{3} \\ 0 & 0 & -\frac{5}{3} \end{bmatrix} \begin{matrix} \text{II} \times -\frac{1}{3} \\ \text{III} + 11\ \text{II} \end{matrix}.$$

All the columns are pivot columns, so $A \sim I$ and the result follows from Proposition 3.

One result about homogeneous systems turns up quite often in practice.

Proposition 4 *A homogeneous system of m equations in n unknowns, with $n > m$, has a nonzero solution.*

Proof. The number of pivot columns cannot exceed m because each requires its own row for the leading entry. There are at least $n - m$ free variables, and if we choose a nonzero value for one of these, we get our nonzero solution.

There is a simple relationship between the solution of a given inhomogeneous system

$$A\boldsymbol{x} = \boldsymbol{b}$$

and the homogeneous system with the same coefficient matrix,

$$A\boldsymbol{x} = \boldsymbol{0}.$$

We can illustrate this by looking back at Example 10. The general solution of $A\boldsymbol{x} = \boldsymbol{b}$ is

$$(3,4,0,0,8) + x_3(-5,1,1,0,0) + x_4(2,3,0,1,0) \quad (x_3, x_4 \text{ free}).$$

The general solution of $A\boldsymbol{x} = \boldsymbol{0}$ is

$$x_3(-5,1,1,0,0) + x_4(2,3,0,1,0).$$

In other words, for the general solution of $A\boldsymbol{x} = \boldsymbol{b}$ we take the general solution of $A\boldsymbol{x} = \boldsymbol{0}$ and add a particular solution of $A\boldsymbol{x} = \boldsymbol{b}$.

Proposition 5 *Let $A\boldsymbol{x} = \boldsymbol{b}$ be a linear system, which has a solution $\boldsymbol{x} = \boldsymbol{d}$. The general solution is $\boldsymbol{d} + \boldsymbol{y}$, where $\boldsymbol{y}$ is the general solution of the system*

$$A\boldsymbol{x} = \boldsymbol{0}.$$

Proof. Since $A\boldsymbol{d} = \boldsymbol{b}$ and

$$A(\boldsymbol{y} + \boldsymbol{d}) = A\boldsymbol{y} + A\boldsymbol{d} = A\boldsymbol{y} + \boldsymbol{b},$$

the equation $A\boldsymbol{y} = \boldsymbol{0}$ is equivalent to the equation $A(\boldsymbol{y}+\boldsymbol{d}) = \boldsymbol{b}$. This is just a rephrasing of the result that we are to prove.

6 Electric circuits

Figure 1 provides information about an electric circuit. Voltage sources are shown like this: ||. The short line indicates the negative terminal. A voltage source drives electric current away from the positive terminal (the long line). The **resistance** of a piece of a circuit is shown next to a symbol ⩘⩘. The notation $h\Omega$ indicates a resistance of h ohms, which means that a voltage source of k volts would send k/h amperes (amps) of current through that piece of circuit considered by itself. The I_j are the currents in the loop inside which they are written. A positive I_j indicates current flowing anticlockwise; a negative I_j thus means a clockwise current. Some loops share a section of circuit, and we interpret this as follows in Figure 1, for example. The current $I_1 - I_2$ flows from left to right in the section common to I_1 and I_2. The current $I_4 - I_3$ flows downwards in the section common to I_3 and I_4. It was determined by Kirchoff that in a given loop, the **algebraic sum of 'resistance × current' products equals the algebraic sum of voltage sources**. The term 'algebraic sum' means that we use the sign convention above for currents; similarly, we consider voltage sources that drive current anticlockwise as positive.

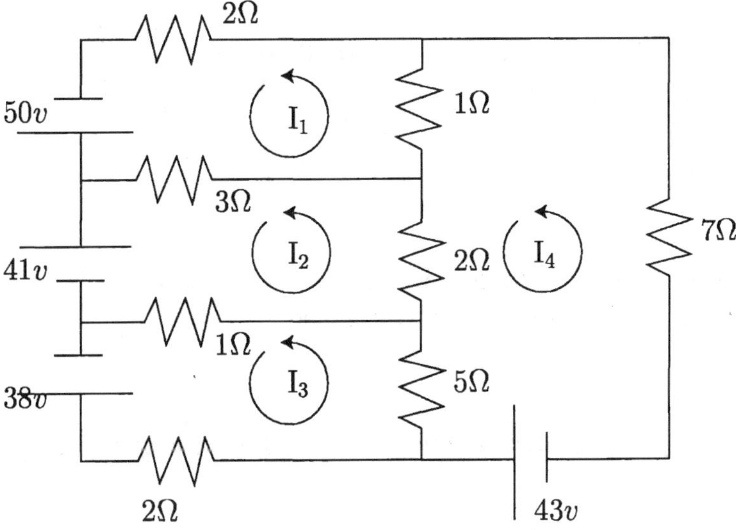

Figure 1. The circuit for Example 12.

Example 12 Consider the circuit in Figure 1. With a little thought we reach the following linear system for $\boldsymbol{I} = (I_1, I_2, I_3, I_4)$.

$$\begin{aligned}
6I_1 - 3I_2 \qquad\quad - \quad I_4 &= 50 \\
-3I_1 + 6I_2 - \ I_3 - \ 2I_4 &= -41 \\
-I_2 + 8I_3 - \ 5I_4 &= 38 \\
-I_1 - 2I_2 - 5I_3 + 15I_4 &= -43.
\end{aligned}$$

The augmented matrix is

$$\begin{bmatrix} 6 & -3 & 0 & -1 & 50 \\ -3 & 6 & -1 & -2 & -41 \\ 0 & -1 & 8 & -5 & 38 \\ -1 & -2 & -5 & 15 & -43 \end{bmatrix} \sim \begin{bmatrix} 1 & 2 & 5 & -15 & 43 \\ 0 & 12 & 14 & -47 & 88 \\ 0 & 1 & -8 & 5 & -38 \\ 0 & -15 & -30 & 89 & -208 \end{bmatrix} \begin{matrix} \text{I} \leftrightarrow \text{IV} \\ \text{I} \times -1 \\ \text{II} + 3\text{I} \\ \text{IV} - 6\text{I} \\ \text{III} \times -1 \end{matrix}$$

$$\sim \begin{bmatrix} 1 & 2 & 5 & -15 & 43 \\ 0 & 1 & -8 & 5 & -38 \\ 0 & 0 & 110 & -107 & 544 \\ 0 & 0 & -150 & 164 & -778 \end{bmatrix} \begin{matrix} \text{II} \leftrightarrow \text{III} \\ \text{III} - 12\text{II} \\ \text{IV} + 15\text{II} \end{matrix} \sim \begin{bmatrix} 1 & 2 & 5 & -15 & 43 \\ 0 & 1 & -8 & 5 & -38 \\ 0 & 0 & 110 & -107 & 544 \\ 0 & 0 & 0 & 1 & -2 \end{bmatrix} \begin{matrix} \text{IV} + \frac{15}{11}\text{III} \\ \text{IV} \times \frac{11}{199} \end{matrix}$$

$$\sim \begin{bmatrix} 1 & 2 & 5 & 0 & 13 \\ 0 & 1 & -8 & 0 & -28 \\ 0 & 0 & 110 & 0 & 330 \\ 0 & 0 & 0 & 1 & -2 \end{bmatrix} \begin{matrix} \text{I} + 15\text{IV} \\ \text{II} - 5\text{IV} \\ \text{III} + 107\text{IV} \end{matrix} \sim \begin{bmatrix} 1 & 0 & 0 & 0 & 6 \\ 0 & 1 & 0 & 0 & -4 \\ 0 & 0 & 1 & 0 & 3 \\ 0 & 0 & 0 & 1 & -2 \end{bmatrix} \begin{matrix} \text{III} \times \frac{1}{110} \\ \text{II} + 8\text{III} \\ \text{I} - 5\text{III} \\ \text{I} - 2\text{II} \end{matrix}$$

so that $\boldsymbol{I} = (6, -4, 3, -2)$. Our physical intuition certainly leads us to expect a unique solution, and this has been confirmed.

Example 13 The augmented matrix associated with Figure 2 is

$$\begin{bmatrix} 9 & -4 & -2 & 70 \\ -4 & 5 & -1 & -26 \\ -2 & -1 & 10 & -27 \end{bmatrix} \sim \begin{bmatrix} 1 & \frac{1}{2} & -5 & \frac{27}{2} \\ 0 & 7 & -21 & 28 \\ 0 & -\frac{17}{2} & 43 & -\frac{103}{2} \end{bmatrix} \begin{matrix} \text{I} \leftrightarrow \text{III} \\ \text{I} \times -\frac{1}{2} \\ \text{II} + 4\text{I} \\ \text{III} - 9\text{I} \end{matrix}$$

$$\sim \begin{bmatrix} 1 & \frac{1}{2} & -5 & \frac{27}{2} \\ 0 & 1 & -3 & 4 \\ 0 & 0 & \frac{35}{2} & -\frac{35}{2} \end{bmatrix} \begin{matrix} \text{II} \times \frac{1}{7} \\ \text{III} + \frac{17}{2}\text{II} \end{matrix} \sim \begin{bmatrix} 1 & 0 & 0 & 8 \\ 0 & 1 & 0 & 1 \\ 0 & 0 & 1 & -1 \end{bmatrix} \begin{matrix} \text{III} \times \frac{2}{35} \\ \text{II} + 3\text{III} \\ \text{I} + 5\text{III} \\ \text{I} - \frac{1}{2}\text{II} \end{matrix}$$

so that $\boldsymbol{I} = (8, 1, -1)$.

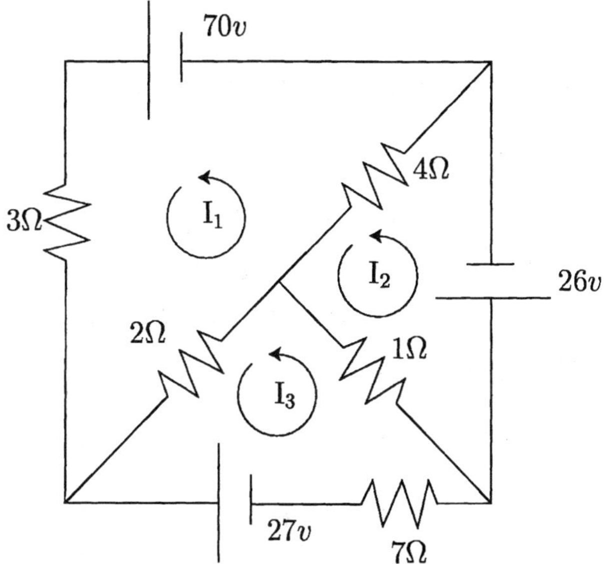

Figure 2. The circuit for Example 13.

Exercises for Chapter 2

Reminder: do Exercises a.b after reading Section a.

1.1 Write the vector equation

$$x_1 \begin{bmatrix} -2 \\ 0 \\ 4 \end{bmatrix} + x_2 \begin{bmatrix} 3 \\ 3 \\ 5 \end{bmatrix} + x_3 \begin{bmatrix} 6 \\ 0 \\ -1 \end{bmatrix} = \begin{bmatrix} -1 \\ 2 \\ 1 \end{bmatrix}$$

as a system of linear equations.

1.2 Write the system of equations

$$\begin{aligned} -x_1 - x_2 + x_3 + 2x_5 &= 0 \\ x_1 + 2x_2 + 4x_3 + x_4 + x_5 &= 0 \end{aligned}$$

as a vector equation.

In Exercises 2.1–2.3, find the augmented matrix for the system of equations.

2.1
$$\begin{aligned} x_1 - 2x_2 &= 3 \\ x_1 + 2x_2 &= 4 \end{aligned}$$

2.2
$$\begin{aligned} x_1 - x_2 - x_3 &= 7 \\ 2x_1 - x_2 + x_3 &= 4 \\ 3x_1 - x_2 &= -1 \end{aligned}$$

2.3
$$\begin{aligned} x_1 - x_2 &= 4 \\ x_2 - x_3 &= 1 \\ 2x_1 - x_2 - x_3 - x_4 &= 0 \end{aligned}$$

In Exercises 3.1–3.6, determine which matrices are in reduced echelon form.

3.1 $\begin{bmatrix} 1 & 1 \\ 0 & 0 \end{bmatrix}$

3.2 $\begin{bmatrix} 0 & 1 \\ 0 & 0 \end{bmatrix}$

3.3 $\begin{bmatrix} 0 & 3 & 0 \\ 0 & 0 & 1 \end{bmatrix}$

3.4 $\begin{bmatrix} 1 & 1 & 0 & 0 & 0 \\ 0 & 2 & 3 & 0 & 0 \\ 0 & 0 & 0 & 1 & 0 \end{bmatrix}$

3.5 $\begin{bmatrix} 1 & 0 & 0 & 5 & 0 & 4 & 0 \\ 0 & 0 & 1 & 0 & -1 & 3 & 0 \\ 0 & 0 & 0 & 0 & 0 & 0 & 1 \\ 0 & 0 & 0 & 0 & 0 & 0 & 0 \end{bmatrix}$

3.6 $\begin{bmatrix} 1 & 0 \\ 0 & 1 \\ 0 & 0 \\ 0 & 0 \end{bmatrix}$

3.7 Let A be a matrix whose rows are $\boldsymbol{a}, \boldsymbol{b}, \boldsymbol{c}, \boldsymbol{d}$. Here $\boldsymbol{a} = [a_1 \cdots a_n], \ldots,$ $\boldsymbol{d} = [d_1 \cdots d_n]$. So

$$A = \begin{bmatrix} \boldsymbol{a} \\ \boldsymbol{b} \\ \boldsymbol{c} \\ \boldsymbol{d} \end{bmatrix}.$$

Let $B - \begin{bmatrix} \boldsymbol{d} \\ \boldsymbol{a} \\ \boldsymbol{b} \\ \boldsymbol{c} \end{bmatrix}$. Is it true that A and B have the same row reduced echelon form?

In Exercises 4.1–4.8, the matrices are augmented matrices for systems of linear equations. In each case, find the reduced echelon form and solve the linear system.

4.1 $\begin{bmatrix} 2 & 4 & 6 & 0 \\ -1 & 2 & 2 & -4 \\ 2 & 8 & 17 & -4 \end{bmatrix}$

4.2 $\begin{bmatrix} 1 & 2 & 5 & 1 \\ 3 & 1 & 0 & 7 \\ 1 & 7 & 20 & -3 \end{bmatrix}$

4.3 $\begin{bmatrix} -1 & 2 & 2 & 4 & 5 \\ 2 & 6 & 0 & 9 & 1 \\ 7 & -4 & -10 & -11 & -24 \end{bmatrix}$

4.4 $\begin{bmatrix} 1 & 2 & 2 & 2 & 11 \\ 5 & 13 & 12 & 15 & 76 \\ -4 & -14 & 5 & 0 & 19 \\ 2 & 4 & 4 & 5 & 25 \end{bmatrix}$

4.5 $\begin{bmatrix} -1 & -3 & 5 & 7 & 2 \\ 1 & 4 & 6 & 0 & 3 \end{bmatrix}$

4.6 $\begin{bmatrix} 1 & -1 & -1 & 0 \\ 1 & 1 & -1 & 0 \\ -1 & 1 & 1 & 0 \end{bmatrix}$

4.7 $\begin{bmatrix} 1 & 4 & 9 \\ 2 & 5 & 7 \\ 1 & 7 & 16 \end{bmatrix}$

4.8 $\begin{bmatrix} 1 & 1 & 1 & 1 & 1 \\ -1 & -2 & -2 & 1 & 0 \\ 3 & 5 & 5 & -1 & 0 \\ 0 & 1 & 1 & -2 & -1 \end{bmatrix}$

In Exercises 4.9–4.10 determine for what values of b the system with augmented matrix B is inconsistent.

4.9 $B = \begin{bmatrix} 5 & 10 & 25+b & 13 \\ 2 & 3 & 10-b & 2 \\ 4 & 8 & 20+b & 11 \end{bmatrix}$

4.10 $B = \begin{bmatrix} 1 & 1 & 2 & 2 \\ 3 & -5 & 1 & 7 \\ 11 & -13 & 82 & b \end{bmatrix}$

4.11 Let A be a 3×3 matrix whose columns are $\boldsymbol{u}, \boldsymbol{v}, \boldsymbol{w}$.

$$A = [\boldsymbol{u}\ \boldsymbol{v}\ \boldsymbol{w}].$$

Let $B = [\boldsymbol{u}\ \boldsymbol{w}\ \boldsymbol{v}]$. If the system $A\boldsymbol{x} = \boldsymbol{b}$ has a unique solution, is it true that the system $B\boldsymbol{x} = \boldsymbol{b}$ has a unique solution?

4.12 Let $\boldsymbol{x}_1$ and $\boldsymbol{x}_2$ be two solutions of the linear system

$$A\boldsymbol{x} = \boldsymbol{b}.$$

Show that every point on the straight line through $\boldsymbol{x}_1$ and $\boldsymbol{x}_2$ is a solution.

4.13 Every circle in the plane has an equation of the form

$$x_1^2 + x_2^2 + ax_1 + bx_2 + c = 0.$$

Find the equation of the circle that passes through (2, 1), (3, 3), (4, 6).

Hint: Set up a linear system whose solution is (a, b, c).

4.14 Find a hyperplane in $\mathbb{R}^4$ that passes through the four points $(1, -1, 0, 7)$, $(0, 1, 2, -5)$, $(2, 1, 2, -9)$, $(3, 0, 0, 3)$. Is it the only such hyperplane?

4.15 A sphere (spherical surface) in $\mathbb{R}^3$ can be written in the form

$$x_1^2 + x_2^2 + x_3^2 + ax_1 + bx_2 + cx_3 = d.$$

Find a sphere that passes through the three points $(1, 0, 3), (1, 2, 5)$, $(-1, 2, 3)$. Is it the only such sphere?

5.1 Suppose that x_1, x_2, x_3 are real numbers and each of x_1, x_2 is the average of the other two numbers. Show that $x_1 = x_2 = x_3$.

5.2 There is an analog of the result of question 5.1 for sets of n real numbers $x_1, \dots, x_n$. State it and give a proof. (You might want to try $n = 4, n = 5$ before the general case.)

Questions 6.1, 6.2; find the currents in each loop of the circuit. All currents are an integer number of amps.

6.1

3Ω
39v
I_1
4Ω
I_2
38v
19v
I_3
2Ω
1v
I_4
1Ω
4Ω

6.2

3v
1Ω
2Ω
6v
1Ω
I_1
I_4
I_2
2Ω
2Ω
I_3
9v
1Ω
1Ω
2Ω
3v

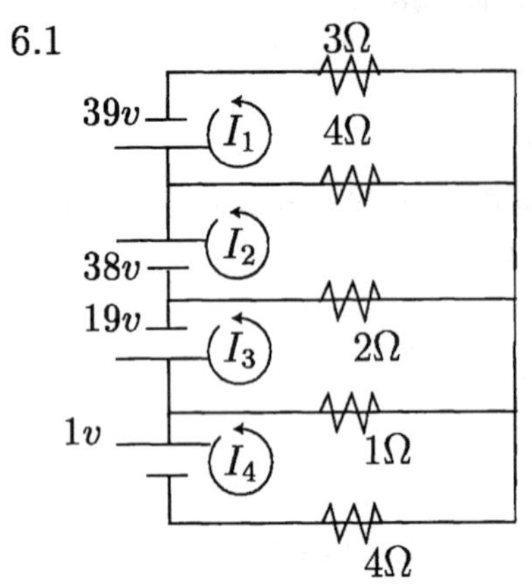

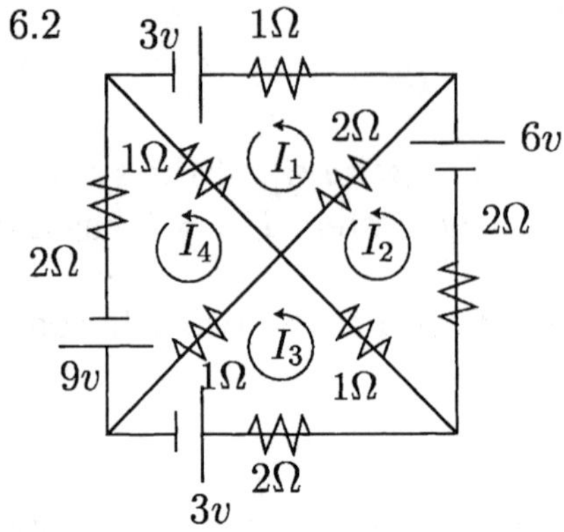

Chapter 3

Linear independence. Subspaces

1 Σ notation

The sign Σ (capital Greek sigma) is a useful shorthand. Given numbers $a_1, \dots, a_n$, we have written their sum up to now as $a_1 + \cdots + a_n$. The Σ notation is

$$a_1 + \cdots + a_n = \sum_{i=1}^{n} a_i.$$

Similarly, if $\boldsymbol{v}_1, \dots, \boldsymbol{v}_r$ are vectors in $\mathbb{R}^m$, we write

$$\boldsymbol{v}_1 + \cdots + \boldsymbol{v}_r = \sum_{j=1}^{r} \boldsymbol{v}_j.$$

The symbols i, j here are in effect integer variables and could be changed at will. For instance,

$$\sum_{j=1}^{r} \boldsymbol{v}_j = \sum_{k=1}^{r} \boldsymbol{v}_k.$$

The Σ notation is particularly useful when there is more than one integer variable to consider. For instance, the i-th row of the matrix $A = [a_{ij}]_{1 \le i \le m, 1 \le j \le n}$ is $(a_{i1}, a_{i2}, \dots, a_{in})$. The sum of the entries on the i-th row is

$$\sum_{j=1}^{n} a_{ij}.$$

Notice that instead of j we could use k, or ℓ; but not i, which has 'already been used' in this context. The sum of the entries on column j is

$$\sum_{k=1}^{m} a_{kj}.$$

Example 1 If A is an $n \times n$ matrix, we recall that the diagonal consisting of the entries on row i, column i is called the **main diagonal**. The sum of these entries is called the **trace** of A, written $\mathrm{Tr}A$:

$$\mathrm{Tr}A = \sum_{i=1}^{n} a_{ii}.$$

Interchange of summation is the equation

$$\sum_{i=1}^{p}\left(\sum_{j=1}^{q} \boldsymbol{v}_{ij}\right) = \sum_{j=1}^{q}\left(\sum_{i=1}^{p} \boldsymbol{v}_{ij}\right) \tag{1}$$

for any vectors $\boldsymbol{v}_{ij} (i = 1, \ldots, p, j = 1, \ldots, q)$. On both sides we have the sum of all pq vectors $\boldsymbol{v}_{11}, \ldots, \boldsymbol{v}_{pq}$. Naturally a similar result is true with real numbers in place of vectors.

Try writing out (1) in full for $p = 2, q = 3$ to see how the two sides are rearrangements of each other.

Example 2 Show that

$$\sum_{i=1}^{n} c(a_i + b_i) = c\sum_{i=1}^{n} a_i + c\sum_{i=1}^{n} b_i.$$

Proof. The left side is the sum of $ca_1 + cb_1, ca_2 + cb_2, \ldots, ca_n + cb_n$, so it is the sum of all $2n$ numbers $ca_1, ca_2, ca_3, \ldots, ca_n, cb_1, cb_2, \ldots, cb_n$. The right side is also seen to be the sum of $ca_1, ca_2, \ldots, ca_n, cb_1, \ldots, cb_n$.

2 Linear span and linear independence

The definitions in this chapter will be used constantly throughout the rest of the book.

Definition 1 Let $\boldsymbol{v}_1, \dots, \boldsymbol{v}_k$ be vectors in $\mathbb{R}^n$. A **linear combination** of $\boldsymbol{v}_1, \dots, \boldsymbol{v}_k$ is a vector of the form

$$a_1\boldsymbol{v}_1 + \cdots + a_k\boldsymbol{v}_k,$$

where the a_i are real numbers.

The **linear span** of $\boldsymbol{v}_1, \dots, \boldsymbol{v}_k$ is the set of all possible linear combinations of $\boldsymbol{v}_1, \dots, \boldsymbol{v}_k$. We write this set $\text{Span}\{\boldsymbol{v}_1, \dots, \boldsymbol{v}_k\}$.

For the sake of brevity, I often drop the word linear and refer to a **combination** of $\boldsymbol{v}_1, \dots, \boldsymbol{v}_k$ and the **span** of $\boldsymbol{v}_1, \dots, \boldsymbol{v}_k$.

Example 3 Let $\boldsymbol{v}_1 = (1, 2, 3), \boldsymbol{v}_2 = (0, 1, -2)$. Then

$$6\boldsymbol{v}_1 - 3\boldsymbol{v}_2 = (6, 12, 18) - (0, 3, -6) = (6, 9, 24)$$

is a **combination** of $\boldsymbol{v}_1$ and $\boldsymbol{v}_2$.

The **span** of $\boldsymbol{v}_1$ and $\boldsymbol{v}_2$ is the set of all $a_1\boldsymbol{v}_1 + a_2\boldsymbol{v}_2$ where a_1 and a_2 are real. We saw in Chapter 1 (since $\boldsymbol{v}_2$ is not proportional to $\boldsymbol{v}_1$) that this is, in fact, a plane through $\mathbf{0}$. To find its equation, let $\boldsymbol{w} = (w_1, w_2, w_3)$ be a normal to $\text{Span}\{\boldsymbol{v}_1, \boldsymbol{v}_2\}$. Then

$$\begin{aligned} w_1 + 2w_2 + 3w_3 &= 0 \\ w_2 - 2w_3 &= 0. \end{aligned}$$

Take $w_3 - 1$, then $w_2 = 2$ and $w_1 = (-2)2 - 3 = -7$. We conclude that the equation of $\text{Span}\{\boldsymbol{v}_1, \boldsymbol{v}_2\}$ is

$$-7w_1 + 2w_2 + w_3 = 0.$$

It is not easy to visualize the linear span in $\mathbb{R}^4$ of $\boldsymbol{v}_1 = (3, 7, 1, -2)$, $\boldsymbol{v}_2 = (2, 0, 1, 4), \boldsymbol{v}_3 = (7, 0, 1, 6)$ (for instance). Actually this linear span will turn out to a hyperplane; see the end of Section 5. The general point of $\text{Span}\{\boldsymbol{v}_1, \boldsymbol{v}_2, \boldsymbol{v}_3\}$ is

$$a_1\boldsymbol{v}_1 + a_2\boldsymbol{v}_2 + a_3\boldsymbol{v}_3 = (3a_1 + 2a_2 + 7a_3, 7a_1, a_1 + a_2 + a_3, -2a_1 + 4a_2 + 6a_3).$$

Clearly, any linear span $\text{Span}\{\boldsymbol{v}_1, \dots, \boldsymbol{v}_k\}$ contains $\mathbf{0}$, because we have the option of choosing all a_i to be 0 in

$$a_1\boldsymbol{v}_1 + \cdots + a_k\boldsymbol{v}_k.$$

Example 4 For any $\boldsymbol{v}_1$, Span$\{\boldsymbol{v}_1\}$ is the line through $\mathbf{0}$ and $\boldsymbol{v}_1$.

In fact Span$\{\boldsymbol{v}_1\}$ is the set of all $a_1\boldsymbol{v}_1$ (a_1 real), which fits the definition of a line (Chapter 1). As a specific example, let $n = 2$ and $\boldsymbol{v}_1 = (2, 7)$. Then Span$\{\boldsymbol{v}_1\}$ is the line with equation

$$2x_2 - 7x_1 = 0.$$

Example 5 $\boldsymbol{v}_1, \boldsymbol{v}_2, \boldsymbol{v}_3$ are nonzero vectors in $\mathbb{R}^3$; $\boldsymbol{v}_2$ is not a multiple of $\boldsymbol{v}_1$; $\boldsymbol{v}_3$ is not in the plane Span$\{\boldsymbol{v}_1, \boldsymbol{v}_2\}$. Then

$$\text{Span}\{\boldsymbol{v}_1, \boldsymbol{v}_2, \boldsymbol{v}_3\} = \mathbb{R}^3. \tag{2}$$

The simplest way to see this is to take any point $\boldsymbol{x}$ in $\mathbb{R}^3$ and subtract a multiple $a_3\boldsymbol{v}_3$, chosen so that $\boldsymbol{x} - a_3\boldsymbol{v}_3$ lies in the plane Span$\{\boldsymbol{v}_1, \boldsymbol{v}_2\}$ (Figure 1). This is possible because $\boldsymbol{v}_3$ is not parallel to the plane. Now

$$\boldsymbol{x} - a_3\boldsymbol{v}_3 \text{ is in Span}\{\boldsymbol{v}_1, \boldsymbol{v}_2\};$$
$$\boldsymbol{x} - a_3\boldsymbol{v}_3 = a_1\boldsymbol{v}_1 + a_2\boldsymbol{v}_2.$$

Shifting $a_3\boldsymbol{v}_3$ to the other side in the last equation, $\boldsymbol{x}$ is in Span$\{\boldsymbol{v}_1, \boldsymbol{v}_2, \boldsymbol{v}_3\}$. Since Span$\{\boldsymbol{v}_1, \boldsymbol{v}_2, \boldsymbol{v}_3\}$ includes any given $\boldsymbol{x}$ in $\mathbb{R}^3$, we must have (2).

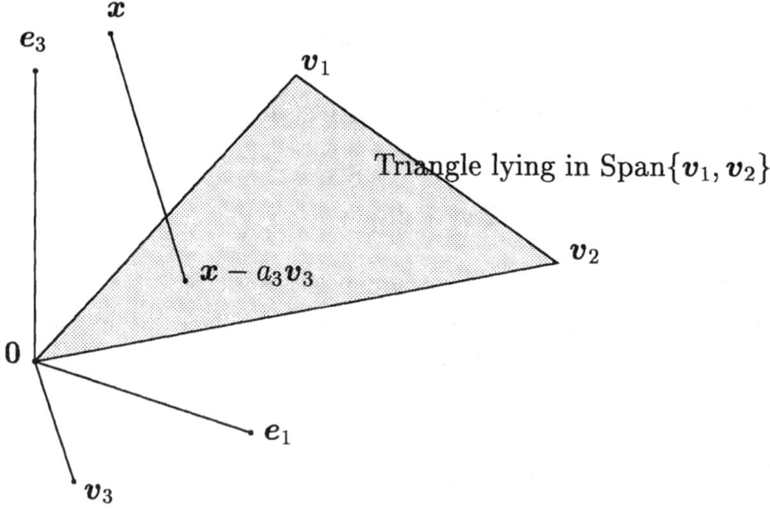

Figure 1. Span$\{\boldsymbol{v}_1, \boldsymbol{v}_2, \boldsymbol{v}_3\}$ is $\mathbb{R}^3$ if Span$\{\boldsymbol{v}_1, \boldsymbol{v}_2\}$ is a plane that misses $\boldsymbol{v}_3$.

Definition 2 Let $\boldsymbol{v}_1, \ldots, \boldsymbol{v}_k$ be vectors in $\mathbb{R}^n$. We say that $\boldsymbol{v}_1, \ldots, \boldsymbol{v}_k$ is a **linearly independent** set if there is no nonzero solution of the vector equation

$$x_1\boldsymbol{v}_1 + x_2\boldsymbol{v}_2 + \cdots + x_k\boldsymbol{v}_k = \mathbf{0}. \tag{3}$$

A set that is not linearly independent is said to be **linearly dependent.** We also say for brevity, '$\boldsymbol{v}_1, \ldots, \boldsymbol{v}_k$ are dependent' (or independent, as the case may be).

Example 6 The set $\boldsymbol{v}_1 = \mathbf{0}, \boldsymbol{v}_2 = (1, 2)$ is *dependent.* To see this, choose $x_1 = 1, x_2 = 0$,

$$x_1\boldsymbol{v}_1 + x_2\boldsymbol{v}_2 = 1(0,0) + 0(1,2) = \mathbf{0}.$$

We have a nonzero solution of the vector equation

$$x_1\boldsymbol{v}_1 + x_2\boldsymbol{v}_2 = \mathbf{0}.$$

We can use the same idea to show that any set in $\mathbb{R}^n$ containing the vector $\mathbf{0}$ is dependent. Say $\boldsymbol{v}_1 = \mathbf{0}, \boldsymbol{v}_2, \ldots, \boldsymbol{v}_k$ is the given set. Then

$$1\boldsymbol{v}_1 + 0\boldsymbol{v}_2 + 0\boldsymbol{v}_3 + \cdots + 0\boldsymbol{v}_k = \mathbf{0}.$$

We have a nonzero solution of (3). This may seem a trite observation, but it is useful.

Example 7 The set $\boldsymbol{v}_1 = (1, 4, 1), \boldsymbol{v}_2 = (1, 5, 2)$ is independent. To see this, if x_1, x_2 satisfy

$$x_1\boldsymbol{v}_1 + x_2\boldsymbol{v}_2 = \mathbf{0},$$

and one of x_1 or x_2 is *not* zero (say, x_2) then

$$\boldsymbol{v}_2 = -\frac{x_1}{x_2}\,\boldsymbol{v}_1$$

and $\boldsymbol{v}_2$ is a multiple of $\boldsymbol{v}_1$. This is absurd, so $\boldsymbol{v}_1, \boldsymbol{v}_2$ is an independent set.

We can use the above argument verbatim to show that a set of two nonzero vectors $\boldsymbol{v}_1, \boldsymbol{v}_2$ in $\mathbb{R}^n$, $\boldsymbol{v}_2$ not a multiple of $\boldsymbol{v}_1$, is a linearly independent set. (Of course, $\boldsymbol{v}_1$ is not a multiple of $\boldsymbol{v}_2$ either.) We describe $\boldsymbol{v}_1, \boldsymbol{v}_2$ as a pair of **non-proportional** vectors.

A **proportional** pair (say $\boldsymbol{v}_1, \boldsymbol{v}_2$ with $\boldsymbol{v}_2 = c\boldsymbol{v}_1$) is dependent since

$$c\boldsymbol{v}_1 - 1\boldsymbol{v}_2 = \mathbf{0}.$$

Example 8 Let $\boldsymbol{v}_1, \boldsymbol{v}_2$ be nonproportional vectors in $\mathbb{R}^3$ and suppose $\boldsymbol{v}_3$ is in the plane through $\mathbf{0}, \boldsymbol{v}_1, \boldsymbol{v}_2$. Then

$$\boldsymbol{v}_3 = t\boldsymbol{v}_1 + u\boldsymbol{v}_2$$

for some real t, u (Chapter 1). Now $\boldsymbol{v}_1, \boldsymbol{v}_2, \boldsymbol{v}_3$ is dependent:

$$t\boldsymbol{v}_1 + u\boldsymbol{v}_2 - \boldsymbol{v}_3 = \mathbf{0}.$$

I think of a set $\boldsymbol{v}_1, \dots, \boldsymbol{v}_k$ as being independent if $\boldsymbol{v}_1 \neq \mathbf{0}, \boldsymbol{v}_2$ is not in the line of $\boldsymbol{v}_1$, $\boldsymbol{v}_3$ is not in the plane of $\boldsymbol{v}_1, \boldsymbol{v}_2; \dots ; \boldsymbol{v}_k$ is not in the 'plane' of $\boldsymbol{v}_1, \dots, \boldsymbol{v}_{k-1}$. We formalize this as follows.

Proposition 1 *The following assertions about a set $\boldsymbol{v}_1, \dots, \boldsymbol{v}_k$ of nonzero vectors in $\mathbb{R}^n$ are equivalent.*

(i) No $\boldsymbol{v}_j$ is a linear combination of $\boldsymbol{v}_1, \dots \boldsymbol{v}_{j-1}$.

(ii) $\boldsymbol{v}_1, \dots, \boldsymbol{v}_k$ are linearly independent.

Proof. Suppose (ii) holds. We cannot have

$$\boldsymbol{v}_j = a_1\boldsymbol{v}_1 + \cdots + a_{j-1}\boldsymbol{v}_{j-1}$$

because then

$$a_1\boldsymbol{v}_1 + \cdots + a_{j-1}\boldsymbol{v}_{j-1} - 1\boldsymbol{v}_j + 0\boldsymbol{v}_{j+1} + \cdots + 0\boldsymbol{v}_k = \mathbf{0}$$

which is a nonzero solution of (3). So (i) holds.

Suppose (i) holds. Let $x_1, \dots, x_k$ be a solution of

$$x_1\boldsymbol{v}_1 + \cdots + x_k\boldsymbol{v}_k = \mathbf{0}.$$

If *some* x_j is not 0, we can choose j *as large as possible* with $x_j \neq 0$. Now $j > 1$ since $\boldsymbol{v}_1 \neq \mathbf{0}$. Since $x_{j+1} = 0, \dots, x_k = 0$, we have

$$\boldsymbol{v}_j = \left(-\frac{x_1}{x_j}\right) \boldsymbol{v}_1 + \cdots + \left(-\frac{x_{j-1}}{x_j}\right) \boldsymbol{v}_{j-1}.$$

We know that this does not occur. So the only solution of (3) is the zero solution and (ii) holds.

It is worth noticing that any subset of a linearly independent set is independent. (Otherwise, we could give a nonzero solution of (3) with some of the terms omitted.)

There is a simple computation to decide whether a given set $\boldsymbol{a}_1, \ldots, \boldsymbol{a}_k$ is independent. Form the matrix with columns $\boldsymbol{a}_1, \ldots, \boldsymbol{a}_k$:

$$A = [\boldsymbol{a}_1 \ldots \boldsymbol{a}_k].$$

Reduce A to echelon form by row operations to find out which columns of A are pivot columns. If **all columns of A are pivot columns,** $\boldsymbol{a}_1, \ldots, \boldsymbol{a}_k$ are independent. To see this, consider the reduced echelon form of A,

$$B = [\boldsymbol{b}_1 \ldots \boldsymbol{b}_k].$$

The vector equation

$$x_1\boldsymbol{b}_1 + \cdots + x_k\boldsymbol{b}_k = \boldsymbol{0}$$

includes the equations $x_1 = 0, x_2 = 0, \ldots, x_k = 0$ (the first k equations). Therefore (3) has only the zero solution. The equivalent vector equation

$$x_1\boldsymbol{a}_1 + \cdots + x_k\boldsymbol{a}_k = \boldsymbol{0} \tag{4}$$

has only the zero solution. This proves independence of $\boldsymbol{a}_1, \ldots, \boldsymbol{a}_k$.

On the other hand, if A has a nonpivot column, there is a free variable in the general solution of (4), which provides nonzero solutions. In this case, $\boldsymbol{a}_1, \ldots, \boldsymbol{a}_k$ are *dependent.*

Example 9 Determine dependence or independence for the set

$$\boldsymbol{a}_1 = (1, 2, 1, 3), \boldsymbol{a}_2 = (3, -1, 2, 4), \boldsymbol{a}_3 = (6, 2, 2, 5).$$

Solution.

$$A = \begin{bmatrix} 1 & 3 & 6 \\ 2 & -1 & 2 \\ 1 & 2 & 2 \\ 3 & 4 & 5 \end{bmatrix} \sim \begin{bmatrix} 1 & 3 & 6 \\ 0 & -7 & -10 \\ 0 & -1 & -4 \\ 0 & -5 & -13 \end{bmatrix} \begin{matrix} \text{II} - 2\text{I} \\ \text{III} - \text{I} \\ \text{IV} - 3\text{I} \\ \\ \end{matrix}$$

$$\sim \begin{bmatrix} 1 & 3 & 6 \\ 0 & 1 & 4 \\ 0 & 0 & 18 \\ 0 & 0 & 7 \end{bmatrix} \begin{matrix} \text{II} \leftrightarrow \text{III} \\ \text{II} \times -1 \\ \text{III} + 7\text{II} \\ \text{IV} + 5\text{II} \\ \end{matrix} .$$

All columns are pivot columns; $\boldsymbol{a}_1, \boldsymbol{a}_2, \boldsymbol{a}_3$ are independent.

The following result is helpful in developing the notions of **basis** and **dimension** in Section 3.

Proposition 2 *Let $\boldsymbol{u}_1, \ldots, \boldsymbol{u}_k$ be vectors in $\mathbb{R}^n$. We cannot find $k+1$ linearly independent vectors in* $\mathrm{Span}\{\boldsymbol{u}_1, \ldots, \boldsymbol{u}_k\}$.

For instance, in $\mathbb{R}^3$ you will not find four independent vectors. In a plane in $\mathbb{R}^3$ you will not find three independent vectors.

Proof. Take any $\boldsymbol{v}_1, \ldots, \boldsymbol{v}_{k+1}$ in $\mathrm{Span}\{\boldsymbol{u}_1, \ldots, \boldsymbol{u}_k\}$,

$$\begin{aligned} \boldsymbol{v}_1 &= a_{11}\boldsymbol{u}_1 + \cdots + a_{1k}\boldsymbol{u}_k \\ \boldsymbol{v}_2 &= a_{21}\boldsymbol{u}_1 + \cdots + a_{2k}\boldsymbol{u}_k \\ &\cdots \\ \boldsymbol{v}_{k+1} &= a_{k+1,1}\boldsymbol{u}_1 + \cdots + a_{k+1,k}\boldsymbol{u}_k. \end{aligned}$$

Now there is a nonzero solution $x_1, \ldots, x_{k+1}$ of

$$(5) \qquad x_1 \begin{bmatrix} a_{11} \\ \vdots \\ a_{1k} \end{bmatrix} + \cdots + x_{k+1} \begin{bmatrix} a_{k+1,1} \\ \vdots \\ a_{k+1,k} \end{bmatrix} = \mathbf{0}$$

(Chapter 2, Proposition 4). We have

$$x_1\boldsymbol{v}_1 + \cdots + x_{k+1}\boldsymbol{v}_{k+1} = \sum_{j=1}^{k+1} x_j \left(\sum_{i=1}^{k} a_{ji}\boldsymbol{u}_i \right) = \sum_{i=1}^{k} \left(\sum_{j=1}^{k+1} a_{ji}x_j \right) \boldsymbol{u}_i.$$

The coefficient of $\boldsymbol{u}_i$ is the i-th entry of the vector in (5), which is 0. So

$$x_1\boldsymbol{v}_1 + \cdots + x_{k+1}\boldsymbol{v}_{k+1} = \mathbf{0}$$

and $\boldsymbol{v}_1, \ldots, \boldsymbol{v}_{k+1}$ are dependent. This proves Proposition 2.

Example 10 The span of the following set of n vectors is $\mathbb{R}^n$:

$$(6) \qquad \boldsymbol{e}_1 = (1,0,0,\ldots,0), \boldsymbol{e}_2 = (0,1,0,\ldots,0), \ldots, \boldsymbol{e}_n = (0,0,\ldots,0,1).$$

After all, given $\boldsymbol{x} = (x_1, \ldots, x_n)$, we can write $\boldsymbol{x}$ as

$$x_1\boldsymbol{e}_1 + x_2\boldsymbol{e}_2 + \cdots + x_n\boldsymbol{e}_n.$$

We will always use the notation $\boldsymbol{e}_i$ to denote the vectors in (6).

Consequently, we cannot find $n+1$ independent vectors in $\mathbb{R}^n$.

3 Subspaces of $\mathbb{R}^n$

Definition 3 Let V be a set of vectors in $\mathbb{R}^n$, *not* the empty set. V is a **subspace** of $\mathbb{R}^n$ if

(7) $au + bv$ is in V for all $\boldsymbol{u}$ and $\boldsymbol{v}$ in V and real a, b.

We call the property (7) the **closure property** .

The set consisting of $\mathbf{0}$ alone is a subspace. We call it the **zero subspace.** We write $\mathbf{0}$ for this subspace. This is not good 'grammar' since $\mathbf{0}$ is a vector rather than a set of vectors, but it is convenient.

A subspace in general turns out to be the linear span of some set.

Proposition 3 *Let V be a set of vectors in $\mathbb{R}^n$. The following two assertions are equivalent.*

(i) V is a subspace of $\mathbb{R}^n$.

(ii) $V = Span\{\boldsymbol{v}_1, \ldots, \boldsymbol{v}_k\}$ for some $\boldsymbol{v}_1, \ldots, \boldsymbol{v}_k$ in $\mathbb{R}^n$.

Proof. Suppose that (i) holds. If V is the zero subspace, then $V = \text{Span}\{\mathbf{0}\}$ and we get (ii). If V is not the zero subspace, pick $\boldsymbol{v}_1 \neq \mathbf{0}$ in V. Of course $\text{Span}\{\boldsymbol{v}_1\}$ is a subset of V because of (7) (with $\boldsymbol{v}_1 = \boldsymbol{u}, \boldsymbol{v}_2 = \mathbf{0}$). If there are vectors in V, not in $\text{Span}\{\boldsymbol{v}_1\}$, pick $\boldsymbol{v}_2$ in V, $\boldsymbol{v}_2$ not in $\text{Span}\{\boldsymbol{v}_1\}$.

We continue this game in the following way. Once we have got k vectors $\boldsymbol{v}_1, \ldots, \boldsymbol{v}_k$ in V, pick $\boldsymbol{v}_{k+1}$ in V but not in $\text{Span}\{\boldsymbol{v}_1, \ldots, \boldsymbol{v}_k\}$ *if possible.* If this is not possible, stop the game. The game must stop after at most n steps, because Proposition 1 shows that $\boldsymbol{v}_1, \ldots, \boldsymbol{v}_k$ are independent, which is impossible with $k = n + 1$. When we stop with $\boldsymbol{v}_1, \ldots, \boldsymbol{v}_k$, then $\text{Span}\{\boldsymbol{v}_1, \ldots, \boldsymbol{v}_k\}$ is a subset of V because of (7). $\text{Span}\{\boldsymbol{v}_1, \ldots, \boldsymbol{v}_k\}$ is, in fact, equal to V, otherwise the game would continue. Thus (ii) holds.

Now suppose that (ii) holds. Take any $\boldsymbol{u}$ and $\boldsymbol{v}$ in V and real a, b. Then

$$\boldsymbol{u} = a_1\boldsymbol{v}_1 + \cdots + a_k\boldsymbol{v}_k, \boldsymbol{v} = b_1\boldsymbol{v}_1 + \cdots + b_k\boldsymbol{v}_k$$

for some real a_i and b_i;

$$a\boldsymbol{u} + b\boldsymbol{v} = (aa_1 + bb_1)\boldsymbol{v}_1 + \cdots + (aa_k + bb_k)\boldsymbol{v}_k.$$

So $a\boldsymbol{u} + b\boldsymbol{v}$ is in V, and (i) holds.

For the subspace in the following example, the closure property is easier to perceive than the property (ii) of Proposition 3.

Example 11 The solution set U of a homogeneous linear system with $m \times n$ coefficient matrix A,

$$a_{11}x_1 + \cdots + a_{1n}x_n = 0$$
$$\cdots$$
$$a_{m1}x_1 + \cdots + a_{mn}x_n = 0$$

is a subspace of $\mathbb{R}^n$. We say that U is the **null space** of A, and write,

$$U = \text{Nul } A.$$

To see that U is a subspace, take $\boldsymbol{x}$ and $\boldsymbol{y}$ in U. Then $\boldsymbol{a}_i \cdot \boldsymbol{x} = 0 = \boldsymbol{a}_i \cdot \boldsymbol{y}$, where $\boldsymbol{a}_i = (a_{i1}, \ldots, a_{in})$. Consequently $\boldsymbol{a}_i \cdot (a\boldsymbol{x} + b\boldsymbol{y}) = 0$. Here i is any of $1, \ldots, m$. This shows that $a\boldsymbol{x} + b\boldsymbol{y}$ is in U, and U is a subspace.

We do, in fact, know how to write down $\boldsymbol{v}_1, \ldots, \boldsymbol{v}_k$ in U which satisfy $U = \text{Span}\{\boldsymbol{v}_1, \ldots, \boldsymbol{v}_k\}$. In Chapter 2 we gave the general solution of the system in the form

$$x_{m(1)}\boldsymbol{v}_1 + \cdots + x_{m(k)}\boldsymbol{v}_k$$

where $x_{m(1)}, \ldots, x_{m(k)}$ are free variables. This is equivalent to

$$U = \text{Span}\{\boldsymbol{v}_1, \ldots, \boldsymbol{v}_k\}.$$

For example, just after Proposition 4 we noted that with A as in Example 10, the general solution of $A\boldsymbol{x} = \boldsymbol{0}$ is

$$x_3\boldsymbol{v}_1 + x_4\boldsymbol{v}_2, \boldsymbol{v}_1 = (-5, 1, 1, 0, 0), \boldsymbol{v}_2 = (2, 3, 0, 1, 0).$$

Here the null space is $U = \text{Span}\{\boldsymbol{v}_1, \boldsymbol{v}_2\}$.

The only subspaces of $\mathbb{R}^3$ are the zero subspace, lines through $\boldsymbol{0}$, planes through $\boldsymbol{0}$, and $\mathbb{R}^3$ itself. The last three alternatives correspond to the game in Proposition 3 stopping when $k = 1, 2, 3$ respectively.

The game we used in the proof of Proposition 3 generates a set $\boldsymbol{v}_1, \ldots, \boldsymbol{v}_k$ that is a **basis**, as defined below.

Definition 4 Let V be a subspace of $\mathbb{R}^n$, not the zero subspace. A **basis** of V is a linearly independent set $\boldsymbol{v}_1, \ldots, \boldsymbol{v}_k$ such that

$$V = \text{Span}\{\boldsymbol{v}_1, \ldots, \boldsymbol{v}_k\}.$$

For any nonzero subspace V of $\mathbb{R}^n$, the set $\boldsymbol{v}_1, \ldots, \boldsymbol{v}_k$ obtained in the proof of Proposition 3 is linearly independent. This follows from Proposition 1, since $\boldsymbol{v}_j$ is not a combination of $\boldsymbol{v}_1, \ldots, \boldsymbol{v}_{j-1}$. Accordingly this set $\boldsymbol{v}_1, \ldots, \boldsymbol{v}_k$ is a basis of V.

Of course, there are infinitely many choices of basis. For example, if V is a plane in $\mathbb{R}^3$, the last paragraph assures us that *any* nonproportional pair of vectors $\boldsymbol{v}_1, \boldsymbol{v}_2$ in V could be chosen as a basis of V.

A crucial observation is that any two bases of V have the same number of vectors. ('Bases,' pronounced bayseas, is the plural of basis.)

Proposition 4 *Let V be a subspace of $\mathbb{R}^n$. Any two bases of V have the same number of vectors.*

Proof. Let $\boldsymbol{u}_1, \ldots, \boldsymbol{u}_k$ and $\boldsymbol{v}_1, \ldots, \boldsymbol{v}_j$ be two bases of V. We have j linearly independent elements $\boldsymbol{v}_1, \ldots, \boldsymbol{v}_j$ in $V = \operatorname{Span}\{\boldsymbol{u}_1, \ldots, \boldsymbol{u}_k\}$. Proposition 2 assures us that $j \leq k$. Reversing roles, we get $k \leq j$. So $k = j$.

Proposition 4 enables us to define dimension.

Definition 5 The **dimension** of a subspace V of $\mathbb{R}^n$ is the number of vectors in any basis of V. We write $\dim V$ for the dimension of V.

Naturally $\dim V \leq n$, since $\mathbb{R}^n$ cannot contain $n+1$ linearly independent vectors.

For completeness, the zero subspace is assigned dimension 0. The zero subspace does not have a basis.

We can now deduce that the dimension of $\mathbb{R}^n$ is n. For the set $\boldsymbol{e}_1, \ldots, \boldsymbol{e}_n$ in Example 4 is independent. The vector equation $x_1\boldsymbol{e}_1 + \cdots + x_n\boldsymbol{e}_n = \boldsymbol{0}$ reads $x_1 = 0, x_2 = 0, \ldots, x_n = 0$. So $\boldsymbol{e}_1, \ldots, \boldsymbol{e}_n$ is one basis of $\mathbb{R}^n$; we call it the **standard basis** . Now we know, of course, that

$$\dim \mathbb{R}^n = n.$$

The dimension of a plane V through $\boldsymbol{0}$ in $\mathbb{R}^3$ is 2; we noted a few paragraphs ago that we can find (many) bases of V with 2 elements. The dimension of a line through $\boldsymbol{0}$ in $\mathbb{R}^n$ is 1.

It seems reasonable that the dimension of a hyperplane W in $\mathbb{R}^n$, with equation

$$a_1x_1 + a_2x_2 + \cdots + a_nx_n = 0, \tag{8}$$

should be $n-1$. The easiest demonstration is to provide a basis. Suppose, for instance, that $a_1 \neq 0$. The general solution of the linear system (8) is

$$x_1 = \left(-\frac{a_2}{a_1}\right) x_2 + \cdots + \left(-\frac{a_n}{a_1}\right) x_n$$

or

$$\text{(9)} \qquad \boldsymbol{x} = x_2 \begin{bmatrix} -a_2/a_1 \\ 1 \\ 0 \\ \vdots \\ 0 \end{bmatrix} + \cdots + x_n \begin{bmatrix} -a_n/a_1 \\ 0 \\ \vdots \\ 0 \\ 1 \end{bmatrix} = x_2\boldsymbol{v}_2 + \cdots + x_n\boldsymbol{v}_n,$$

say. Here $x_2, \ldots, x_n$ are free. The hyperplane W is Span$\{\boldsymbol{v}_2, \ldots, \boldsymbol{v}_n\}$. Now $\boldsymbol{v}_2, \ldots, \boldsymbol{v}_n$ are independent. To see this, set $x_2\boldsymbol{v}_2 + \cdots + x_n\boldsymbol{v}_n = \boldsymbol{0}$ in (9). We get a linear system which includes the equations $x_2 = 0, x_3 = 0, \ldots, x_n = 0$. Therefore $\boldsymbol{v}_2, \ldots, \boldsymbol{v}_n$ is a basis of W, and

$$\dim W = n - 1.$$

This suggests a general result on the dimension of the space Nul A. The general solution to

$$A\boldsymbol{x} = \boldsymbol{0}$$

is

$$\text{(10)} \qquad \boldsymbol{v} = x_{m(1)}\boldsymbol{v}_1 + \cdots + x_{m(k)}\boldsymbol{v}_k \qquad (x_{m(1)}, \ldots, x_{m(k)} \text{ free})$$

as we recalled in Example 5. Now k of the coordinates of $\boldsymbol{v}$ are actually $x_{m(1)}, \ldots, x_{m(k)}$. For instance, let $n = 5$, and suppose that the general solution is

$$\boldsymbol{v} = \begin{bmatrix} -2x_2 - 3x_3 - x_5 \\ x_2 \\ x_3 \\ 4x_2 + 6x_3 + 2x_5 \\ x_5 \end{bmatrix} = x_2\boldsymbol{v}_1 + x_3\boldsymbol{v}_2 + x_5\boldsymbol{v}_3$$

(x_2, x_3, x_5 free). Here $\boldsymbol{v}_1 = (-2, 1, 0, 4, 0)$, $\boldsymbol{v}_2 = (-3, 0, 1, 6, 0)$, $\boldsymbol{v}_3 = (-1, 0, 0, 2, 1)$. Three coordinates of $\boldsymbol{v}$ are x_2, x_3, x_5. If $x_2\boldsymbol{v}_1 + x_3\boldsymbol{v}_2 + x_5\boldsymbol{v}_3$ is $\boldsymbol{0}$, then x_2, x_3, x_5 are 0. In the general case, if we set $\boldsymbol{v}$ equal to $\boldsymbol{0}$ in (10), then $x_{m(1)}, \ldots, x_{m(k)}$ are 0. It follows that $\boldsymbol{v}_1, \ldots, \boldsymbol{v}_k$ not only have span equal to Nul A, but are linearly independent. Since $\boldsymbol{v}_1, \ldots, \boldsymbol{v}_k$ is a basis of Nul A, we have established:

Proposition 5 dim(*Nul* A) *is the number of non-pivot columns of* A.

Here we use the simple fact that there is a free variable for each non-pivot column.

The **nullity** of A, or nullity A, is another expression used for dim(Nul A).

Example 12 Find a basis of Nul A, and determine the nullity of A, where

$$A = \begin{bmatrix} 1 & 0 & 1 & 2 & 0 & 0 \\ -4 & 1 & 1 & -11 & 0 & -1 \\ -7 & 0 & -7 & -14 & 1 & -2 \\ 6 & 0 & 6 & 12 & 0 & 1 \end{bmatrix}.$$

Solution. You will find that the reduced echelon form of A is

$$B = \begin{bmatrix} 1 & 0 & 1 & 2 & 0 & 0 \\ 0 & 1 & 5 & -3 & 0 & 0 \\ 0 & 0 & 0 & 0 & 1 & 0 \\ 0 & 0 & 0 & 0 & 0 & 1 \end{bmatrix}.$$

The general solution of $A\boldsymbol{x} = \mathbf{0}$, in which x_3 and x_4 are free variables, is

$$\begin{bmatrix} -x_3 - 2x_4 \\ -5x_3 + 3x_4 \\ x_3 \\ x_4 \\ 0 \\ 0 \end{bmatrix} = x_3 \begin{bmatrix} -1 \\ -5 \\ 1 \\ 0 \\ 0 \\ 0 \end{bmatrix} + x_4 \begin{bmatrix} -2 \\ 3 \\ 0 \\ 1 \\ 0 \\ 0 \end{bmatrix} = x_3\boldsymbol{v}_1 + x_4\boldsymbol{v}_2, \text{ say.}$$

Now $\boldsymbol{v}_1, \boldsymbol{v}_2$ is a basis of Nul A. The linear independence of $\boldsymbol{v}_1, \boldsymbol{v}_2$ is obvious. By definition, the nullity of A is 2.

4 General propositions about bases

The following propositions are very general and useful.

Proposition 6 *Let* $\boldsymbol{v}_1, \dots, \boldsymbol{v}_j$ *be linearly independent vectors in a subspace* V *of* $\mathbb{R}^n$. *Suppose* $\dim V = r$. *We can find* $r - j$ *vectors* $\boldsymbol{v}_{j+1}, \dots \boldsymbol{v}_r$ *to obtain a basis* $\boldsymbol{v}_1, \dots, \boldsymbol{v}_j, \boldsymbol{v}_{j+1}, \dots, \boldsymbol{v}_r$ *of* V.

Proof. This is simply the proof of Proposition 3 (i) in which $\boldsymbol{v}_1, \ldots, \boldsymbol{v}_j$ have already been selected. (Note $\boldsymbol{v}_i$ is not in $\text{Span}\{\boldsymbol{v}_1, \ldots, \boldsymbol{v}_{i-1}\}$ for $i \leq j$.) Pick $\boldsymbol{v}_{j+1}$ not in $\text{Span}\{\boldsymbol{v}_1, \ldots, \boldsymbol{v}_j\}$ (if possible), and continue in this way until we have a basis $\boldsymbol{v}_1, \ldots, \boldsymbol{v}_k$ of V. The fact that the procedure provides a basis of V was noted after Definition 4. Finally, we must have $k = r$ because every basis of V has r members.

The special case $j = r$ yields:

Proposition 7 *Let V be a subspace of $\mathbb{R}^n$, $\dim V = r$. Any r linearly independent vectors in V are a basis of V.*

As a kind of mirror image of Proposition 7 we have:

Proposition 8 *Let V be a subspace of $\mathbb{R}^n$, $\dim V = r$. Suppose that $\boldsymbol{v}_1, \ldots, \boldsymbol{v}_r$ are vectors in V with $\text{Span}\{\boldsymbol{v}_1, \ldots, \boldsymbol{v}_r\} = V$. Then $\boldsymbol{v}_1, \ldots, \boldsymbol{v}_r$ is a basis of V.*

Proof. If $\boldsymbol{v}_1, \ldots, \boldsymbol{v}_r$ is a dependent set, some $\boldsymbol{v}_i$ is *either* $\mathbf{0}$ *or* a combination of $\boldsymbol{v}_1, \ldots, \boldsymbol{v}_{i-1}$. This implies

$$\text{Span}\{\boldsymbol{v}_1, \ldots, \boldsymbol{v}_{i-1}, \boldsymbol{v}_{i+1}, \ldots, \boldsymbol{v}_r\} = \text{Span}\{\boldsymbol{v}_1, \ldots, \boldsymbol{v}_r\}.$$

(The vector $c_i\boldsymbol{v}_i$ can be replaced by a combination of $\boldsymbol{v}_1, \ldots, \boldsymbol{v}_{i-1}$ in any combination $c_1\boldsymbol{v}_1 + \cdots + c_r\boldsymbol{v}_r$.) We now have r independent vectors (any basis of V) in the linear span of $r-1$ vectors $\boldsymbol{v}_1, \ldots, \boldsymbol{v}_{i-1}, \boldsymbol{v}_{i+1}, \ldots, \boldsymbol{v}_r$. This is impossible, and we conclude that the vectors $\boldsymbol{v}_1, \ldots, \boldsymbol{v}_r$ are independent. Since $\boldsymbol{v}_1, \ldots, \boldsymbol{v}_r$ span V, they form a basis of V.

Proposition 9 *Let V and W be subspaces of $\mathbb{R}^n$ such that W is contained in V and* $\dim W = \dim V$. *Then $W = V$.*

Proof. Let $\boldsymbol{v}_1, \ldots, \boldsymbol{v}_r$ be a basis of W. Apply Proposition 7 to see that $\boldsymbol{v}_1, \ldots, \boldsymbol{v}_r$ is a basis of V. Hence W and V are both equal to $\text{Span}\{\boldsymbol{v}_1, \ldots, \boldsymbol{v}_r\}$.

5 Column space. Rank

Given a set of vectors $\boldsymbol{a}_1, \ldots, \boldsymbol{a}_n$ in $\mathbb{R}^m$, there is a simple way to pick out from them a basis of $V = \text{Span}\{\boldsymbol{a}_1, \ldots, \boldsymbol{a}_n\}$. It is a variant of our 'pivot column test' for independence. Write A for the matrix with columns $\boldsymbol{a}_1, \ldots, \boldsymbol{a}_n$,

$$A = [\boldsymbol{a}_1 \ldots \boldsymbol{a}_n].$$

Definition 6 The **column space** of an $m \times n$ matrix $A = [\boldsymbol{a}_1 \dots \boldsymbol{a}_n]$ is the subspace Span$\{\boldsymbol{a}_1, \dots, \boldsymbol{a}_n\}$ of $\mathbb{R}^m$. We write Col A for the column space.

So far, then, all we have done is write Span$\{\boldsymbol{a}_1, \dots, \boldsymbol{a}_n\}$ as Col A.

Proposition 10 *Let A be an $m \times n$ matrix. The pivot columns of A are a basis of Col A. Moreover,*

$$\dim(\text{Col } A) + \dim(\text{Nul } A) = n. \tag{11}$$

Before proving the proposition we introduce another technical term.

Definition 7 The matrix A is said to have **rank** r if $\dim(\text{Col } A) = r$.

The equation (11) can be written more succinctly as

$$\text{rank } A + \text{nullity } A = n$$

Proof of Proposition 10 Let B be the reduced echelon form of A, with columns $\boldsymbol{b}_1, \dots, \boldsymbol{b}_n$:

$$B = [\boldsymbol{b}_1 \dots \boldsymbol{b}_n].$$

Suppose B has r nonzero rows. The pivot columns of B are, let us say,

$$\boldsymbol{b}_{h(1)}, \dots, \boldsymbol{b}_{h(r)}.$$

In fact, this last list of vectors is just $\boldsymbol{e}_1, \dots, \boldsymbol{e}_r$, the first r members of the standard basis of $\mathbb{R}^m$. So $\boldsymbol{b}_{h(1)}, \dots, \boldsymbol{b}_{h(r)}$ are independent. It is obvious that the span of $\boldsymbol{e}_1, \dots, \boldsymbol{e}_r$ is Col B, because every vector $\boldsymbol{y}$ in Col B has $y_{r+1} = \dots = y_m = 0$. This establishes that $\boldsymbol{b}_{h(1)}, \dots, \boldsymbol{b}_{h(r)}$ has span Col B and is a basis of Col B.

We can 'transfer' this fact to A. Firstly, if

$$x_1\boldsymbol{a}_{h(1)} + \cdots + x_r\boldsymbol{a}_{h(r)} = \boldsymbol{0},$$

we carry out the row operations that take A into B and get

$$x_1\boldsymbol{b}_{h(1)} + \cdots + x_r\boldsymbol{b}_{h(r)} = \boldsymbol{0}.$$

This has only the zero solution, and we have proved the pivot columns $\boldsymbol{a}_{h(1)}, \dots, \boldsymbol{a}_{h(r)}$ of A are linearly independent.

Given any column $\boldsymbol{a}_i$, we can write

$$\boldsymbol{a}_i = z_1\boldsymbol{a}_{h(1)} + \cdots + z_r\boldsymbol{a}_{h(r)} \quad (\text{some real } z_1, \dots, z_r). \tag{12}$$

To do this, simply write

$$b_i = z_1 b_{h(1)} + \cdots + z_r b_{h(r)}. \tag{13}$$

Just as above, if the equation (13) is true, then the equation (12) must be true. This proves that every a_i is in the span of the pivot columns. In turn, every **combination** of $a_1, \ldots, a_n$ is in the span of the pivot columns. This completes the proof that the set of pivot columns is a basis of Col A. As for the equation (11), we recall Proposition 5. The total number of pivot columns and non-pivot columns is n.

Just a slight change in the above argument yields the following piece of unfinished business from Chapter 2.

Proposition 11 *A matrix A has only one reduced echelon form.*

Proof. Suppose B, C are reduced echelon forms of A. Certainly A, B, C have the same zero columns. The pivot columns of B and C are the nonzero columns that are not linear combinations of columns to their left. Now let $A = [a_1 \cdots a_n], B = [b_1 \cdots b_n], C = [c_1 \cdots c_n]$. The equation

$$a_j = x_1 a_1 + \cdots + x_{j-1} a_{j-1}$$

is equivalent to either of the equations

$$b_j = x_1 b_1 + \cdots + x_{j-1} b_{j-1}, \tag{14}$$

$$c_j = x_1 c_1 + \cdots + x_{j-1} c_{j-1}. \tag{15}$$

Thus the pivot columns of B and C are in the same places as the nonzero columns of A that are not combinations of columns to their left. Since both sets of pivot columns are $e_1, e_2, \ldots$ reading from left to right, the pivot columns of B and C are the same vectors, in the same positions.

Now a *non*-pivot, nonzero column b_j *does* satisfy a relation (14). As above, we have (15) with the same numbers $x_1, \ldots, x_{j-1}$. Working from left to right, we get $b_j = c_j$ for non pivot columns also. This completes the proof.

Example 13 Find a basis of the column space of

$$A = \begin{bmatrix} 1 & 2 & 8 & 7 \\ 3 & 0 & 6 & 3 \\ 5 & -2 & 4 & 4 \\ 7 & 1 & 17 & 2 \end{bmatrix} = \begin{bmatrix} a_1 & a_2 & a_3 & a_4 \end{bmatrix}.$$

Solution. $A \sim \begin{bmatrix} 1 & 2 & 8 & 7 \\ 0 & -6 & -18 & -18 \\ 0 & -12 & -36 & -31 \\ 0 & -13 & -39 & -47 \end{bmatrix} \begin{matrix} \text{II} - 3\text{I} \\ \text{III} - 5\text{I} \\ \text{IV} - 7\text{I} \\ \\ \end{matrix} \sim \begin{bmatrix} 1 & 2 & 8 & 7 \\ 0 & 1 & 3 & 3 \\ 0 & 0 & 0 & 5 \\ 0 & 0 & 0 & -8 \end{bmatrix} \begin{matrix} \text{II} \times -\frac{1}{6} \\ \text{II} + 12\text{II} \\ \text{IV} + 13\text{II} \\ \\ \end{matrix}$

$$\sim \begin{bmatrix} 1 & 2 & 8 & 7 \\ 0 & 1 & 3 & 3 \\ 0 & 0 & 0 & 5 \\ 0 & 0 & 0 & 0 \end{bmatrix} \text{IV} + \tfrac{8}{5}\,\text{II} \sim \begin{bmatrix} 1 & 0 & 2 & 1 \\ 0 & 1 & 3 & 3 \\ 0 & 0 & 0 & 5 \\ 0 & 0 & 0 & 0 \end{bmatrix} \text{I} - 2\text{II} \quad = B.$$

The pivot columns are 1, 2 and 4. Hence $\boldsymbol{a}_1, \boldsymbol{a}_2, \boldsymbol{a}_4$ is a basis of Col A.

Detecting hidden relations. It is clear from a glance at B that $\boldsymbol{b}_3 = 2\boldsymbol{b}_1 + 3\boldsymbol{b}_2$, and therefore $\boldsymbol{a}_3 = 2\boldsymbol{a}_1 + 3\boldsymbol{a}_2$ (check!) This 'hidden relation' explains why $\boldsymbol{a}_3$ could be discarded without changing the span of the set of column vectors.

A common student error. If $A \sim B$ and B is in echelon form it is not necessarily true that the pivot columns of B are a basis of Col A. It is just the **numbers** of the pivot column (such as: columns 2, 4, 6, 7) that we can transfer to A to find a set of columns that is a basis of Col A.

For example, let $A = \begin{bmatrix} 1 & 3 \\ 1 & 3 \end{bmatrix}$. A basis of Col A is $\begin{bmatrix} 1 \\ 1 \end{bmatrix}$, so the column space is the line

$$x_1 = x_2.$$

The reduced echelon form of A is $B = \begin{bmatrix} 1 & 3 \\ 0 & 0 \end{bmatrix}$; the column space of B is the linear span of the first column, namely the line

$$x_2 = 0.$$

It is useful to note that the **row space** of A (the linear span of the set of row vectors of A) has the same dimension as the column space. This might be deemed surprising if A is 89×164, for instance.

Proposition 12 *The dimension of the row space of A is equal to the rank of A.*

Proof. A row operation on a matrix A gives a matrix B with the same row space as A. To see this, let $\boldsymbol{v}_1, \ldots, \boldsymbol{v}_m$ be the rows of A. If we use Operation 1, let us say adding c times row 2 to row 1, the rows of B are

$$\boldsymbol{v}_1 + c\boldsymbol{v}_2, \boldsymbol{v}_2, \boldsymbol{v}_3, \ldots, \boldsymbol{v}_m.$$

Every vector in $\operatorname{Span}\{\boldsymbol{v}_1 + c\boldsymbol{v}_2, \boldsymbol{v}_2, \ldots, \boldsymbol{v}_m\}$ is obviously in $\operatorname{Span}\{\boldsymbol{v}_1, \boldsymbol{v}_2, \ldots, \boldsymbol{v}_m\}$.

Now A can be obtained from B by Operation 1. So every vector in $\operatorname{Span}\{\boldsymbol{v}_1, \boldsymbol{v}_2, \ldots, \boldsymbol{v}_m\}$ is in $\operatorname{Span}\{\boldsymbol{v}_1 + c\boldsymbol{v}_2, \boldsymbol{v}_2, \ldots, \boldsymbol{v}_m\}$. This proves that A and B have the same row space.

The argument is similar if Operation 2 is used. The rows of B are, let us say,

$$c\boldsymbol{v}_1, \boldsymbol{v}_2, \boldsymbol{v}_3, \ldots, \boldsymbol{v}_m$$

where $c \neq 0$. Every vector in $\operatorname{Span}\{\boldsymbol{v}_1, c\boldsymbol{v}_2, \boldsymbol{v}_3, \ldots, \boldsymbol{v}_m\}$ is obviously in $\operatorname{Span}\{\boldsymbol{v}_1, \boldsymbol{v}_2, \ldots, \boldsymbol{v}_m\}$. The rest of the proof that A and B have the same row space is similar to the last paragraph, since A can be obtained from B by Operation 2.

It follows that the reduced echelon form C of A has the same row space V as A. We claim that the nonzero rows $\boldsymbol{u}_1, \ldots, \boldsymbol{u}_r$ of C are a basis of V. Obviously the span of $\boldsymbol{u}_1, \ldots, \boldsymbol{u}_r$ is V. Now $\boldsymbol{u}_1, \ldots, \boldsymbol{u}_r$ are linearly independent because each has a 1 in a column that otherwise consists of zeros. For instance, the first three rows $\boldsymbol{u}_1, \boldsymbol{u}_2, \boldsymbol{u}_3$ of

$$\begin{bmatrix} 1 & 0 & 0 & * & * & 0 \\ 0 & 0 & 1 & * & * & 0 \\ 0 & 0 & 0 & 0 & 0 & 1 \\ 0 & 0 & 0 & 0 & 0 & 0 \end{bmatrix}$$

are linearly independent because the linear system $x_1\boldsymbol{u}_1 + x_2\boldsymbol{u}_2 + x_3\boldsymbol{u}_3 = 0$ includes the equations $x_1 = 0, x_2 = 0$ and $x_3 = 0$.

We conclude that $\dim V$ is the number of nonzero rows of C, which is equal to the number of pivot columns of C. We are now able to see that $\dim V = \dim(\text{Col } A) = \text{rank of } A$.

As an example, we can use Example 13 above. A basis of the row space V is $(1,0,2,0), (0,1,3,0), (0,0,0,1)$. The row space has dimension 3, as does Col A.

Let us finish this section by returning to the subspace $\operatorname{Span}\{\boldsymbol{v}_1, \boldsymbol{v}_2, \boldsymbol{v}_3\}$ where $\boldsymbol{v}_1 = (3,7,1,-2), \boldsymbol{v}_2 = (2,0,1,4), \boldsymbol{v}_3 = (7,0,1,6)$, just after Example

3. The matrix $A = [\boldsymbol{v}_1\ \boldsymbol{v}_2\ \boldsymbol{v}_3]$ satisfies

$$A = \begin{bmatrix} 3 & 2 & 7 \\ 7 & 0 & 0 \\ 1 & 1 & 1 \\ -2 & 4 & 6 \end{bmatrix} \sim \begin{bmatrix} 1 & 1 & 1 \\ 0 & -7 & -7 \\ 0 & -1 & 4 \\ 0 & 6 & 8 \end{bmatrix} \begin{matrix} \text{I} \leftrightarrow \text{III} \\ \text{II} - 7\text{I} \\ \text{III} - 3\text{I} \\ \text{IV} + 2\text{I} \end{matrix}$$

$$\sim \begin{bmatrix} 1 & 1 & 1 \\ 0 & 1 & 1 \\ 0 & 0 & 5 \\ 0 & 0 & 0 \end{bmatrix} \begin{matrix} \text{II} \times -\frac{1}{7} \\ \text{III} + \text{II} \\ \text{IV} - 6\text{II} \\ \text{IV} - \frac{2}{5}\,\text{III} \end{matrix} .$$

Clearly $\operatorname{Span}\{\boldsymbol{v}_1, \boldsymbol{v}_2, \boldsymbol{v}_3\}$ has dimension 3. Let us find a hyperplane H that contains $\boldsymbol{v}_1, \boldsymbol{v}_2, \boldsymbol{v}_3$. That is, we look for a hyperplane

$$\boldsymbol{u} \cdot \boldsymbol{x} = 0$$

where $\boldsymbol{u} \cdot \boldsymbol{v}_1 = 0, \boldsymbol{u} \cdot \boldsymbol{v}_2 = 0, \boldsymbol{u} \cdot \boldsymbol{v}_3 = 0$. The linear system for $\boldsymbol{u}$ has coefficient matrix

$$\begin{bmatrix} 3 & 7 & 1 & -2 \\ 2 & 0 & 1 & 4 \\ 7 & 0 & 1 & 6 \end{bmatrix} \sim \begin{bmatrix} 1 & 0 & \frac{1}{2} & 2 \\ 0 & 7 & -\frac{1}{2} & -8 \\ 0 & 0 & -\frac{5}{2} & -8 \end{bmatrix} \begin{matrix} \text{I} \leftrightarrow \text{II} \\ \text{I} \times \frac{1}{2} \\ \text{II} - 3\text{I} \\ \text{III} - 7\text{I} \end{matrix}$$

$$\sim \begin{bmatrix} 1 & 0 & 0 & \frac{2}{5} \\ 0 & 7 & 0 & -\frac{32}{5} \\ 0 & 0 & 1 & \frac{16}{5} \end{bmatrix} \begin{matrix} \text{III} \times -\frac{2}{5} \\ \text{II} + \frac{1}{2}\,\text{III} \\ \text{I} - \frac{1}{2}\,\text{III} \end{matrix} .$$

A solution is $\left(-\frac{2}{5}, \frac{32}{35}, -\frac{16}{5}, 1\right)$ or $(-14, 32, -112, 35)$, so H has equation

$$-14x_1 + 32x_2 - 112x_3 + 35x_4 = 0.$$

Since H has dimension 3, we deduce from Proposition 9 that $H = \operatorname{Span}\{\boldsymbol{v}_1, \boldsymbol{v}_2, \boldsymbol{v}_3\}$.

6 Coordinate vectors. Change of coordinates

Given a basis $\boldsymbol{v}_1, \ldots, \boldsymbol{v}_r$ of a subspace V of $\mathbb{R}^n$ and a vector $\boldsymbol{v}$ in V, we can certainly write $\boldsymbol{v}$ in at least one way as

$$\boldsymbol{v} = a_1\boldsymbol{v}_1 + \cdots + a_r\boldsymbol{v}_r.$$

In fact, there is only one choice of $\boldsymbol{a} = (a_1, \dots, a_r)$, since if

$$\boldsymbol{v} = b_1\boldsymbol{v}_1 + \cdots + b_r\boldsymbol{v}_r$$

then

$$(b_1 - a_1)\boldsymbol{v}_1 + \cdots + (b_r - a_r)\boldsymbol{v}_r = \mathbf{0}.$$

In view of the linear independence of $\boldsymbol{v}_1, \dots, \boldsymbol{v}_r$, we get $b_1 = a_1, b_2 = a_2, \dots, b_r = a_r$.

We think of $\boldsymbol{v}_1, \dots, \boldsymbol{v}_r$ as providing a coordinate system for V and call $\boldsymbol{a}$ the **coordinate vector** of $\boldsymbol{v}$ for the basis $\boldsymbol{v}_1, \dots, \boldsymbol{v}_r$. In Figure 2, we illustrate the case $n = 3, r = 2$.

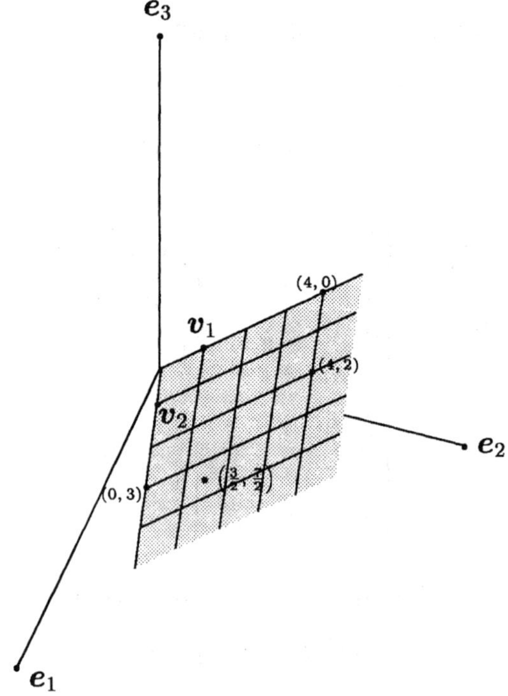

Figure 2. Coordinate pairs for the basis $\boldsymbol{v}_1, \boldsymbol{v}_2$ of the shaded plane.

Example 14 The plane

$$x_1 + 3x_2 + 7x_3 = 0$$

has a basis $\boldsymbol{v}_1, \boldsymbol{v}_2$ where $\boldsymbol{v}_1 = (7, -14, 5), \boldsymbol{v}_2 = (0, 7, -3)$. Find the coordinate vector $\boldsymbol{a}$ of $\boldsymbol{v} = (1, 2, -1)$ for the basis $\boldsymbol{v}_1, \boldsymbol{v}_2$.

Solution. We must solve $(1, 2, -1) = a_1\boldsymbol{v}_1 + a_2\boldsymbol{v}_2$. The augmented matrix is

$$\begin{bmatrix} 7 & 0 & 1 \\ -14 & 7 & 2 \\ 5 & -3 & -1 \end{bmatrix} \sim \begin{bmatrix} 1 & 0 & \frac{1}{7} \\ 0 & 7 & 4 \\ 0 & -3 & -\frac{12}{7} \end{bmatrix} \begin{matrix} \text{II} + 2\text{I} \\ \text{I} \times \frac{1}{7} \\ \text{III} - 5\text{I} \end{matrix}$$

$$\sim \begin{bmatrix} 1 & 0 & \frac{1}{7} \\ 0 & 1 & \frac{4}{7} \\ 0 & 0 & 0 \end{bmatrix} \begin{matrix} \text{II} \times \frac{1}{7} \\ \text{III} + 3\text{II} \end{matrix}.$$

Thus $\boldsymbol{a} = \left(\frac{1}{7}, \frac{4}{7}\right)$.

It is very useful to switch from a given basis of a subspace to a more convenient basis, as we shall see in our work on eigenvalues. We now derive the formula for the change of coordinates when we switch bases.

Suppose that $\boldsymbol{v}_1, \dots, \boldsymbol{v}_r$ and $\boldsymbol{w}_1, \dots, \boldsymbol{w}_r$ are two bases of the subspace V of $\mathbb{R}^n$. Then

$$\boldsymbol{w}_j = p_{1j}\boldsymbol{v}_1 + \cdots + p_{rj}\boldsymbol{v}_r \tag{16}$$

for certain real numbers p_{ij} $(i, j = 1, \dots, r)$. Let $P = [p_{ij}]_{1 \le i,j \le r}$. Let $\boldsymbol{y}$ be the coordinate vector of a given point $\boldsymbol{v}$ for $\boldsymbol{v}_1, \dots \boldsymbol{v}_r$, and $\boldsymbol{z}$ the coordinate vector of $\boldsymbol{v}$ for $\boldsymbol{w}_1, \dots, \boldsymbol{w}_r$:

$$\begin{aligned} \boldsymbol{v} = y_1\boldsymbol{v}_1 + \cdots + y_r\boldsymbol{v}_r &= \sum_{j=1}^{r} z_j\boldsymbol{w}_j \\ &= \sum_{j=1}^{r} z_j \left(\sum_{i=1}^{r} p_{ij}\boldsymbol{v}_i \right) = \sum_{i=1}^{r} \left(\sum_{j=1}^{r} p_{ij}z_j \right) \boldsymbol{v}_i. \end{aligned}$$

Thus $y_i = \sum_{j=1}^{r} p_{ij}z_j$, which can be expressed succinctly as

$$\boldsymbol{y} = P\boldsymbol{z}. \tag{17}$$

Proposition 13 *Let $\boldsymbol{v}_1, \dots, \boldsymbol{v}_r$ and $\boldsymbol{w}_1, \dots, \boldsymbol{w}_r$ be bases of the subspace V,*

$$\boldsymbol{w}_j = p_{1j}\boldsymbol{v}_1 + \cdots + p_{rj}\boldsymbol{v}_r.$$

The coordinate vectors $\boldsymbol{y}$ and $\boldsymbol{z}$ for $\boldsymbol{v}_1, \dots, \boldsymbol{v}_r$ and $\boldsymbol{w}_1, \dots, \boldsymbol{w}_r$ satisfy (17).

If $r = n$ and $\boldsymbol{y}$ denotes coordinate vector for the standard basis, then

$$P = [\boldsymbol{w}_1 \cdots \boldsymbol{w}_n]. \tag{18}$$

The simplest way to get (18) is to note that (16) reads: $\boldsymbol{w}_j$ is column j of P.

Example 15 Find the coordinates of (z_1, z_2) for the basis $\boldsymbol{v}_1 = (1,6), \boldsymbol{v}_2 = (1,7)$ of $\mathbb{R}^2$.

Solution. Here $\boldsymbol{w}_1 = \boldsymbol{e}_1, \boldsymbol{w}_2 = \boldsymbol{e}_2$. Thus the coordinate vector of $\boldsymbol{z}$ for $(1,6),(1,7)$ is

$$\begin{bmatrix} y_1 \\ y_2 \end{bmatrix} = P \begin{bmatrix} z_1 \\ z_2 \end{bmatrix} = \begin{bmatrix} p_{11} & p_{12} \\ p_{21} & p_{22} \end{bmatrix} \cdot \begin{bmatrix} z_1 \\ z_2 \end{bmatrix}$$

where the p_{ij} are found from (16):

$$\begin{aligned} (1,0) = \boldsymbol{e}_1 = p_{11}\boldsymbol{v}_1 + p_{21}\boldsymbol{v}_2 = (p_{11} + p_{21}, 6p_{11} + 7p_{21}) \\ (0,1) = \boldsymbol{e}_2 = p_{12}\boldsymbol{v}_1 + p_{22}\boldsymbol{v}_2 = (p_{12} + p_{22}, 6p_{12} + 7p_{22}). \end{aligned}$$

We now solve the equations to get $p_{21} = 1 - p_{11}$,

$$6p_{11} + 7(1 - p_{11}) = 0, p_{11} = 7, p_{21} = -6.$$

Similarly, $p_{22} = -p_{12}$,

$$6p_{12} + 7(-p_{12}) = 1, p_{12} = -1, p_{22} = 1.$$

The change of coordinates is

$$\begin{bmatrix} y_1 \\ y_2 \end{bmatrix} = \begin{bmatrix} 7 & -1 \\ -6 & 1 \end{bmatrix} \begin{bmatrix} z_1 \\ z_2 \end{bmatrix}.$$

As a simple check we take $\boldsymbol{z} = (1,6)$ and get $\boldsymbol{y} = (1,0)$; likewise $\boldsymbol{z} = (1,7)$ gives $\boldsymbol{y} = (0,1)$.

Proposition 14 *Let V be an r-dimensional subspace of $\mathbb{R}^n$ and let P be the $r \times r$ matrix of a change of coordinates as in (17). Then P has rank r.*

Proof. We need only show that P has zero nullity, because of Proposition 10. If $P\boldsymbol{z} = \boldsymbol{0}$ for some $\boldsymbol{z}$, then the coordinate vector of $z_1\boldsymbol{w}_1 + \cdots + z_r\boldsymbol{w}_r$ for $\boldsymbol{v}_1, \ldots, \boldsymbol{v}_r$ is $\boldsymbol{0}$. (We have labeled the two bases involved just as in (16).) Hence $z_1\boldsymbol{w}_1 + \cdots + z_r\boldsymbol{w}_r = \boldsymbol{0}$. Independence of the $\boldsymbol{w}_i$ gives $\boldsymbol{z} = \boldsymbol{0}$. This proves that P has zero nullity.

Exercises for Chapter 3

Attempt exercises a.b after reading Section a.

2.1 Let $\boldsymbol{w}_1, \boldsymbol{w}_2, \boldsymbol{w}_3$ be vectors in $\mathbb{R}^n$ with

$$\boldsymbol{w}_3 = 6\boldsymbol{w}_1 - 9\boldsymbol{w}_2.$$

Write the linear combination $a_1\boldsymbol{w}_1 + a_2\boldsymbol{w}_2 + a_3\boldsymbol{w}_3$ in the form

(i) $b_1\boldsymbol{w}_1 + b_2\boldsymbol{w}_2$,

(ii) $c_1\boldsymbol{w}_1 + c_2\boldsymbol{w}_3$

(iii) $d_1\boldsymbol{w}_2 + d_2\boldsymbol{w}_3$.

2.2 Find three linearly dependent vectors in $\mathbb{R}^3$ such that none of them is a multiple of another.

2.3 Decide whether each of the following statements is true or false.

(i) A subset of a linearly dependent set is linearly dependent.

(ii) A set that contains a linearly dependent set is linearly dependent.

(iii) A set that contains a linearly independent set is linearly independent.

2.4 An **upper triangular matrix** is a square matrix A having zeros below the main diagonal ($a_{ij} = 0$ if $i > j$). If the entries a_{ii} on the main diagonal of an upper triangular matrix A are nonzero, prove that the column vectors of A are linearly independent.

In each of Exercises 2.5–2.7, determine whether the column vectors of A are linearly independent.

2.5 $A = \begin{bmatrix} 1 & 2 & 3 & 4 \\ 3 & 2 & 1 & 0 \\ 0 & -1 & -3 & -5 \\ 2 & 4 & 6 & 9 \end{bmatrix}$

2.6 $A = \begin{bmatrix} 1 & 2 & 3 \\ 0 & 0 & 1 \end{bmatrix}$

2.7 $\begin{bmatrix} 1 & 1 & -1 & 3 & 4 \\ -1 & 1 & -7 & 6 & 7 \\ 0 & 2 & -8 & 2 & 4 \\ 3 & 7 & -19 & 1 & 8 \end{bmatrix}$

3.1 Find a basis of Nul A, A as in Exercise 2.5.

3.2 Find a basis of Nul A, A as in Exercise 2.6.

3.3 Find a basis of Nul A, A as in Exercise 2.7.

3.4 Let A be an $m \times n$ matrix with $n > m$. Show that dim(Nul A) $\geq n - m$.

3.5 Let $\boldsymbol{u}_1, \ldots, \boldsymbol{u}_r$ be a linearly independent set in $\mathbb{R}^n$ and let $\boldsymbol{u}$ be a vector in $\mathbb{R}^n$. Show that the following two statements are equivalent.

(i) Span$\{\boldsymbol{u}_1, \ldots, \boldsymbol{u}_r, \boldsymbol{u}\}$ has dimension $r + 1$.

(ii) $\boldsymbol{u}$ is not in Span$\{\boldsymbol{u}_1, \ldots, \boldsymbol{u}_r\}$.

3.6 Let V and W be subspaces of $\mathbb{R}^n$. Then $V \cap W$ (the **intersection** of V and W) denotes the set of vectors belonging to both V and W. Show that $V \cap W$ is a subspace of $\mathbb{R}^n$. (**Hint:** show that $V \cap W$ has the closure property.)

3.7 Give examples of subspaces V and W of $\mathbb{R}^3$ with the following properties.

(i) $\dim V = 2, \dim W = 2, \dim V \cap W = 2$.

(ii) $\dim V = 2, \dim W = 2, \dim V \cap W = 1$.

(iii) $\dim V = 2, \dim W = 1, \dim V \cap W = 1$.

(iv) $\dim V = 2, \dim W = 1, \dim V \cap W = 0$.

3.8 Let V and W be subspaces of $\mathbb{R}^n$. Then $V + W$ (the **sum** of V and W) denotes the set of all possible sums $\boldsymbol{v} + \boldsymbol{w}$ with $\boldsymbol{v}$ in V, $\boldsymbol{w}$ in W. Show that $V + W$ is a subspace of $\mathbb{R}^n$. **Hint:** Show that $V + W$ has the closure property.

3.9 If V is a plane through $\mathbf{0}$ in $\mathbb{R}^3$ and W is a straight line through $\mathbf{0}$ in $\mathbb{R}^3$, show that

(i) $V + W = \mathbb{R}^3$ if W does not lie in V.

(ii) $V + W = V$ if W lies in V.

3.10 If V and W are planes through $\mathbf{0}$ in $\mathbb{R}^3$, show that

(i) $V + W = \mathbb{R}^3$ if $W \neq V$.

(ii) $V + W = V$ if $W = V$.

4.1 Let $\mathrm{Span}\{\boldsymbol{u}_1, \ldots, \boldsymbol{u}_r\} = V$ and $\mathrm{Span}\{\boldsymbol{w}_1, \ldots, \boldsymbol{w}_s\} = W$. Show that

$$V + W = \mathrm{Span}\{\boldsymbol{u}_1, \ldots, \boldsymbol{u}_r, \boldsymbol{w}_1, \ldots, \boldsymbol{w}_s\}.$$

4.2 Let V, W be subspaces of $\mathbb{R}^n$. Let $\boldsymbol{u}_1, \ldots, \boldsymbol{u}_k$ be a basis of $V \cap W$. Let $\boldsymbol{u}_1, \ldots, \boldsymbol{u}_k, \boldsymbol{v}_1, \ldots, \boldsymbol{v}_r$ be a basis of V and $\boldsymbol{u}_1, \ldots, \boldsymbol{u}_k, \boldsymbol{w}_1, \ldots, \boldsymbol{w}_s$ a basis of W.

(i) Show that $\boldsymbol{u}_1, \ldots, \boldsymbol{u}_k, \boldsymbol{v}_1, \ldots, \boldsymbol{u}_r, \boldsymbol{w}_1, \ldots, \boldsymbol{w}_s$ is independent.
Hint: Suppose that $\boldsymbol{v} = a_1\boldsymbol{v}_1 + \cdots + a_r\boldsymbol{v}_r = b_1\boldsymbol{u}_1 + \cdots + b_k\boldsymbol{u}_k + c_1\boldsymbol{w}_1 + \cdots + c_s\boldsymbol{w}_s$. Show that $\boldsymbol{v}$ is in both V and W.

(ii) Show that $\mathrm{Span}\{\boldsymbol{u}_1, \ldots, \boldsymbol{u}_k, \boldsymbol{v}_1, \ldots, \boldsymbol{v}_r, \boldsymbol{w}_1, \ldots, \boldsymbol{w}_s\} = V + W$.

(iii) Show that

$$\dim(V + W) + \dim(V \cap W) = \dim V + \dim W.$$

4.3 In each of the examples you found in Exercise 3.7, find $V + W$ and show that the value of $\dim(V+W) + \dim(V \cap W)$ conforms with Exercise 4.2.

In each of questions 4.4–4.6, verify that $\boldsymbol{v}_1, \boldsymbol{v}_2, \boldsymbol{v}_3$ is a basis of $\mathbb{R}^3$.

4.4 $\boldsymbol{v}_1 = (1, 9, 0), \boldsymbol{v}_2 = (0, 1, -1), \boldsymbol{v}_3 = (2, 7, 0)$.

4.5 $\boldsymbol{v}_1 = (2, 3, 2), \boldsymbol{v}_2 = (2, 2, 3), \boldsymbol{v}_3 = (3, 2, 2)$.

4.6 $\boldsymbol{v}_1 = (1, 5, 6), \boldsymbol{v}_2 = (4, 1, 2), \boldsymbol{v}_3 = (-1, 1, 0)$.

5.1 Find a basis of Col A, with A as in Exercise 2.5.

5.2 Find a basis of Col A, with A as in Exercise 2.6.

5.3 Find a basis of Col A, with A as in Exercise 2.7.

5.4 Show that the following two statements are equivalent.

(i) $\boldsymbol{b}$ is in Col A.

(ii) The linear system $A\boldsymbol{x} = \boldsymbol{b}$ is consistent.

5.5 Let $\boldsymbol{w}_1, \dots, \boldsymbol{w}_{n-1}$ be linearly independent vectors in $\mathbb{R}^n$. Show that the general solution of the linear system

$$\boldsymbol{w}_1 \cdot \boldsymbol{x} = 0, \dots, \boldsymbol{w}_{n-1} \cdot \boldsymbol{x} = 0$$

is a straight line through $\mathbf{0}$. (**Hint:** rank + nullity = n.)

6.1 Find the coordinate vector of (1, 0, 0) for the basis $\boldsymbol{v}_1, \boldsymbol{v}_2, \boldsymbol{v}_3$ of Exercise 4.4.

6.2 Find the coordinate vector of (1, 0, 0) for the basis $\boldsymbol{v}_1, \boldsymbol{v}_2, \boldsymbol{v}_3$ of Exercise 4.5.

6.3 Find the coordinate vector of (1, 0, 0) for the basis $\boldsymbol{v}_1, \boldsymbol{v}_2, \boldsymbol{v}_3$ of Exercise 4.6.

Chapter 4

Linear mappings. Matrix algebra

1 Linear mappings

A **mapping** $f : X \to Y$ is a rule that assigns a unique $f(x)$ in Y to every member x of the set X. For example, $f(x) = \sin x$ defines a mapping $f : \mathbb{R} \to \mathbb{R}$.

For any mapping $f : X \to Y$, we say that $f(x)$ is the **image** of x. We also say that f **maps** X **into** Y.

Let m, n be positive integers.

Definition 1 Let $T : \mathbb{R}^n \to \mathbb{R}^m$ be a mapping which has the properties

$$T(\boldsymbol{x} + \boldsymbol{z}) = T(\boldsymbol{x}) + T(\boldsymbol{z}), \tag{1}$$

$$T(c\boldsymbol{x}) = cT(\boldsymbol{x}) \tag{2}$$

for any $\boldsymbol{x}, \boldsymbol{z}$ in $\mathbb{R}^n$ and any real c. We say that T is a **linear** mapping.

Briefly, we say 'T is linear'. We shall write $T\boldsymbol{x}$ instead of $T(\boldsymbol{x})$; however, expressions such as $T(c\boldsymbol{x}), T(\boldsymbol{x} + \boldsymbol{z})$ need brackets.

We note right away from (2) that $T\mathbf{0} = \mathbf{0}$, since $T\mathbf{0} = T(0\mathbf{0}) = 0T\mathbf{0} = \mathbf{0}$. But what do linear mappings look like? The following examples should help.

Example 1 Let a be a given angle. For $\boldsymbol{x}$ in $\mathbb{R}^2$ let $T\boldsymbol{x}$ be the vector obtained by rotating $\boldsymbol{x}$ anticlockwise through a. Then $T : \mathbb{R}^2 \to \mathbb{R}^2$ is linear. To get (1), (2), rotate the parallellogram with vertices $\mathbf{0}, \boldsymbol{x}, \boldsymbol{z}, \boldsymbol{x} + \boldsymbol{z}$ through

a to get a parallellogram with vertices $\mathbf{0}, T\boldsymbol{x}, T\boldsymbol{z}, T(\boldsymbol{x}+\boldsymbol{z})$. Thus (1) holds. Similarly, rotate $c\boldsymbol{x}$ to see that $T(c\boldsymbol{x}) = cT\boldsymbol{x}$ (Figure 1).

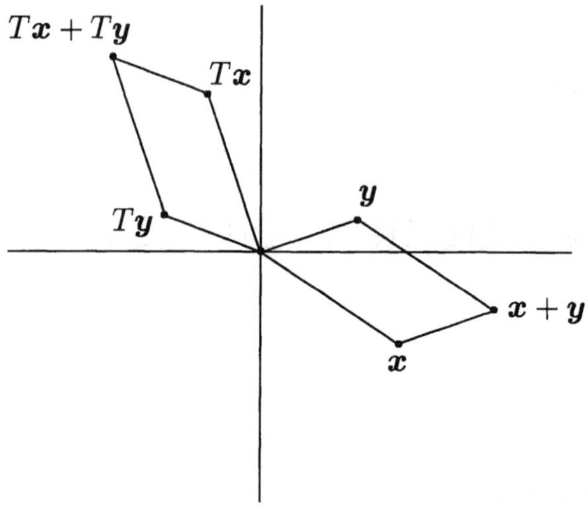

Figure 1. Add first, then rotate; or rotate first, then add.

Example 2 Let $\boldsymbol{b}$ be a nonzero vector in $\mathbb{R}^n$ and let $S\boldsymbol{x}$ denote the projection of $\boldsymbol{x}$ onto $\boldsymbol{b}$, that is, the foot of the perpendicular from $\boldsymbol{x}$ to the line through $\mathbf{0}$ and $\boldsymbol{b}$. The formula for $S\boldsymbol{x}$ is

$$S\boldsymbol{x} = \frac{\boldsymbol{x}\cdot\boldsymbol{b}}{|\boldsymbol{b}|^2}\,\boldsymbol{b}.$$

To see that $S\boldsymbol{x}$ is the foot of the perpendicular, verify that $(\boldsymbol{x} - S\boldsymbol{x})\cdot\boldsymbol{b} = 0$. See Figure 24 of Chapter 1.

Now S is a linear mapping, $S : \mathbb{R}^n \to \mathbb{R}^n$. To verify this,

$$\begin{aligned} S(\boldsymbol{x}+\boldsymbol{y}) &= \frac{(\boldsymbol{x}+\boldsymbol{y})\cdot\boldsymbol{b}}{|\boldsymbol{b}|^2}\,\boldsymbol{b} \\ &= \frac{\boldsymbol{x}\cdot\boldsymbol{b}}{|\boldsymbol{b}|^2}\,\boldsymbol{b} + \frac{\boldsymbol{y}\cdot\boldsymbol{b}}{|\boldsymbol{b}|^2}\,\boldsymbol{b} = S\boldsymbol{x} + S\boldsymbol{y}; \end{aligned}$$

$$S(c\boldsymbol{x}) = \frac{(c\boldsymbol{x})\cdot\boldsymbol{b}}{|\boldsymbol{b}|^2}\,\boldsymbol{b} = c\frac{\boldsymbol{x}\cdot\boldsymbol{b}}{|\boldsymbol{b}|^2}\,\boldsymbol{b} = cS\boldsymbol{x}.$$

See Figure 2.

As a simple example, let $n = 2, \boldsymbol{b} = \boldsymbol{e}_1$. Then

$$S\boldsymbol{x} = (\boldsymbol{x}\cdot\boldsymbol{e}_1)\boldsymbol{e}_1 = x_1\boldsymbol{e}_1.$$

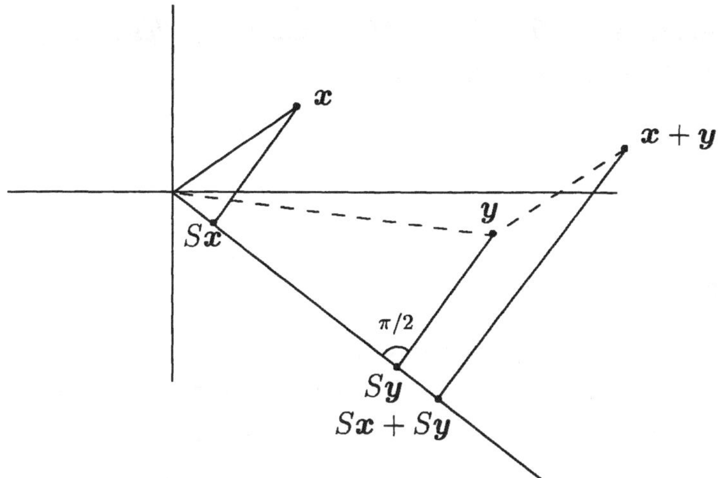

Figure 2. The projection of $\boldsymbol{x}+\boldsymbol{y}$ is the sum of the projections of $\boldsymbol{x}$ and $\boldsymbol{y}$.

Example 3 Let $T:\mathbb{R}^3 \to \mathbb{R}^3$ be defined by

$$T\boldsymbol{x} = (x_1 + x_2 - x_3, x_2 + x_3, 4x_3).$$

Then

$$\begin{aligned} T(\boldsymbol{x}+\boldsymbol{y}) &= (x_1 + y_1 + x_2 + y_2 - x_3 - y_3, x_2 + y_2 + x_3 + y_3, 4x_3 + 4y_3) \\ &= T\boldsymbol{x} + T\boldsymbol{y}, \\ T(c\boldsymbol{x}) &= (cx_1 + cx_2 - cx_3, cx_2 + cx_3, 4cx_3) \\ &= cT\boldsymbol{x}. \end{aligned}$$

So T is linear.

We will use these examples to illustrate the concepts introduced in this chapter. The following mapping acts in a neutral way on vectors, rather like 1 in multiplication of numbers.

Example 4 Let $I:\mathbb{R}^n \to \mathbb{R}^n, I\boldsymbol{x} = \boldsymbol{x}$. We call I the **identity** map . You can easily check that I is linear.

Any linear mapping T can be written in the form

$$T\boldsymbol{x} = A\boldsymbol{x} \tag{3}$$

where A is $m \times n$ if T maps $\mathbb{R}^n$ to $\mathbb{R}^m$. Let us record this in

Proposition 1 *Let $T : \mathbb{R}^n \to \mathbb{R}^m$ be linear and let A be the $m \times n$ matrix with columns $T\boldsymbol{e}_1, \ldots, T\boldsymbol{e}_n$,*

$$A = [T\boldsymbol{e}_1 \ldots T\boldsymbol{e}_n]. \tag{4}$$

Then

$$T\boldsymbol{x} = A\boldsymbol{x}$$

for all $\boldsymbol{x}$.

Proof. $T\boldsymbol{x} = T(x_1\boldsymbol{e}_1 + \cdots + x_n\boldsymbol{e}_n)$
$= x_1 T\boldsymbol{e}_1 + \cdots + x_n T\boldsymbol{e}_n$ by repeated use of (1), (2)
$= A\boldsymbol{x}$.

We say that A is the matrix of T when (4) holds; or, T has matrix A. Of course we could start with an $m \times n$ matrix A and use (3) to define a linear mapping $T : \mathbb{R}^n \to \mathbb{R}^m$.

Example 5 (i) The matrix of the rotation in Example 1 is

$$\begin{bmatrix} \cos a & -\sin a \\ \sin a & \cos a \end{bmatrix}$$

since $T(\boldsymbol{e}_1) = (\cos a, \sin a), T(\boldsymbol{e}_2) = (-\sin a, \cos a)$ (verify via a sketch!).

(ii) The matrix of the projection $S : \mathbb{R}^2 \to \mathbb{R}^2$ in Example 2 (case $n = 2$) is

$$\frac{1}{b_1^2 + b_2^2} \begin{bmatrix} b_1^2 & b_1 b_2 \\ b_1 b_2 & b_2^2 \end{bmatrix}$$

since $S(\boldsymbol{e}_j) = \frac{b_j \boldsymbol{b}}{|\boldsymbol{b}|^2}$. (The scalar product of a matrix and a real number occurs here. See Definition 2 for a formal definition.)

(iii) The matrix of the mapping T in Example 3 is

$$\begin{bmatrix} 1 & 1 & -1 \\ 0 & 1 & 1 \\ 0 & 0 & 4 \end{bmatrix}$$

since

$$T\boldsymbol{x} = \begin{bmatrix} x_1 + x_2 - x_3 \\ x_2 + x_3 \\ 4x_3 \end{bmatrix} = \begin{bmatrix} 1 & 1 & -1 \\ 0 & 1 & 1 \\ 0 & 0 & 4 \end{bmatrix} \begin{bmatrix} x_1 \\ x_2 \\ x_3 \end{bmatrix}.$$

The matrix of I in Example 4 is the matrix

$$I = \begin{bmatrix} 1 & & & & \\ & \cdot & & & \\ & & \cdot & & \\ & & & \cdot & \\ & & & & 1 \end{bmatrix}$$

encountered in Chapter 2 (blank spaces denote zeros). Note the use of the same symbol for the mapping I and its matrix.

Definition 2 The **sum** $S+T$ of two mappings $S : \mathbb{R}^n \to \mathbb{R}^m, T : \mathbb{R}^n \to \mathbb{R}^m$ is defined by

$$(S+T)\boldsymbol{x} = S\boldsymbol{x} + T\boldsymbol{x}.$$

The **scalar product** cS is defined by

$$(cS)\boldsymbol{x} = c(S\boldsymbol{x}).$$

The **sum** $A+B$ of two $m \times n$ matrices $A = [a_{ij}], B = [b_{ij}]$ and the scalar product cA are defined by

$$A + B = [a_{ij} + b_{ij}], \quad cA = [ca_{ij}].$$

For example,

$$\begin{bmatrix} a_{11} & a_{12} \\ a_{21} & a_{22} \end{bmatrix} + \begin{bmatrix} b_{11} & b_{12} \\ b_{21} & b_{22} \end{bmatrix} = \begin{bmatrix} a_{11}+b_{11} & a_{12}+b_{12} \\ a_{21}+b_{21} & a_{22}+b_{22} \end{bmatrix},$$

$$2\begin{bmatrix} 0 & 1 \\ -3 & 4 \end{bmatrix} = \begin{bmatrix} 0 & 2 \\ -6 & 8 \end{bmatrix}.$$

It is a simple observation, but an important one, that if S has matrix A and T has matrix B, then $S+T$ is a linear mapping with matrix $A+B$. For

$$(A+B)\boldsymbol{x} = A\boldsymbol{x} + B\boldsymbol{x} = (S+T)\boldsymbol{x} \tag{5}$$

by definition of $S+T$. Similarly cS is linear and has matrix cA.

Example 6 Let T, S be as in Examples 1 and 2 with $a = \frac{\pi}{2}, \boldsymbol{b} = (2,3)$. Find the matrix of $S+T$.

Solution. Since $\cos\frac{\pi}{2} = 0, \sin\frac{\pi}{2} = 1$, the matrix of T is

$$\begin{bmatrix} 0 & -1 \\ 1 & 0 \end{bmatrix}.$$

The matrix of S is

$$\frac{1}{13}\begin{bmatrix} 4 & 6 \\ 6 & 9 \end{bmatrix}$$

(Example 5). So $S + T$ has matrix

$$\begin{bmatrix} \frac{4}{13} & -\frac{7}{13} \\ \frac{19}{13} & \frac{9}{13} \end{bmatrix}.$$

Now for multiplication (or **composition**) of mappings. We say that linear mappings T, S are **compatible** if

$$S : \mathbb{R}^n \to \mathbb{R}^m, T : \mathbb{R}^m \to \mathbb{R}^p$$

for some m, n and p.

Definition 3 Suppose T, S is a compatible pair. The mapping TS is defined by

$$(TS)\boldsymbol{x} = T(S\boldsymbol{x}). \tag{6}$$

That is, to map $\boldsymbol{x}$ under TS we apply S first to $\boldsymbol{x}$, then apply T to $S\boldsymbol{x}$. Notice that TS maps $\mathbb{R}^n$ into $\mathbb{R}^p$.

Example 7 Find the matrices of TS and ST where T, S are as in Example 4.

Solution. Here $m = n = p = 2$. We have

$$\begin{aligned} (TS)\boldsymbol{x} &= T\left(\frac{4x_1 + 6x_2}{13}, \frac{6x_1 + 9x_2}{13}\right) \\ &= \left(\frac{-6x_1 - 9x_2}{13}, \frac{4x_1 + 6x_2}{13}\right) \end{aligned}$$

so TS has matrix

$$\frac{1}{13}\begin{bmatrix} -6 & -9 \\ 4 & 6 \end{bmatrix}.$$

Now

$$\begin{aligned}(ST)\boldsymbol{x} &= S(-x_2, x_1)\\ &= \left(\frac{6x_1 - 4x_2}{13}, \frac{9x_1 - 6x_2}{13}\right).\end{aligned}$$

Thus ST has matrix

$$\frac{1}{13}\begin{bmatrix}6 & -4\\ 9 & -6\end{bmatrix}.$$

Notice that ST and TS differ! But at least we can say that both are linear.

Proposition 2 *The product of compatible linear mappings, T, S is linear.*

Proof.

$$(TS)(\boldsymbol{x} + \boldsymbol{y}) = T(S\boldsymbol{x} + S\boldsymbol{y}) = (TS)\boldsymbol{x} + (TS)\boldsymbol{y}$$

on using (1) twice;

$$(TS)(c\boldsymbol{x}) = T(cS\boldsymbol{x}) = c(TS)\boldsymbol{x}$$

on using (2) twice.

Suppose

$$S : \mathbb{R}^n \to \mathbb{R}^m, T : \mathbb{R}^m \to \mathbb{R}^p, U : \mathbb{R}^p \to \mathbb{R}^q.$$

Then

$$U(TS) = (UT)S$$

(**associative law** for mappings). To see this, both mappings are carried out in the same three steps: map S under $\boldsymbol{x}$, then $S\boldsymbol{x}$ under T, then $T(S\boldsymbol{x})$ under U.

We can multiply the matrices of compatible mappings; we defer matrix multiplication to Section 3.

2 One-to-one mappings, onto mappings and invertible mappings

A mapping $T : \mathbb{R}^n \to \mathbb{R}^m$ is said to be **one-to-one** if, for each $\boldsymbol{y}$ in $\mathbb{R}^m$, at most one $\boldsymbol{x}$ maps to $\boldsymbol{y}$. Equivalently, the equation

$$T\boldsymbol{u} = T\boldsymbol{v} \tag{7}$$

implies $\boldsymbol{u} = \boldsymbol{v}$.

Example 8 A rotation is one-to-one, since distinct vectors have distinct images. A projection S onto $\boldsymbol{b}$ is not one-to-one, since $S\boldsymbol{c} = \boldsymbol{0}$ for any $\boldsymbol{c}$ perpendicular to $\boldsymbol{b}$.

The mapping T in Example 3 is one-to-one. To see this, let

$$A = \begin{bmatrix} 1 & 1 & -1 \\ 0 & 1 & 1 \\ 0 & 0 & 4 \end{bmatrix} = [\boldsymbol{a}_1\ \boldsymbol{a}_2\ \boldsymbol{a}_3]$$

be the matrix of A. Then (7) reads

$$u_1\boldsymbol{a}_1 + u_2\boldsymbol{a}_2 + u_3\boldsymbol{a}_3 = v_1\boldsymbol{a}_1 + v_2\boldsymbol{a}_2 + v_3\boldsymbol{a}_3.$$

Since $\boldsymbol{a}_1, \boldsymbol{a}_2, \boldsymbol{a}_3$ are independent, it follows that $\boldsymbol{u} = \boldsymbol{v}$.

Proposition 3 *Let $T : \mathbb{R}^n \to \mathbb{R}^m$ be linear. The following statements are equivalent.*

(i) T is one-to-one.

(ii) The matrix of T has linearly independent columns.

Proof. Let $A = [\boldsymbol{a}_1 \ldots \boldsymbol{a}_n]$ be the matrix of T. Suppose that (i) holds. Then the only vector with $T\boldsymbol{x} = \boldsymbol{0}$ is $\boldsymbol{x} = \boldsymbol{0}$. But

$$(8) \qquad T\boldsymbol{x} = A\boldsymbol{x} = x_1\boldsymbol{a}_1 + \cdots + x_n\boldsymbol{a}_n.$$

So the only solution of $x_1\boldsymbol{a}_1 + \cdots + x_n\boldsymbol{x}_n = \boldsymbol{0}$ is the zero solution, giving (ii).

Now suppose that (ii) holds. If $T\boldsymbol{u} = T\boldsymbol{v}$, then using (8),

$$(u_1 - v_1)\boldsymbol{a}_1 + \cdots + (u_n - v_n)\boldsymbol{a}_n = \boldsymbol{0}.$$

Since the $\boldsymbol{a}_i$ are independent, we deduce that $\boldsymbol{u} = \boldsymbol{v}$.

Definition 4 Let $T : \mathbb{R}^n \to \mathbb{R}^m$ be a linear mapping. The **kernel** of T, written Ker T, is the set of $\boldsymbol{x}$ for which $T\boldsymbol{x} = \boldsymbol{0}$. The **image** of T, written Im T, is the set of all $T\boldsymbol{x}$ as $\boldsymbol{x}$ varies over $\mathbb{R}^n$.

These are familiar objects, in fact. If $A = [\boldsymbol{a}_1 \ldots \boldsymbol{a}_n]$ is the matrix of T, then Ker T = Nul A is obviously true. Moreover, Im T = Col A, since both Im T and Col A consist of all

$$x_1\boldsymbol{a}_1 + \cdots + x_n\boldsymbol{a}_n$$

with $x_1, \ldots, x_n$ real. The **rank** of T is defined to be rank A and the **nullity** of T is defined to be nullity A.

Definition 5 The linear map $T : \mathbb{R}^n \to \mathbb{R}^m$ is said to be *onto* if $\text{Im}\, T = \mathbb{R}^m$.

In other words, if T is an onto map, then for every $\boldsymbol{y}$ in $\mathbb{R}^m$ there is at least one $\boldsymbol{x}$ with $T\boldsymbol{x} = \boldsymbol{y}$.

Example 9 A rotation T is onto, since given $\boldsymbol{y}$ we can always find $\boldsymbol{x}$ such that $T\boldsymbol{x} = \boldsymbol{y}$. A projection $S : \mathbb{R}^n \to \mathbb{R}^n$ onto $\boldsymbol{b}$, with $n > 1$, is not onto, since for $\boldsymbol{y}$ not on the line $\text{Span}\{\boldsymbol{b}\}$, we cannot find $\boldsymbol{x}$ with $S\boldsymbol{x} = \boldsymbol{y}$.

The mapping T in Example 3 is onto, since if

$$A = \begin{bmatrix} 1 & 1 & -1 \\ 0 & 1 & 1 \\ 0 & 0 & 4 \end{bmatrix}$$

and $\boldsymbol{y}$ is given in $\mathbb{R}^3$, we will be able to solve the linear system $A\boldsymbol{x} = \boldsymbol{y}$.

Proposition 4 *Let $T : \mathbb{R}^n \to \mathbb{R}^m$ be linear with matrix A. The following statements are equivalent.*

(i) T is onto.

(ii) A has rank m.

Proof. If (i) holds, then $\text{Im}\, T = \text{Col}\, A = \mathbb{R}^m$. So $\text{rank}\, A = \dim(\text{Col}\, A) = m$.

If (ii) holds, $\text{Col}\, A$ is a subspace of $\mathbb{R}^m$ of dimension m. Thus $\text{Col}\, A$ must be $\mathbb{R}^m$ (Proposition 9 of Chapter 3). Since $\text{Im}\, T = \mathbb{R}^m$, T is onto.

Notice, for instance, that the matrix of S in Example 5 has rank 1 since the columns are proportional.

A linear mapping $T : \mathbb{R}^n \to \mathbb{R}^m$ that is one-to-one and onto is said to be **invertible** . In this case, m and n must be equal, as we now show.

Proposition 5 *Let $T : \mathbb{R}^n \to \mathbb{R}^m$ be invertible. Then $m = n$.*

Proof. The matrix A of T has rank m (Proposition 4). Since A has n columns, we must have $m \leq n$. Moreover, these n columns are linearly independent vectors in $\mathbb{R}^m$ (Proposition 3). So $n \leq m$, and in fact $n = m$.

If $T : \mathbb{R}^n \to \mathbb{R}^n$ is invertible, the equation $\boldsymbol{y} = T\boldsymbol{x}$ has *exactly* one solution $\boldsymbol{x}$ for any given $\boldsymbol{y}$ in $\mathbb{R}^n$. We write T^{-1} for the **inverse** mapping that sends $\boldsymbol{y}$ to $\boldsymbol{x}$.

Example 10 (i) Let $T : \mathbb{R}^2 \to \mathbb{R}^2$ be a rotation through angle a. Then T^{-1} is the rotation through angle $-a$.

(ii) It is not hard to compute T^{-1} for the mapping T in Example 3. Given $\boldsymbol{y}$, we simply solve the system $A\boldsymbol{x} = \boldsymbol{y}$ for $\boldsymbol{x}$. The augmented matrix is

$$\begin{bmatrix} 1 & 1 & -1 & y_1 \\ 0 & 1 & 1 & y_2 \\ 0 & 0 & 4 & y_3 \end{bmatrix} \sim \begin{bmatrix} 1 & 1 & 0 & y_1 + \frac{y_3}{4} \\ 0 & 1 & 0 & y_2 - \frac{y_3}{4} \\ 0 & 0 & 1 & \frac{y_3}{4} \end{bmatrix} \begin{matrix} \text{III} \times \frac{1}{4} \\ \text{II} - \text{III} \\ \text{I} + \text{III} \end{matrix}$$

$$\sim \begin{bmatrix} 1 & 0 & 0 & y_1 - y_2 + \frac{y_3}{2} \\ 0 & 1 & 0 & y_2 - \frac{y_3}{4} \\ 0 & 0 & 1 & \frac{y_3}{4} \end{bmatrix} \text{I} - \text{II} \; .$$

Thus $T^{-1}\boldsymbol{y} = \left(y_1 - y_2 + \frac{y_3}{2}, y_2 - \frac{y_3}{4}, \frac{y_3}{4}\right)$. The matrix of T^{-1} is

$$\begin{bmatrix} 1 & -1 & \frac{1}{2} \\ 0 & 1 & -\frac{1}{4} \\ 0 & 0 & \frac{1}{4} \end{bmatrix}.$$

Proposition 6 *The invertible linear mappings $T : \mathbb{R}^n \to \mathbb{R}^n$ are the linear mappings of rank n. The mapping T^{-1} is linear, and satisfies $T^{-1}T = TT^{-1} = I$.*

Proof. If T is invertible, then T has rank n from Proposition 4. If T has rank n, then T is onto by Proposition 4. Moreover the matrix of T has n linearly independent columns. So T is one-to-one (Proposition 3) and in fact invertible.

To see that T^{-1} is linear, we start with the augmented matrix $[A|\boldsymbol{y}]$ of the system whose solution is $T^{-1}\boldsymbol{y}$. Row operations change the entries of the last column into linear forms in $y_1, \ldots, y_n$, and we end up with, let us say,

$$\left[\begin{array}{c|c} & b_{11}y_1 + \cdots + b_{1n}y_n \\ I & \vdots \\ & b_{n1}y_1 + \cdots + b_{nn}y_n \end{array}\right].$$

Compare Example 10, above. Now $T^{-1}\boldsymbol{y} = B\boldsymbol{y}$ and T^{-1} is linear.

Finally $T^{-1}(T\boldsymbol{x}) = \boldsymbol{x}$ from the definition of T^{-1}. Similarly $T(T^{-1}\boldsymbol{x}) = \boldsymbol{x}$. This proves that $T^{-1}T = TT^{-1} = I$. We note a useful result.

Proposition 7 *Let $T : \mathbb{R}^n \to \mathbb{R}^n, S : \mathbb{R}^n \to \mathbb{R}^n$ be linear mappings such that $TS = I$. Then T is invertible and $S = T^{-1}$.*

Proof. T is onto, since for a given $\boldsymbol{y}$, $T(S\boldsymbol{y}) = \boldsymbol{y}$ provides the solution of $T\boldsymbol{x} = \boldsymbol{y}$. Now T has rank n (Proposition 4) so T is invertible (Proposition 6). Also, $T(T^{-1}\boldsymbol{x}) = T(S\boldsymbol{x})$ for any $\boldsymbol{x}$. (Both sides are $\boldsymbol{x}$.) Since T is one-to-one,

$$T^{-1}\boldsymbol{x} = S\boldsymbol{x}.$$

This proves that $S = T^{-1}$.

Example 11 Compute Ker S and Im S for the projection S onto $\boldsymbol{b}$. Verify that rank S + nullity $S = n$.

Solution. Obviously the image Im S consists of Span$\{\boldsymbol{b}\}$ since every $S\boldsymbol{x}$ is in Span$\{\boldsymbol{b}\}$, and $S(t\boldsymbol{b}) = t\boldsymbol{b}$ for a given t. We find that rank $S = 1$.

The kernel Ker S is the hyperplane with equation

$$b_1x_1 + \cdots + b_nx_n = 0, \tag{9}$$

since $S\boldsymbol{x} = \boldsymbol{0}$ is equivalent to $\boldsymbol{x} \cdot \boldsymbol{b} = 0$. We recall that the hyperplane has dimension $n - 1$, giving rank S + nullity $S = n$.

It is helpful to have a little geometric insight into the effect that an invertible mapping T has on given objects or figures in $\mathbb{R}^n$. The **image** of a set of vectors E in $\mathbb{R}^n$ under T is the set $T(E)$ consisting of all $T\boldsymbol{x}$ with $\boldsymbol{x}$ in E. For simplicity, we take $n = 2$. We see below that the image of a square is a parallellogram. Since any figure can be made up of 'little squares', this gives a clear idea of what happens to an arbitrary plane figure E when we compute the image of E.

Proposition 8 *Let $T : \mathbb{R}^2 \to \mathbb{R}^2$ be an invertible linear mapping. The image of a square under T is a parallellogram.*

Proof. Let Q be a square with three adjacent vertices $\boldsymbol{a}+\boldsymbol{b}, \boldsymbol{a}, \boldsymbol{a}+\boldsymbol{c}$; $\boldsymbol{b}$ and $\boldsymbol{c}$ are perpendicular (Figure 3). Then Q consists of all

$$\boldsymbol{a} + t\boldsymbol{b} + u\boldsymbol{c} \quad (0 \le t \le 1, 0 \le u \le 1); \tag{10}$$

(observe that $t\boldsymbol{b}, u\boldsymbol{c}$ are obtained by shrinking $\boldsymbol{b}, \boldsymbol{c}$ by a chosen amount). Now the image of Q is the set of all

$$T(\boldsymbol{a} + t\boldsymbol{b} + u\boldsymbol{c}) = T\boldsymbol{a} + tT\boldsymbol{b} + uT\boldsymbol{c}, \tag{11}$$

with $0 \leq t \leq 1, 0 \leq u \leq 1$. Much as before, this is a parallellogram with adjacent vertices $T\boldsymbol{a} + T\boldsymbol{b}, T\boldsymbol{a}, T\boldsymbol{a} + T\boldsymbol{c}$ (Figure 4). (Of course, $T\boldsymbol{b}$ cannot be a multiple $kT\boldsymbol{c}$, since $T(\boldsymbol{b} - k\boldsymbol{c}) \neq \boldsymbol{0}$.)

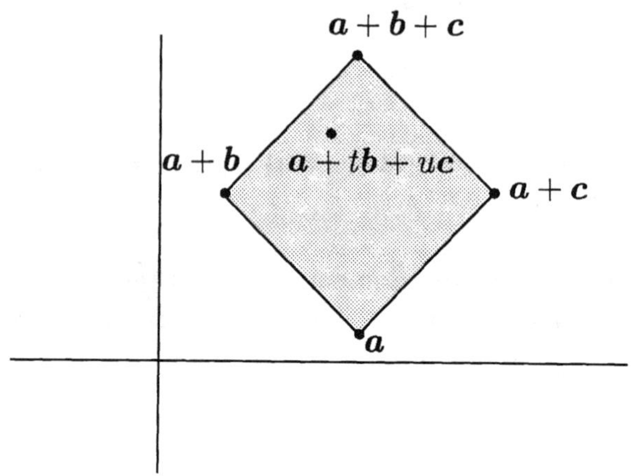

Figure 3. Parametric form of a square.

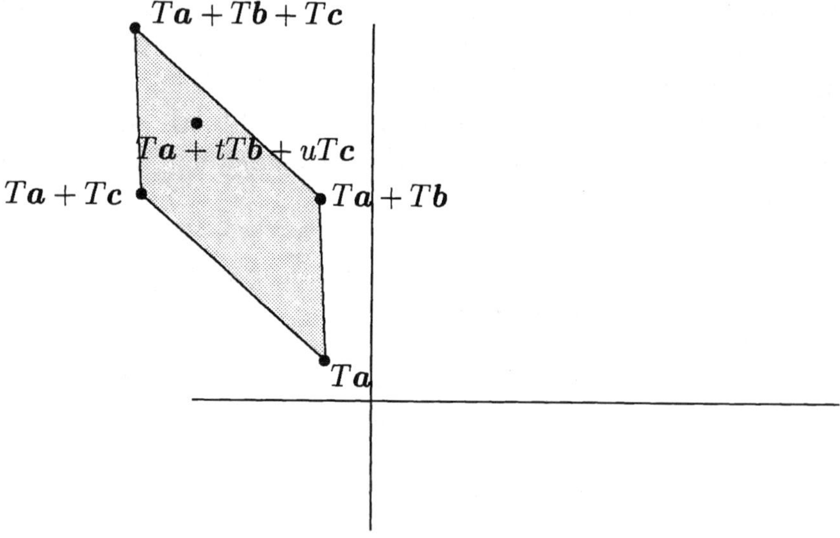

Figure 4. Image of a square under T.

Similarly, for an invertible $T : \mathbb{R}^3 \to \mathbb{R}^3$, the image of a cube is a parallellepiped.

With a little more effort, one can show that there is a constant k, which depends on T, such that the image of a plane figure E has area $k\times$ (area of E). For an invertible $T : \mathbb{R}^3 \to \mathbb{R}^3$, volumes are transformed in the same way, and one can even define n-dimensional volume and show that the image of a body E in $\mathbb{R}^n$ has n-dimensional volume $k\times$ (volume of E) under an invertible mapping T. Again, k depends on T. This is used in multivariable calculus. The idea is to approximate (for fixed $\boldsymbol{x}_0$) the differences $\boldsymbol{f}(\boldsymbol{x}_0+\boldsymbol{x})-\boldsymbol{f}(\boldsymbol{x}_0)$ of a vector function $\boldsymbol{f}$ by a linear mapping $T\boldsymbol{x}$, when the length of $\boldsymbol{x}$ is small. See Further Reading.

Example 12 Let $T : \mathbb{R}^4 \to \mathbb{R}^4$,

$$T(\boldsymbol{x}) = (x_1 + x_2 - x_3, x_1 + 2x_2 + 4x_4, x_3, x_2 + x_4).$$

Show that T is invertible.

Proof. The matrix of T is

$$\begin{bmatrix} 1 & 1 & -1 & 0 \\ 1 & 2 & 0 & 4 \\ 0 & 0 & 1 & 0 \\ 0 & 1 & 0 & 1 \end{bmatrix} \sim \begin{bmatrix} 1 & 1 & -1 & 0 \\ 0 & 1 & 1 & 4 \\ 0 & 0 & 1 & 0 \\ 0 & 0 & -1 & -3 \end{bmatrix} \begin{matrix} \text{II} - \text{I} \\ \text{IV} - \text{II} \\ \\ \\ \end{matrix} \sim \begin{bmatrix} 1 & 1 & -1 & 0 \\ 0 & 1 & 1 & 4 \\ 0 & 0 & 1 & 0 \\ 0 & 0 & 0 & -3 \end{bmatrix} \begin{matrix} \text{IV} + \text{III} \\ \\ \\ \\ \end{matrix},$$

which has rank 4. Thus T is invertible.

3 Algebraic rules of linear mappings. Matrix algebra

The following are the algebraic rules for linear mappings.

1° $S + T = T + S$ if S and T map $\mathbb{R}^n$ into $\mathbb{R}^m$.

2° $(S + T) + U = S + (T + U)$ if S, T and U map $\mathbb{R}^n$ into $\mathbb{R}^m$.

3° There is a mapping $0 : \mathbb{R}^n \to \mathbb{R}^m$ such that $S + 0 = 0 + S = S$ whenever S maps $\mathbb{R}^n$ into $\mathbb{R}^m$.

4° For each S there is a mapping $-S$ such that $S + (-S) = 0$.

5° $a(bS) = (ab)S$ for any S and real a, b.

6° $a(S+T) = aS + aT$ if S and T map $\mathbb{R}^n$ into $\mathbb{R}^m$.

7° $(a+b)S = aS + bS$ for any S and real a, b.

8° $1S = S$ for any S.

9° $(UT)S = U(TS)$ if U, T and T, S are compatible pairs.

10° There is a mapping $I : \mathbb{R}^n \to \mathbb{R}^n$ such that $SI = S$ if S maps $\mathbb{R}^n$ into $\mathbb{R}^m$ and $IT = T$ if T maps $\mathbb{R}^m$ into $\mathbb{R}^n$.

11° $(S+T)U = SU + TU$ if $U : \mathbb{R}^m \to \mathbb{R}^p$ and S, T map $\mathbb{R}^p$ into $\mathbb{R}^q$.

12° $U(S+T) = US + UT$ if $U : \mathbb{R}^p$ and S, T map $\mathbb{R}^q$ into $\mathbb{R}^m$.

It would be tedious to write out the proofs (we already proved 9°). The mapping 0 is defined by $0\boldsymbol{x} = \mathbf{0}$. The mapping $-S$ is $(-1)S$. Let us prove 12°:

$$\begin{aligned} U(S+T)\boldsymbol{x} &= U(S\boldsymbol{x} + T\boldsymbol{x}) \\ &= U(S\boldsymbol{x}) + U(T\boldsymbol{x}) = (US)\boldsymbol{x} + (UT)\boldsymbol{x} \end{aligned}$$

which gives the equation in 12°.

Example 13 Let $T : \mathbb{R}^n \to \mathbb{R}^m, S : \mathbb{R}^p \to \mathbb{R}^m$ be linear. Show that

$$\text{nullity } TS \leq \text{ nullity } T + \text{ nullity } S.$$

Solution. Let $\boldsymbol{v}_1, \ldots, \boldsymbol{v}_k$ be a basis of Ker TS. Extend $\boldsymbol{v}_1, \ldots, \boldsymbol{v}_k$ to a basis $\boldsymbol{v}_1, \ldots, \boldsymbol{v}_p$ of $\mathbb{R}^p$. Let $A = [S\boldsymbol{v}_1 \cdots S\boldsymbol{v}_p]$. Obviously Col $A =$ Im S and A has h non-pivot columns, $h =$ nullity S. So $B = [S\boldsymbol{v}_1 \cdots S\boldsymbol{v}_k]$ has at least $k - h$ pivot columns. Every column of B is in Ker T, so $k - h \leq$ nullity T. That is, nullity $TS-$ nullity $S \leq$ nullity T.

In practice, the easiest way to multiply linear mappings is the process of **matrix multiplication** , which we now define.

If A is an $m \times n$ matrix and

$$\boldsymbol{b} = \begin{bmatrix} b_1 \\ \vdots \\ b_n \end{bmatrix}$$

we recall that

$$A\boldsymbol{b} = \begin{bmatrix} \boldsymbol{r}_1 \cdot \boldsymbol{b} \\ \vdots \\ \boldsymbol{r}_m \cdot \boldsymbol{b} \end{bmatrix}$$

where $\boldsymbol{r}_i$ is the i-th row of A. We use this to define AB where B is an $n \times p$ matrix, with columns $\boldsymbol{b}_1, \dots, \boldsymbol{b}_p$,

$$B = [\boldsymbol{b}_1 \dots \boldsymbol{b}_p].$$

We define

$$AB = [A\boldsymbol{b}_1 \; A\boldsymbol{b}_2 \dots A\boldsymbol{b}_p]. \tag{12}$$

We say that A, B is a **compatible** pair of matrices if A is $m \times n$ and B is $n \times p$ for some m, n, p. Notice that AB is $m \times p$.

There is a matching between matrix multiplication and multiplication of mappings in the following sense.

Proposition 9 *Let T, S be a compatible pair of linear mappings. If the matrix of T is A and the matrix of S is B, then the matrix of TS is AB.*

Proof. We have $T : \mathbb{R}^n \to \mathbb{R}^m$ and $S : \mathbb{R}^p \to \mathbb{R}^n$ for some m, n, p. Now A is $m \times n$ and B is $n \times p$, so A, B is a compatible pair. Take $\boldsymbol{y}$ in $\mathbb{R}^p$. Then

$$\begin{aligned} S\boldsymbol{y} &= B\boldsymbol{y} = y_1\boldsymbol{b}_1 + \cdots + y_p\boldsymbol{b}_p \ (\boldsymbol{b}_j = \text{ column } j \text{ of } B), \\ (TS)\boldsymbol{y} &= T(y_1\boldsymbol{b}_1 + \cdots + y_p\boldsymbol{b}_p) \\ &= y_1T\boldsymbol{b}_1 + \cdots + y_pT\boldsymbol{b}_p \\ &= y_1A\boldsymbol{b}_1 + \cdots + y_pA\boldsymbol{b}_p = AB\boldsymbol{y} \end{aligned}$$

from (12).

As an example of the matching explained above, we recall T, S in Example 7. We found that the matrix of ST is

$$\begin{bmatrix} \frac{6}{13} & -\frac{4}{13} \\ \frac{9}{13} & -\frac{6}{13} \end{bmatrix}.$$

The matrix of S is

$$A = \begin{bmatrix} \frac{4}{13} & \frac{6}{13} \\ \frac{6}{13} & \frac{9}{13} \end{bmatrix}$$

and the matrix of T is

$$B = \begin{bmatrix} 0 & -1 \\ 1 & 0 \end{bmatrix}.$$

We have

$$\begin{aligned} AB &= \begin{bmatrix} \frac{4}{13} & \frac{6}{13} \\ \frac{6}{13} & \frac{9}{13} \end{bmatrix} \begin{bmatrix} 0 & -1 \\ 1 & 0 \end{bmatrix} \\ &= \begin{bmatrix} \frac{6}{13} & -\frac{4}{13} \\ \frac{9}{13} & -\frac{6}{13} \end{bmatrix}. \end{aligned}$$

Thus we can obtain the matrix of ST as (matrix of S) $\times$ (matrix of T).

Because of the matching in Proposition 9, the rules of matrix algebra are the same as those for linear mappings:

1° $B + A = A + B$ if A, B are $m \times n$ matrices.

2° $(A + B) + C = A + (B + C)$ if A, B, C are $m \times n$ matrices.

3° There is an $m \times n$ matrix 0 such that $A + 0 = 0 + A$ whenever A is $m \times n$.

4° For any A there is a matrix $-A$ such that $A + (-A) = 0$.

5° $a(bB) = (ab)B$ for any B and real a, b.

6° $a(A + B) = aA + aB$ if A and B are $m \times n$ and a is real.

7° $(a + b)B = aB + bB$ for any B and real a, b.

8° $1B = B$ for any B.

9° $(AB)C = A(BC)$ whenever A, B and B, C are compatible pairs.

10° Let I be the $n \times n$ identity matrix. If B is $m \times n$, then $BI = B$. If C is $n \times m$, then $IC = C$.

11° $(A + B)C = AC + BC$ if C is $p \times m$ and A, B are $q \times p$.

12° $C(B + A) = CB + CA$ if C is $p \times m$ and B, A are $m \times q$.

Let us just prove 9°. Let A be $m \times n, B$ be $n \times p$ and C be $p \times q$. Let U, T, S be the mappings whose matrices are A, B, C respectively.

U maps $\mathbb{R}^n$ into $\mathbb{R}^m$, T maps $\mathbb{R}^p$ into $\mathbb{R}^n$, S maps $\mathbb{R}^q$ into $\mathbb{R}^p$. So UT, $(UT)S$, TS and $U(TS)$ all exist and

$$(UT)S = U(TS) \tag{13}$$

by associativity for mappings. Now UT has matrix AB, so $(UT)S$ has matrix $(AB)C$. It also has matrix $A(BC)$ by (13), and the rule 9° follows.

Naturally we could prove the rules 1°–12° just from the equation (12). A useful version of (12) for this type of work is:

if $A = [a_{ik}]$ and $B = [b_{kj}]$ are compatible, the i, j entry of AB is

$$\sum_{k=1}^{n} a_{ik}b_{kj} \tag{14}$$

(n = number of columns of A).

To see this, the i, j entry of AB is the i-th entry of $A\boldsymbol{b}_j$, from (12). This is $(a_{i1}, \ldots, a_{in}) \cdot (b_{1j}, \ldots, b_{nj})$, and we obtain (14).

Example 14 In the case of two 2×2 matrices,

$$\begin{bmatrix} a_{11} & a_{12} \\ a_{21} & a_{22} \end{bmatrix} \begin{bmatrix} b_{11} & b_{12} \\ b_{21} & b_{22} \end{bmatrix} = \begin{bmatrix} a_{11}b_{11} + a_{12}b_{21} & a_{11}b_{12} + a_{12}b_{22} \\ a_{21}b_{11} + a_{22}b_{21} & a_{21}b_{12} + a_{22}b_{22} \end{bmatrix}.$$

We say that two $n \times n$ matrices A and B **commute** (or, A **commutes with** B) if $AB = BA$. It is clear from Example 7 that A, B may fail to commute.

Example 15 Let $B = \begin{bmatrix} 1 & 2 \\ 3 & 4 \end{bmatrix}$. Which matrices commute with B?

Solution. Let $C = \begin{bmatrix} c_{11} & c_{12} \\ c_{21} & c_{22} \end{bmatrix}$. Then

$$BC = \begin{bmatrix} c_{11} + 2c_{21} & c_{12} + 2c_{22} \\ 3c_{11} + 4c_{21} & 3c_{12} + 4c_{22} \end{bmatrix},$$

$$CB = \begin{bmatrix} c_{11} + 3c_{12} & 2c_{11} + 4c_{12} \\ c_{21} + 3c_{22} & 2c_{21} + 4c_{22} \end{bmatrix}.$$

The equation $BC = CB$ is equivalent to a linear system in $(c_{11}, c_{12}, c_{21}, c_{22})$:

$$\begin{aligned} c_{11} + 2c_{21} &= c_{11} + 3c_{12} \\ c_{12} + 2c_{22} &= 2c_{11} + 4c_{12} \\ 3c_{11} + 4c_{21} &= c_{21} + 3c_{22} \\ 3c_{12} + 4c_{22} &= 2c_{21} + 4c_{22} \end{aligned}$$

with coefficient matrix

$$\begin{bmatrix} 0 & -3 & 2 & 0 \\ 2 & 3 & 0 & -2 \\ 3 & 0 & 3 & -3 \\ 0 & 3 & -2 & 0 \end{bmatrix} \sim \begin{bmatrix} 1 & 0 & 1 & -1 \\ 0 & 1 & -\frac{2}{3} & 0 \\ 2 & 3 & 0 & -2 \\ 0 & 0 & 0 & 0 \end{bmatrix} \begin{matrix} \text{IV} + \text{I} \\ \text{III} \times \frac{1}{3} \\ \text{I} \leftrightarrow \text{II} \\ \text{I} \leftrightarrow \text{III} \\ \text{II} \times -\frac{1}{3} \end{matrix} \sim \begin{bmatrix} 1 & 0 & 1 & -1 \\ 0 & 1 & -\frac{2}{3} & 0 \\ 0 & 0 & 0 & 0 \\ 0 & 0 & 0 & 0 \end{bmatrix} \begin{matrix} \text{III - 2I} \\ \text{III - 3II} \end{matrix}$$

There are two free variables, and

$$C = \begin{bmatrix} -c_{21} + c_{22} & \frac{2c_{21}}{3} \\ c_{21} & c_{22} \end{bmatrix}.$$

Any linear combination $uI + vB$ of I and B obviously commutes with B. Let us show **no other matrix C commutes with B** by writing C in this form. Comparing c_{21} with the bottom left entry of $uI + vB$, we get $3v = c_{21}$. Comparing c_{22} now yields $c_{22} = u + \frac{4}{3}c_{21}$. You should check that

$$C = \left(-\frac{4c_{21}}{3} + c_{22}\right) I + \frac{c_{21}}{3} B.$$

4 Powers and inverses

Let $T : \mathbb{R}^n \to \mathbb{R}^n$ be linear. We define $T^r = TT \ldots T$ (r factors). It is also convenient to write $T^0 = I$, if $T \neq 0$. If T is invertible, then T^{-r} is, by definition, $(T^{-1})^r$.

Example 16 Show that if $T : \mathbb{R}^n \to \mathbb{R}^n$ is invertible, then

$$T^4 T^{-17} = T^{-13}.$$

Solution. Writing this out as $T \ldots TT^{-1} \ldots T^{-1}$, we cancel four pairs TT^{-1} and get T^{-13}.

This argument of course gives $T^aT^b = T^{a+b}$ for any a and b, whether positive, negative, or zero.

In exactly the same way, we can define $A^r = A \ldots A$ (r factors) if A is an $n \times n$ matrix. It is surprising that every $n \times n$ matrix A satisfies a polynomial equation

$$A^n + c_1 A^{n-1} + \cdots + c_{n-1}A + c_0 I = 0.$$

This is the Cayley-Hamilton theorem, which is Proposition 6 of Chapter 6. Naturally the mapping T, where $T\boldsymbol{x} = A\boldsymbol{x}$, satisfies the same polynomial equation.

We shall define the inverse of an $n \times n$ matrix in the following way.

Definition 6 Let A be an $n \times n$ matrix. We say that A has an inverse B (or, A is **invertible**) if the $n \times n$ matrix B satisfies

$$BA = AB = I.$$

In this case, we write $B = A^{-1}$.

This corresponds exactly to the existence of T^{-1} such that $TT^{-1} = I = T^{-1}T$, where A is the matrix of T. So the invertible matrices are just the matrices of invertible mappings $T : \mathbb{R}^n \to \mathbb{R}^n$. Proposition 6 gives us the following result.

Proposition 10 *The invertible $n \times n$ matrices are those $n \times n$ matrices whose rank is n.*

It should be observed that $(A^{-1})^{-1} = A$ for an invertible matrix A. This follows directly from Definition 6.

If $AB = I$, it follows that B commutes with A, as we see in the next result.

Proposition 11 *Let A and B be $n \times n$ matrices such that $AB = I$. Then A is invertible and $B = A^{-1}$.*

Proof. This is just a rewording of Proposition 7 in terms of matrix multiplication.

We need a practical method of finding the inverse of an $n \times n$ matrix A of rank n. This is easy enough if A is 2×2,

$$A = \begin{bmatrix} a_{11} & a_{12} \\ a_{21} & a_{22} \end{bmatrix}.$$

The **determinant** of A is the number $\det A = a_{11}a_{22} - a_{12}a_{21}$.

Proposition 12 *Let A be a 2×2 matrix, $A = (a_{ij})$. The following statements are equivalent.*

(i) A has rank 2.

(ii) $\det A \neq 0$.

Proof. Suppose first that (i) holds. If $a_{11} \neq 0$, then $\boldsymbol{a}_2 = \begin{bmatrix} a_{12} \\ a_{22} \end{bmatrix} \neq \frac{a_{12}}{a_{11}} \begin{bmatrix} a_{11} \\ a_{21} \end{bmatrix}$. Hence $a_{22} \neq (a_{12}/a_{11})a_{21}$ and $\det A \neq 0$. If $a_{11} = 0$, then neither a_{12} nor a_{21} is zero since A has no zero row or column. So $\det A = -a_{12}a_{21} \neq 0$.

Now suppose that (ii) holds. Then writing $A = [\boldsymbol{a}_1 \ \boldsymbol{a}_2]$, $\boldsymbol{a}_1$ and $\boldsymbol{a}_2$ are not zero. If $\boldsymbol{a}_2 = c\boldsymbol{a}_1$, then

$$a_{11}a_{22} - a_{12}a_{21} = a_{11}ca_{21} - ca_{11}a_{21} = 0$$

which is absurd. So $\boldsymbol{a}_1, \boldsymbol{a}_2$ is a linearly independent set (Proposition 1 of Chapter 3). So (i) holds.

We can now write down the inverse of any 2×2 matrix of rank 2. It is

$$A^{-1} = \frac{1}{\det A} \begin{bmatrix} a_{22} & -a_{12} \\ -a_{21} & a_{11} \end{bmatrix}. \tag{15}$$

For, if B is the matrix on the right-hand side,

$$AB = \frac{1}{\det A} \begin{bmatrix} a_{11}a_{22} - a_{21}a_{12} & -a_{11}a_{12} + a_{12}a_{11} \\ a_{21}a_{22} - a_{22}a_{21} & -a_{21}a_{12} + a_{22}a_{11} \end{bmatrix} = \begin{bmatrix} 1 & 0 \\ 0 & 1 \end{bmatrix}.$$

We can obtain a formula analogous to (15) for an $n \times n$ matrix A of rank n, where $n > 2$; this requires the theory of $n \times n$ determinants, which is the subject of the next chapter. For the present, we note that the j-th column of A^{-1} is the solution of the linear system

$$A\boldsymbol{x} = \boldsymbol{e}_j. \tag{16$_j$}$$

We can conveniently solve all the linear systems $(16)_1, \ldots, (16)_n$ at the same time by row reduction of the $n \times 2n$ matrix

$$[A|I].$$

Since A has rank n, the row reduced form of A is I. After performing the row operations that take A into I, the matrix $[A|I]$ is transformed into

$$[I|C]$$

for some $n \times n$ matrix C. If we ignore all the last n columns but one, we see that column j of C is the solution of $(16)_j$. That is, $C = A^{-1}$.

Example 17 Find the inverse of

$$A = \begin{bmatrix} 1 & 2 & -1 \\ 2 & 0 & 3 \\ -2 & 4 & 5 \end{bmatrix}$$

Solution.

$$\begin{bmatrix} 1 & 2 & -1 & 1 & 0 & 0 \\ 2 & 0 & 3 & 0 & 1 & 0 \\ -2 & 4 & 5 & 0 & 0 & 1 \end{bmatrix} \sim \begin{bmatrix} 1 & 2 & -1 & 1 & 0 & 0 \\ 0 & -4 & 5 & -2 & 1 & 0 \\ 0 & 8 & 3 & 2 & 0 & 1 \end{bmatrix} \begin{matrix} \text{II} - 2\text{I} \\ \text{III} + 2\text{I} \\ \end{matrix}$$

$$\sim \begin{bmatrix} 1 & 2 & -1 & 1 & 0 & 0 \\ 0 & 1 & -\frac{5}{4} & \frac{1}{2} & -\frac{1}{4} & 0 \\ 0 & 0 & 13 & -2 & 2 & 1 \end{bmatrix} \begin{matrix} \text{III} + 2\text{II} \\ \text{II} \times -\frac{1}{4} \\ \end{matrix} \sim \begin{bmatrix} 1 & 2 & 0 & \frac{11}{13} & \frac{2}{13} & \frac{1}{13} \\ 0 & 1 & 0 & \frac{4}{13} & -\frac{3}{52} & \frac{5}{52} \\ 0 & 0 & 0 & -\frac{2}{13} & \frac{2}{13} & \frac{1}{13} \end{bmatrix} \begin{matrix} \text{III} \times \frac{1}{13} \\ \text{II} + \frac{5}{4}\text{III} \\ \text{I} + \text{III} \end{matrix}$$

$$\sim \begin{bmatrix} 1 & 0 & 0 & \frac{3}{13} & \frac{7}{26} & -\frac{3}{26} \\ 0 & 1 & 0 & \frac{4}{13} & -\frac{3}{52} & \frac{5}{52} \\ 0 & 0 & 1 & -\frac{2}{13} & \frac{2}{13} & \frac{1}{13} \end{bmatrix} \begin{matrix} \text{I} - 2\text{II} \\ \\ \end{matrix}.$$

Hence $A^{-1} = \begin{bmatrix} \frac{3}{13} & \frac{7}{26} & -\frac{3}{26} \\ \frac{4}{13} & -\frac{3}{52} & \frac{5}{52} \\ -\frac{2}{13} & \frac{2}{13} & \frac{1}{13} \end{bmatrix}$ (check!).

This method is essentially the same as the augmented matrix method in Example 10.

If we try to invert an $n \times n$ matrix A of rank less than n by this approach, we obtain a last row $[0\ 0 \cdots 0 * * \cdots *]$ (n zeros). At this point we find out that A is not invertible.

5 Inverse of a product. Transpose

If S and T are invertible linear mappings from $\mathbb{R}^n$ to $\mathbb{R}^n$, then ST is invertible and

$$(ST)^{-1} = T^{-1}S^{-1}.$$

To see this, we observe that

$$T^{-1}S^{-1}ST = T^{-1}T = I,$$
$$ST\,T^{-1}S^{-1} = SS^{-1} = I.$$

In the same way (or by matching linear mappings to matrices) we see that if A and B are invertible $n \times n$ matrices, then AB is invertible and

$$(AB)^{-1} = B^{-1}A^{-1}.$$

Definition 7 The **transpose** of an $m \times n$ matrix $A = [a_{ij}]_{1 \le i \le m, 1 \le j \le n}$ is the $n \times m$ matrix A^t whose i, j entry is a_{ji}:

$$A^t = [a_{ji}]_{1 \le i \le n, 1 \le j \le m}.$$

Thus the j th column vector of A^t is

$$\begin{bmatrix} a_{j1} \\ \vdots \\ a_{jn} \end{bmatrix}$$

which has the same entries as the j-th row of A.

When an expression such as $\boldsymbol{x}^t$ or $\boldsymbol{y}^t$ appears, it will always denote a row vector; for example,

$$\boldsymbol{x} = \begin{bmatrix} x_1 \\ x_2 \\ x_3 \end{bmatrix}, \quad \boldsymbol{x}^t = [x_1\ x_2\ x_3].$$

Example 18 $$\begin{bmatrix} a_{11} & a_{12} & a_{13} \\ a_{21} & a_{22} & a_{23} \end{bmatrix}^t = \begin{bmatrix} a_{11} & a_{21} \\ a_{12} & a_{22} \\ a_{13} & a_{23} \end{bmatrix}.$$

Proposition 13 *Let A be $m \times n$ and let B be $n \times p$. Then*

$$(AB)^t = B^tA^t.$$

Proof. Both $(AB)^t$ and B^tA^t are $p \times m$ matrices. The i, j entry of AB is

$$\sum_{k=1}^{n} a_{ik}b_{kj}.$$

Thus the i, j entry of $(AB)^t$ is $\sum_{k=1}^{n} a_{jk}b_{ki}$. This is the inner product of the i-th row $(b_{1i}, \dots, b_{ni})$ of B^t with the j-th column $(a_{j1}, \dots, a_{jn})$ of A^t, which is the i, j entry of B^tA^t.

Proposition 14 *Let A be an invertible matrix. Then A^t is invertible and*

$$(A^t)^{-1} = (A^{-1})^t.$$

Proof. By Proposition 13,

$$(A^{-1})^tA^t = (AA^{-1})^t = I^t = I,$$

and similarly,

$$A^t(A^{-1})^t = (A^{-1}A)^t = I.$$

This proves that $(A^{-1})^t$ is the inverse of A^t.

6 Change of basis

Let $\boldsymbol{v}_1, \dots, \boldsymbol{v}_n$ and $\boldsymbol{w}_1, \dots, \boldsymbol{w}_n$ be two bases of $\mathbb{R}^n$. Let $T : \mathbb{R}^n \to \mathbb{R}^n$ be linear. If $T\left(\sum_{j=1}^{n} y_j\boldsymbol{v}_j\right)$ has coordinate vector $A\boldsymbol{y}$ for the basis $\boldsymbol{v}_1, \dots, \boldsymbol{v}_n$, we say that **the matrix of** T is A **for the basis** $\boldsymbol{v}_1, \dots, \boldsymbol{v}_n$. 'The matrix of T is A' now becomes a shorthand for 'The matrix of T is A for the standard basis.'

Proposition 15 (Change of basis). *Let T have matrix A for the basis $\boldsymbol{v}_1, \dots, \boldsymbol{v}_n$. Suppose*

$$\boldsymbol{y} = P\boldsymbol{z}, \tag{17}$$

where $\boldsymbol{y}$ and $\boldsymbol{z}$ are coordinate vectors for $\boldsymbol{v}_1, \dots, \boldsymbol{v}_n$ and $\boldsymbol{w}_1, \dots, \boldsymbol{w}_n$ respectively. Then T has matrix $P^{-1}AP$ for the basis $\boldsymbol{w}_1, \dots, \boldsymbol{w}_n$. In particular $P = [\boldsymbol{w}_1 \cdots \boldsymbol{w}_n]$ if $\boldsymbol{v}_1, \dots, \boldsymbol{v}_n$ is the standard basis.

Proof. The change of coordinates can also be written

$$\boldsymbol{z} = P^{-1}\boldsymbol{y}$$

(multiply both sides of (17) by P^{-1}). The coordinate vector of $T\left(\sum_{i=1}^{n} z_i\boldsymbol{w}_i\right)$ is $A(P\boldsymbol{z}) = AP\boldsymbol{z}$ for $\boldsymbol{v}_1, \dots, \boldsymbol{v}_n$. Hence the coordinate vector of $T\left(\sum_{i=1}^{n} z_i\boldsymbol{w}_i\right)$ is $P^{-1}(AP\boldsymbol{z})$ for $\boldsymbol{w}_1, \dots, \boldsymbol{w}_n$. Since

$$P^{-1}(AP\boldsymbol{z}) = (P^{-1}AP)\boldsymbol{z},$$

this proves the proposition. (The final sentence is simply a reminder of a fact from Proposition 13 of Chapter 3.)

In Chapter 6 we shall see that the matrix of T for a suitably chosen basis can have a much simpler appearance than the matrix of T for the standard basis.

Example 19 Find the matrix of the mapping

$$T(\boldsymbol{x}) = (2x_1 - x_2, x_1 - 4x_2)$$

for the basis $\boldsymbol{w}_1 = (1, 2), \boldsymbol{w}_2 = (2, 5)$.

Solution. T has matrix $A = \begin{bmatrix} 2 & -1 \\ 1 & -4 \end{bmatrix}$ for the basis $\boldsymbol{e}_1, \boldsymbol{e}_2$. The matrix B of T for $\boldsymbol{w}_1, \boldsymbol{w}_2$ is $P^{-1}AP$ where $P = [\boldsymbol{w}_1\ \boldsymbol{w}_2]$. So

$$\begin{aligned} B &= \begin{bmatrix} 1 & 2 \\ 2 & 5 \end{bmatrix}^{-1} \begin{bmatrix} 2 & -1 \\ 1 & -4 \end{bmatrix} \begin{bmatrix} 1 & 2 \\ 2 & 5 \end{bmatrix} \\ &= \begin{bmatrix} 5 & -2 \\ -2 & 1 \end{bmatrix} \begin{bmatrix} 2 & -1 \\ 1 & -4 \end{bmatrix} \begin{bmatrix} 1 & 2 \\ 2 & 5 \end{bmatrix} = \begin{bmatrix} 8 & 3 \\ -3 & -2 \end{bmatrix} \begin{bmatrix} 1 & 2 \\ 2 & 5 \end{bmatrix} = \begin{bmatrix} 14 & 31 \\ -7 & -16 \end{bmatrix}. \end{aligned}$$

Thus

$$T(z_1\boldsymbol{w}_1 + z_2\boldsymbol{w}_2) = (14z_1 + 31z_2)\boldsymbol{w}_1 + (-7z_1 - 16z_2)\boldsymbol{w}_2$$

(check this for $\boldsymbol{z} = (1, 0), (0, 1)$!).

7 Orthogonal matrices

Let $P = [\boldsymbol{p}_1 \ldots \boldsymbol{p}_n]$ be a matrix whose columns have length 1 and are perpendicular to each other:

$$\boldsymbol{p}_i \cdot \boldsymbol{p}_j = \begin{cases} 1 & \text{if } i = j \\ 0 & \text{if } i \neq j. \end{cases} \tag{18}$$

Then P is said to be an **orthogonal** matrix.

Example 20 $\begin{bmatrix} \frac{1}{\sqrt{2}} & \frac{1}{\sqrt{2}} \\ -\frac{1}{\sqrt{2}} & \frac{1}{\sqrt{2}} \end{bmatrix}$ is an orthogonal matrix . To check this, compute $\boldsymbol{p}_1 \cdot \boldsymbol{p}_1, \boldsymbol{p}_1 \cdot \boldsymbol{p}_2, \boldsymbol{p}_2 \cdot \boldsymbol{p}_2$. The matrix

$$\begin{bmatrix} \frac{1}{\sqrt{3}} & \frac{1}{\sqrt{2}} & \frac{1}{\sqrt{6}} \\ \frac{1}{\sqrt{3}} & 0 & -\frac{2}{\sqrt{6}} \\ \frac{1}{\sqrt{3}} & -\frac{1}{\sqrt{2}} & \frac{1}{\sqrt{6}} \end{bmatrix}$$

is also orthogonal. To check this, you should compute the six inner products $\boldsymbol{p}_1 \cdot \boldsymbol{p}_1, \boldsymbol{p}_1 \cdot \boldsymbol{p}_2, \boldsymbol{p}_1 \cdot \boldsymbol{p}_3, \boldsymbol{p}_2 \cdot \boldsymbol{p}_2, \boldsymbol{p}_2 \cdot \boldsymbol{p}_3, \boldsymbol{p}_3 \cdot \boldsymbol{p}_3$.

Proposition 16 *Let P be an orthogonal matrix. Then*

$$P^{-1} = P^t. \tag{19}$$

Proof. Let

$$P = [\boldsymbol{p}_1 \cdots \boldsymbol{p}_n].$$

The i, j entry of P^tP is

$$\boldsymbol{p}_i \cdot \boldsymbol{p}_j.$$

Thus P^tP has 1's on the main diagonal ($i = j$) and all other entries zero. This proves that

$$P^tP = I. \tag{20}$$

Now $P^t = P^{-1}$ from Proposition 11.

If P is a matrix satisfying (20), then we can deduce (18) from (20); we find that P is orthogonal.

Proposition 17 *The product of orthogonal matrices P, Q is orthogonal. The inverse of an orthogonal matrix P is orthogonal.*

Proof. $(PQ)^t = Q^tP^t = Q^{-1}P^{-1} = (PQ)^{-1}$, using results from Section 5. Thus PQ is orthogonal.

Moreover, $(P^{-1})^t = (P^t)^{-1} = (P^{-1})^{-1}$, from Proposition 14 and the fact that P is orthogonal. Thus the transpose of P^{-1} is the inverse of P^{-1}, and P^{-1} is orthogonal. (We can simplify $(P^{-1})^{-1}$ to P.)

Example 21 Rotation in $\mathbb{R}^3$.

Let T be a rotation anticlockwise about an axis $\boldsymbol{v}_3$ of length 1 through angle a. To find the matrix of T, compute two perpendicular unit vectors $\boldsymbol{v}_1, \boldsymbol{v}_2$ which are in the plane perpendicular to $\boldsymbol{v}_3$. Thus

$$\boldsymbol{v}_i \cdot \boldsymbol{v}_j = 0 \text{ if } i \neq j, \boldsymbol{v}_i \cdot \boldsymbol{v}_i = 1 \quad (1 \leq i \leq j \leq 3).$$

We also arrange the pair $\boldsymbol{v}_1, \boldsymbol{v}_2$ so that a small anticlockwise rotation about $\boldsymbol{v}_3$ moves $\boldsymbol{v}_1$ towards $\boldsymbol{v}_2$.

By looking at Example 5, it is easy to see that

$$\begin{aligned} T(y_1\boldsymbol{v}_1 + y_2\boldsymbol{v}_2 + y_3\boldsymbol{v}_3) &= (y_1 \cos a - y_2 \sin a)\boldsymbol{v}_1 \\ &\quad + (y_1 \sin a + y_2 \cos a)\boldsymbol{v}_2 + y_3\boldsymbol{v}_3. \end{aligned}$$

The matrix of T for $\boldsymbol{v}_1, \boldsymbol{v}_2, \boldsymbol{v}_3$ is

$$A = \begin{bmatrix} \cos a & -\sin a & 0 \\ \sin a & \cos a & 0 \\ 0 & 0 & 1 \end{bmatrix},$$

which is seen to be an orthogonal matrix.

The change of coordinates $\boldsymbol{y} = P\boldsymbol{x}$ ($\boldsymbol{x}$ = coordinate vector for standard basis) is given by

$$P^{-1} = [\boldsymbol{v}_1\ \boldsymbol{v}_2\ \boldsymbol{v}_3]$$

(Proposition 13 of Chapter 3). Hence the matrix of T for the standard basis is

$$B = P^{-1}AP = [\boldsymbol{v}_1\ \boldsymbol{v}_2\ \boldsymbol{v}_3]A[\boldsymbol{v}_1\ \boldsymbol{v}_2\ \boldsymbol{v}_3]^{-1}$$

which is an orthogonal matrix by Proposition 17.

We get a further simplification by using $P = (P^{-1})^t$:

$$B = [\boldsymbol{v}_1\ \boldsymbol{v}_2\ \boldsymbol{v}_3] \begin{bmatrix} \cos a & -\sin a & 0 \\ \sin a & \cos a & 0 \\ 0 & 0 & 1 \end{bmatrix} [\boldsymbol{v}_1\ \boldsymbol{v}_2\ \boldsymbol{v}_3]^t.$$

Exercises for Chapter 4

Reminder: Attempt Exercise a.b after reading Section a.

In these exercises **all mappings are assumed to be linear**. Triangles, squares, etc. are 'solid' figures, rather than outlines.

Given a set of vectors E in $\mathbb{R}$ and $T : \mathbb{R}^n \to \mathbb{R}^m$, let $T(E)$ be the set of $T\boldsymbol{x}$ ($\boldsymbol{x}$ in E). We call $T(E)$ the **image** of E under T.

1.1 Show that the image of a line segment is a point or a line segment.

1.2 Let $T(\boldsymbol{x}) = (x_1 + x_2, 2x_1 + x_2)$. Draw figures to illustrate the image of the triangle with vertices $(1,0), (2,0), (3,2)$ and the image of the square with vertices $(0,0), (1,0), (0,1), (1,1)$.

1.3 If $T : \mathbb{R}^2 \to \mathbb{R}^2$ is projection onto a vector, show that the image of a square is a line segment. In the case where T is a projection onto $(-1,3)$, find the image of the square with vertices $(1,1), (1,-1), (-1,1), (-1,-1)$.

1.4 Let $T : \mathbb{R}^2 \to \mathbb{R}^2$ be projection onto a vector. Suppose L is a line segment and $T(L)$ is a line segment. Prove that length of $T(L) \leq$ length of L.

1.5 Let T be the mapping $T(x_1, x_2, x_3, x_4) = (x_1 + x_2, x_1 - x_3, x_1 + 2x_4)$ and let S be the mapping

$$S(x_1, x_2, x_3) = (x_1 + x_2 - x_3, x_2 + x_3, x_1 + 3x_3, x_3).$$

Compute TS. Find the matrix A of T, the matrix B of S, and the matrix of TS.

2.1 Let $T : \mathbb{R}^3 \to \mathbb{R}^3, T(x_1, x_2, x_3) = (x_1, x_2, 0)$. ($T$ is the **projection onto the plane** $x_3 = 0$.) Find Im T and Ker T and verify that rank T + nullity $T = 3$.

2.2 Let T be as in Exercise 2.1. Show that the image of a triangle under T is either a triangle or a line segment. Give an example of each case.

2.3 Let T be as in Exercise 2.1. Suppose that K is a triangle and $T(K)$ is a triangle. Show that area of $T(K) \leq$ area of K.

2.4 Let T and S be as in Exercise 1.5. Find the rank and nullity of S, T, ST and TS. Which of these are one-to-one, onto, or both?

2.5 Let T, S be a compatible pair. Show that

$$\text{Im}(TS) = T(\text{Im } S).$$

Deduce that rank $TS \leq \min(\text{rank } S, \text{ rank } T)$.

2.6 In Exercise 2.5 suppose that S is onto. Show that

$$\text{rank } TS = \text{rank } T.$$

2.7 Let $T : \mathbb{R}^n \to \mathbb{R}^n$. Show that $|T\boldsymbol{x}| \leq K|\boldsymbol{x}|$ where $K = |T\boldsymbol{e}_1| + \cdots + |T\boldsymbol{e}_n|$.

Hint: First show $|T\boldsymbol{x}| \leq |x_1|\ |T\boldsymbol{e}_1| + \cdots + |x_n|\ |T\boldsymbol{e}_n|$.

2.8 Let $T : \mathbb{R}^n \to \mathbb{R}^n$ be invertible. Show that $|T\boldsymbol{x}| \geq c|\boldsymbol{x}|$ for a positive constant c.

Hint: Apply Exercise 2.7 to T^{-1}.

2.9 Let $S : \mathbb{R}^n \to \mathbb{R}^n$ and $T : \mathbb{R}^n \to \mathbb{R}^n$. Suppose that TS is invertible. Show that T and S are both invertible.

Hint: Use Exercise 2.5.

2.10 Let $T : \mathbb{R}^n \to \mathbb{R}^m$. Let $\boldsymbol{v}_1, \ldots, \boldsymbol{v}_k$ be vectors in $\mathbb{R}^n$ and suppose that $T\boldsymbol{v}_1, \ldots, T\boldsymbol{v}_k$ are linearly independent. Show that $\boldsymbol{v}_1, \ldots, \boldsymbol{v}_k$ must be linearly independent.

2.11 Let $T : \mathbb{R}^n \to \mathbb{R}^m$ be one-to-one. Let $\boldsymbol{v}_1, \ldots, \boldsymbol{v}_k$ be linearly independent vectors in $\mathbb{R}^n$. Show that $T\boldsymbol{v}_1, \ldots, T\boldsymbol{v}_k$ are linearly independent.

2.12 Let V be a subspace of $\mathbb{R}^n$ and let $T : \mathbb{R}^n \to \mathbb{R}^m$. Show that $T(V)$ is a subspace of $\mathbb{R}^m$.

2.13 In Exercise 2.12 show that $\dim T(V) \leq \dim V$. If T is one-to-one, show that $\dim T(V) = \dim V$.

Hint: Exercises 2.10 and 2.11.

2.14 Let V and W be subspaces of $\mathbb{R}^n$ and let $T : \mathbb{R}^n \to \mathbb{R}^m$ be one-to-one. Show that $T(V \cap W) = T(V) \cap T(W)$.

3.1 Find the matrices B that commute with

$$A = \begin{bmatrix} 1 & 0 & 1 \\ 0 & 1 & 0 \\ 1 & 0 & 1 \end{bmatrix}.$$

Can all these matrices B be written as $x_1 A + x_2 I$?

3.2 Find the matrices B that commute with

$$A = \begin{bmatrix} \cos a & -\sin a \\ \sin a & \cos a \end{bmatrix}$$

for a given angle $a, 0 < a < \pi/2$. Can all these matrices B be written as

$$c \begin{bmatrix} \cos b & -\sin b \\ \sin b & \cos b \end{bmatrix}?$$

3.3 In Exercise 1.5, verify that AB is the matrix of TS.

4.1 Find the value of b for which

$$B = \begin{bmatrix} 3 & 1 & b \\ 2 & 2 & 1 \\ 1 & 0 & b \end{bmatrix}$$

is not invertible (say, $b = b_0$). For $b \neq b_0$, find B^{-1}.

4.2 Find the inverse of

$$\begin{bmatrix} a_{11} & a_{12} & a_{13} \\ 0 & a_{22} & a_{23} \\ 0 & 0 & a_{33} \end{bmatrix}$$

given that $a_{11}a_{22}a_{33} \neq 0$.

4.3 Using Exercise 4.2 and the relation $(A^{-1})^t = (A^t)^{-1}$, write down the inverse of

$$\begin{bmatrix} a_{11} & 0 & 0 \\ a_{21} & a_{22} & 0 \\ a_{31} & a_{32} & a_{33} \end{bmatrix}$$

given that $a_{11}a_{22}a_{33} \neq 0$.

4.4 Find the inverse of the linear mapping $T : \mathbb{R}^2 \to \mathbb{R}^2, T(x_1, x_2) = (3x_1 - 2x_2, x_1 + 3x_2)$ by solving $T(\boldsymbol{x}) = \boldsymbol{y}$ for $\boldsymbol{x}$. Find the matrix A of T. Find A^{-1} and verify that it is the matrix of T^{-1}.

4.5 Find the inverse of

$$\begin{bmatrix} 1 & 0 & 2 & 3 \\ 2 & 0 & 1 & 3 \\ 0 & 1 & 2 & 3 \\ 3 & 0 & 1 & 2 \end{bmatrix}$$

(if it exists).

4.6 Let A be a 2×2 matrix, $A = \begin{bmatrix} a_{11} & a_{12} \\ a_{21} & a_{22} \end{bmatrix}$. Show that

$$A^2 - (a_{11} + a_{22})A + (a_{11}a_{22} - a_{21}a_{12})I = 0.$$

If A has rank 1, show that A^2 is a scalar multiple of A.

4.7 Suppose A is a square matrix that satisfies a cubic equation

$$A^3 + a_1A^2 + a_2A + a_3I = 0,$$

where $a_3 \neq 0$. Show that A is invertible.

5.1 Show that rank (A^t) = rank (A) for any matrix A.

5.2 If $\boldsymbol{u}$ and $\boldsymbol{v}$ are column vectors in $\mathbb{R}^n$ ($n \times 1$ matrices), show that $\boldsymbol{u}^t\boldsymbol{v}$ is a 1×1 matrix whose only entry is $\boldsymbol{u} \cdot \boldsymbol{v}$. (We usually identify $\boldsymbol{u}^t\boldsymbol{v}$ with $\boldsymbol{u} \cdot \boldsymbol{v}$.)

5.3 If the $n \times n$ matrix A satisfies $A = A^t$, we say that A is **symmetric** . Let $\boldsymbol{u}$ be as in Exercise 5.2, with $n = 3$. Show that for any 3×3 matrix B, there is a symmetric 3×3 matrix A such that

$$\boldsymbol{u}^tA\boldsymbol{u} = \boldsymbol{u}^tB\boldsymbol{u} \quad \text{for all } \boldsymbol{u}.$$

6.1 Find the matrix of the projection onto $\boldsymbol{b}$ for the basis $(1, 2), (1, -2)$.

6.2 Find the matrix of the mapping $T : \mathbb{R}^3 \to \mathbb{R}^3, T\boldsymbol{x} = (c_1x_1, c_2x_2, c_3x_3)$ for the basis $(1, 0, 1), (-1, 0, 1), (0, 1, 0)$.

6.3 Let $T : \mathbb{R}^n \to \mathbb{R}^n$. Let $\boldsymbol{v}_1, \ldots, \boldsymbol{v}_n$ be a basis of $\mathbb{R}^n$ and let A be the matrix of T for $\boldsymbol{v}_1, \ldots, \boldsymbol{v}_n$. Show that rank A = rank T.

Hint: Exercise 2.6.

6.4 By multiplying two rotations show that

$$\begin{bmatrix} \cos a & -\sin a \\ \sin a & \cos a \end{bmatrix} \begin{bmatrix} \cos b & -\sin b \\ \sin b & \cos b \end{bmatrix} = \begin{bmatrix} \cos(a+b) & -\sin(a+b) \\ \sin(a+b) & \cos(a+b) \end{bmatrix}.$$

Deduce that

$$\cos(a+b) = \cos a \cos b - \sin a \sin b, \ \sin(a+b) = \sin a \cos b + \sin b \cos a.$$

Chapter 5

Determinants

1 Permutations

Let A be $n \times n$ invertible. What is the denominator in the algebraic formula for A^{-1}? This denominator is the **determinant** of A, $\det A$. We saw in Chapter 4 that $\det A$ is $a_{11}a_{22} - a_{12}a_{21}$ in the 2×2 case. You would be unlikely to guess the formula for $\det A$ from this special case. We first need a little detour. A **permutation** of $\{1, \ldots, n\}$ is an ordered set

$$(1) \qquad p(1), p(2), \ldots, p(n)$$

that is a rearrangement of $1, \ldots, n$. We refer to the permutation as p for brevity. For instance,

$$2, 1, 3$$

is a permutation of 1, 2, 3. Here

$$p(1) = 2, p(2) = 1, p(3) = 3.$$

There are

$$n! = n(n-1) \cdots 3 \cdot 2 \cdot 1$$

permutations of $1, \ldots, n$, since there are n choices for $p(1)$, and, for each of these, $n - 1$ choices for $p(2)$, and so on. The symbol $n!$ is pronounced n **factorial**. The six permutations of 1, 2, 3 are

$$(2) \qquad 1, 2, 3; 1, 3, 2;\ 2, 1, 3;\ 2, 3, 1;\ 3, 1, 2;\ \text{and } 3, 2, 1.$$

The pairs $(p(i), p(j))$ that we can extract from the list (1) for which

$$i < j, \quad p(i) > p(j)$$

are called **disordered** pairs of p. For instance, in the permutation

$$2, 1, 3$$

the only disordered pair is (2, 1). In the permutation

$$1, 6, 4, 2, 5, 3 \tag{3}$$

of 1, 2, 3, 4, 5, 6 we have seven disordered pairs (6, 4), (6, 2), (6, 5), (6, 3), (4, 2), (4, 3), (5, 3).

Definition 1 A permutation p is said to be **even** if the number of its disordered pairs is even. Otherwise the permutation is **odd** .

Example 1 The permutation in (3) is odd because there are 7 disordered pairs. The permutation

$$1, 5, 2, 4, 3, 6$$

is even because there are four disordered pairs.

The **parity** of a permutation means the adjective 'even' or 'odd' that applies to it; 'q has **opposite parity** to p' means exactly one of p, q is even.

It is fairly easy to see that for any $n \geq 2$ there are $n!/2$ permutations of each parity. Just take any permutation

$$p(1), p(2), p(3), \dots, p(n)$$

in the even list and switch $p(1), p(2)$. This either adds 1 or subtracts 1 in the count of disordered pairs, and gives an odd permutation, p' say:

$$p(2), p(1), p(3), \dots, p(n).$$

Of course different p give rise to different p', and every odd permutation is p' for some even p. If there are m even permutations, then $2m = n!$, or $m = n!/2$.

It is useful to note what happens to the parity of a permutation when we switch *any* pair $p(i), p(j)$.

Proposition 1 *Let p' be obtained from p by switching the places of $p(i)$ and $p(j)$. Then p' has opposite parity to p.*

Proof. If $p(i), p(j)$ are adjacent in

$$p(1), \ldots, p(n),$$

the switch adds one disordered pair (if $p(i), p(j)$ is not disordered) and subtracts one disordered pair (if $p(i), p(j)$ is disordered). Thus p' has opposite parity to p.

If there are k integers $a_1, \ldots, a_k$ in between $p(i)$ and $p(j)$ then $(p(i), a_1)$, $(p(i), a_2), \ldots, (p(i), a_k)$ all switch from not disordered to disordered, or vice versa; similarly for $(a_1, p(j)), \ldots, (a_k, p(j))$. With $(p(i), p(j))$, this gives a total of $2k + 1$ changes of ± 1 in the disordered pair count. Now an odd p gives an even p' and an even p gives an odd p'.

Example 2 In (3), switch 4 and 3. The pairs (4, 3), (4, 2), (4, 5), (2, 3), (5, 3) switch places. Now p', the permutation

$$1, 6, 3, 2, 5, 4$$

has six disordered pairs (we lost three and gained two).

We now turn to an $n \times n$ matrix

$$A = \begin{bmatrix} a_{11} & \cdots & a_{1n} \\ \vdots & & \vdots \\ a_{n1} & & a_{nn} \end{bmatrix}.$$

A product

$$a_{1p(1)} a_{2p(2)} \cdots a_{np(n)}$$

where p is a permutation of $1, \ldots, n$ is said to be a **special product** . Special products contain a factor from each row–and a factor from each column. For instance, if

$$A = \begin{bmatrix} a_{11} & a_{12} \\ a_{21} & a_{22} \end{bmatrix},$$

the permutations are 1, 2 and 2, 1. The special products are $a_{11}a_{22}$ and $a_{12}a_{21}$.

In the 3×3 case, take the boxed elements from

$$\begin{bmatrix} a_{11} & \boxed{a_{12}} & a_{13} \\ a_{21} & a_{22} & \boxed{a_{23}} \\ \boxed{a_{31}} & a_{32} & a_{33} \end{bmatrix}$$

to get an example of a special product, $a_{12}\, a_{23}\, a_{31}$.

A **signed product** is a special product $\pm a_{1p(1)} \cdots a_{np(n)}$ with a plus sign if p is even and a minus sign if p is odd.

Definition 2 The **determinant** of A is the sum of the $n!$ signed products. We write $\det A$ for the determinant of A.

We see that the 2×2 matrix above has determinant

$$\det A = a_{11}a_{22} - a_{12}a_{21}$$

because 1, 2 is even and 2, 1 is odd. This agrees with Chapter 4.

Example 3 Let

$$A = \begin{bmatrix} a_{11} & a_{12} & a_{13} \\ a_{21} & a_{22} & a_{23} \\ a_{31} & a_{32} & a_{33} \end{bmatrix}.$$

Taking special products in the same order as in (2),

$$\begin{aligned} \det A = a_{11}a_{22}a_{33} - a_{11}a_{23}a_{32} - a_{12}a_{21}a_{33} + a_{12}a_{23}a_{31} \\ + a_{13}a_{21}a_{32} - a_{13}a_{22}a_{31}. \end{aligned}$$

2 Row reduction

The determinant of a 4×4 matrix is a sum of 24 terms, which we will not write out. Clearly we need a practical method of computing determinants, and we turn to row reduction. It is convenient to discuss interchange of rows first.

Proposition 2 *If B is obtained from A by interchanging two rows i and j, then* $\det B = -\det A$.

Proof. A special product from B,

$$b_{1p(1)}\cdots b_{np(n)},$$

is equal to

$$a_{1p'(1)}\cdots a_{np'(n)}$$

where p' is the permutation obtained from p by switching $p(i)$ and $p(j)$. The reason for this is that

$$b_{jp(j)} = a_{ip(j)}, b_{ip(i)} = a_{jp(i)}$$

and the other factors $b_{\ell p(\ell)}$ of the special product are equal to $a_{\ell p(\ell)}$.

As p runs over the $n!$ possibilities, so too does p'. The sum of all the signed products $\pm b_{1p(1)}\cdots b_{np(n)}$ is the sum of all the signed products $\pm a_{1p'(1)}\cdots a_{np'(n)}$ with the 'wrong' signs attached (because p, p' have opposite parity, Proposition 1). Hence $\det B = -\det A$.

Example 4 The matrix $B = \begin{bmatrix} a_{21} & a_{22} \\ a_{11} & a_{12} \end{bmatrix}$ has determinant

$$a_{21}a_{12} - a_{22}a_{11} = -\det A.$$

From Proposition 2 we get a useful corollary.

Corollary 1 *If A is a matrix with two identical rows, then*

$$\det A = 0.$$

Proof. If the i-th and j-th rows are identical, the matrix B in Proposition 2 is equal to A. So $-\det A = \det A$, and $\det A = 0$.

If a matrix A has one row that is a sum of two given vectors, we can split $\det A$ into a sum of two determinants. For instance

$$\det \begin{bmatrix} a_{11} & a_{12} \\ b_{21}+c_{21} & b_{22}+c_{22} \end{bmatrix} = \det \begin{bmatrix} a_{11} & a_{12} \\ b_{21} & b_{22} \end{bmatrix} + \det \begin{bmatrix} a_{11} & a_{12} \\ c_{21} & c_{22} \end{bmatrix}.$$

This amounts to splitting $a_{11}(b_{22}+c_{22})$ into $a_{11}b_{22}, a_{11}c_{22}$ and splitting $-a_{12}(b_{21}+c_{21})$ into $-a_{12}b_{21}, -a_{12}c_{21}$. The general case is

Proposition 3 *Let A have rows $\boldsymbol{r}_1, \dots, \boldsymbol{r}_n$ where $\boldsymbol{r}_j = \boldsymbol{s} + \boldsymbol{t}$. Then*

$$\det A = \det B + \det C \tag{4}$$

where B has rows $\boldsymbol{r}_1, \dots, \boldsymbol{r}_{j-1}, \boldsymbol{s}, \boldsymbol{r}_{j+1}, \dots, \boldsymbol{r}_n$ and C has rows $\boldsymbol{r}_1, \dots, \boldsymbol{r}_{j-1}$, $\boldsymbol{t}, \boldsymbol{r}_{j+1}, \dots, \boldsymbol{r}_n$.

Notice this does **not** say that $\det(B + C) = \det B + \det C$.

Proof. Let $\boldsymbol{s} = (s_1 \dots, s_n), \boldsymbol{t} = (t_1, \dots, t_n)$. Every signed product

$$\pm a_{1p(1)} \cdots a_{jp(j)} \cdots a_{np(n)} \tag{5}$$

equals the sum of the two signed products

$$\pm a_{1p(1)} \cdots s_{p(j)} \cdots a_{np(n)} \text{ and } \pm a_{1p(1)} \cdots t_{p(j)} \cdots a_{np(n)}$$

differing in the jth factor from (5). If we add up these $n!$ equalities, we get (4).

Example 5 $\det \begin{bmatrix} 1 & 2 & 0 \\ 3 & 1 & 0 \\ 5 & 2 & 2 \end{bmatrix} + \det \begin{bmatrix} 1 & 2 & 0 \\ -3 & 2 & 0 \\ 5 & 2 & 2 \end{bmatrix} = \det \begin{bmatrix} 1 & 2 & 0 \\ 0 & 3 & 0 \\ 5 & 2 & 2 \end{bmatrix}$. If you work out the three determinants, leaving out zero terms, you will get this in the form

$$(2 - 12) + (4 + 12) = 6.$$

We now consider row operations.

Proposition 4 *Let B be obtained from A by multiplying row i by c. Then*

$$\det B = c \det A. \tag{6}$$

Proof. The signed products of B are

$$\pm a_{1p(1)} \cdots c a_{ip(i)} \cdots a_{np(n)},$$

in other words, c times the signed products of A. Adding the $n!$ terms, we get (6).

Taking $c = 0$, a matrix with a zero row has zero determinant.

Proposition 5 *Let B be obtained from A by adding c times row i to row j, where $i \neq j$. Then*

$$\det B = \det A.$$

Proof. Write the rows of A as $\boldsymbol{r}_1, \ldots, \boldsymbol{r}_n$. The jth row of B, $\boldsymbol{r}_j + c\boldsymbol{r}_i$, is the only one differing from the corresponding row of A. By Proposition 3,

$$\det B = \det A + \det C$$

where C has rows $\boldsymbol{r}_1, \ldots, \boldsymbol{r}_{j-1}, c\boldsymbol{r}_i, \boldsymbol{r}_{j+1}, \ldots, \boldsymbol{r}_n$. By Proposition 4 and the Corollary to Proposition 2, $\det C = c0 = 0$. We deduce that

$$\det B = \det A.$$

A nice consequence of Propositions 4 and 5 is that a single number, $\det A$, reveals whether or not a matrix A is invertible.

Proposition 6 *Let A be $n \times n$. The following statements are equivalent.*

(i) A is invertible.

(ii) $\det A \neq 0$.

Proof. Let B be the reduced echelon form of A. Each row operation used is going from A to B produces a new matrix with determinant a nonzero multiple of the previous determinant. (The 'multiplier' is 1 in the case of Operation 1.) So $\det B$ is a nonzero multiple of $\det A$.

Now suppose (i) holds. Then A has rank n and $B = I$ (Proposition 10 of Chapter 4). Clearly $\det B = 1$ (only one signed product is nonzero). So $\det A \neq 0$.

Suppose (ii) holds. Then $\det B \neq 0$. So B has no zero row, and has n pivot columns. So $B = I$, and A has rank n. Again using Proposition 10 of Chapter 4, we see that (i) holds.

In calculating $\det A$, where A is $n \times n$ with $n \geq 3$, a short route is to obtain an upper triangular matrix

$$B = \begin{bmatrix} b_{11} & * & \cdots & * \\ & \cdot & & * \\ & & \cdot & \cdot \\ & & & \cdot \\ & & & b_{nn} \end{bmatrix}$$

by row reduction. The determinant of B is the product of its diagonal entries.

Proposition 7 *The determinant of an upper triangular $n \times n$ matrix $B = [b_{ij}]$ is $b_{11}b_{22}\cdots b_{nn}$.*

Proof. Consider a nonzero special product

$$b_{1p(1)} \dots b_{np(n)}.$$

We must have $p(n) = n$, since $b_{ni} = 0$ for $i < n$. We must have $p(n-1) = n-1$, since $p(n-1) \neq n$ and $b_{n-1,i} = 0$ for $i < n-1$. Continuing in this way,

$$p(n) = n, p(n-1) = n-1, \dots, p(2) = 2, p(1) = 1.$$

Thus $b_{11}b_{22}\dots b_{nn}$ is the only possible nonzero special product. Since $1, 2, 3, \dots, n$ is an even permutation,

$$\det B = b_{11}b_{22}\dots b_{nn}.$$

Example 6 Evaluate $\det A$ where

$$A = \begin{bmatrix} 4 & 1 & 2 & 1 \\ 3 & 6 & 7 & 7 \\ 0 & 2 & 1 & 3 \\ 1 & 5 & 1 & 2 \end{bmatrix}.$$

Solution. We show row reduction in much the same notation as before, but obtain a sequence of equalities between real numbers starting with $\det A$. Propositions 2, 4, 5 are used.

$$\det A = -\det \begin{bmatrix} 1 & 5 & 1 & 2 \\ 3 & 6 & 7 & 7 \\ 0 & 2 & 1 & 3 \\ 4 & 1 & 2 & 1 \end{bmatrix} \begin{matrix} \text{I} \leftrightarrow \text{IV} \\ \\ \\ \\ \end{matrix} = -\det \begin{bmatrix} 1 & 5 & 1 & 2 \\ 0 & -9 & 4 & 1 \\ 0 & 2 & 1 & 3 \\ 0 & -19 & -2 & -7 \end{bmatrix} \begin{matrix} \text{II} - 3\text{I} \\ \text{IV} - 4\text{I} \\ \\ \\ \end{matrix}$$

$$= 2\det \begin{bmatrix} 1 & 5 & 1 & 2 \\ 0 & 1 & \frac{1}{2} & \frac{3}{2} \\ 0 & -9 & 4 & 1 \\ 0 & -19 & -2 & -7 \end{bmatrix} \begin{matrix} \text{III} \leftrightarrow \text{II} \\ \text{Factor 2 from II} \\ \\ \\ \end{matrix} = 2\det \begin{bmatrix} 1 & 5 & 1 & 2 \\ 0 & 1 & \frac{1}{2} & \frac{3}{2} \\ 0 & 0 & \frac{17}{2} & \frac{29}{2} \\ 0 & 0 & \frac{15}{2} & \frac{43}{2} \end{bmatrix} \begin{matrix} \text{III} + 9\text{II} \\ \text{IV} + 19\text{II} \\ \\ \\ \end{matrix}$$

$$= 17\det \begin{bmatrix} 1 & 5 & 1 & 2 \\ 0 & 1 & \frac{1}{2} & \frac{3}{2} \\ 0 & 0 & 1 & \frac{29}{17} \\ 0 & 0 & 0 & \frac{148}{17} \end{bmatrix} \begin{matrix} \text{Factor } 17/2 \text{ from III} \\ \text{IV} - \frac{15}{2}\text{ III} \\ \\ \\ \end{matrix} = 148$$

since $\frac{43}{2} - \frac{15}{2}\frac{29}{17} = \frac{148}{17}$. We used Proposition 7 in the last step.

A result that is both attractive and a time-saver is

Proposition 8 *The determinant of the matrix*

$$B = \left[\begin{array}{c|c} U & V \\ \hline & W \end{array}\right]$$

where U is $n \times n$, V is $n \times m$, and W is $m \times m$, is $\det U \det W$.

Proof. We may reduce B to upper triangular form

$$B' = \left[\begin{array}{c|c} U' & V' \\ \hline & W' \end{array}\right]$$

by using Operation I a number of times on $[U|V]$ and on $[0|W]$, so that

$$\det B' = \det B.$$

Now U' and W' are also in upper triangular form and have been produced from U and W respectively by repeated use of Operation I. This gives

$$\det U' = \det U, \ \det W' = \det W.$$

Finally $\det B' = \det U' \det W'$ (product of the terms on the main diagonal). The result follows on combining these equalities.

If we apply this result to the last matrix but one in Example 6,

$$\det A = 2 \det \begin{bmatrix} 1 & 5 \\ 0 & 1 \end{bmatrix} \det \begin{bmatrix} \frac{17}{2} & \frac{29}{2} \\ \frac{15}{2} & \frac{43}{2} \end{bmatrix}$$

$$= 2\left(\frac{17 \cdot 43 - 29 \cdot 15}{4}\right) = 148.$$

Example 7 Evaluate $\det A$, where

$$A = \begin{bmatrix} 1 & 14 & 3 \\ -5 & 2 & 7 \\ 6 & 4 & 1 \end{bmatrix}.$$

Solution. Proposition 8 gives

$$\det A = \det \begin{bmatrix} 1 & 14 & 3 \\ 0 & 72 & 22 \\ 0 & -80 & -17 \end{bmatrix} \begin{matrix} \text{II} + 5\text{I} \\ \text{III} - 6\text{I} \\ \end{matrix} = 72 \times -17 + 22 \times 80 = 536.$$

The following simple relationship is frequently useful.

Proposition 9 *Let A be $n \times n$. Then*

$$\det A^t = \det A.$$

Proof. The special products for A^t are of the form

$$a_{p(1)1} \cdots a_{p(n)n}. \tag{7}$$

These are all products which contain one term from each row, and one term from each column, of A. Thus the special products for A^t are the same as the special products of A.

Let us rewrite (7) by rearranging the product as

$$a_{1q(1)} \cdots a_{nq(n)} \tag{8}$$

where q is a permutation that depends on p. Now p, q have the same number of disordered pairs. In (8) this is the number of 'subproducts' $a_{iu}a_{jv}$ with $i < j$ and $u > v$. But $i = p(u)$ and $j = p(v)$, so this is equal to the number of pairs v, u with $v < u$ and $p(v) > p(u)$.

Hence the *signed* products for A^t are the same as the signed products for A. Adding, $\det A^t = \det A$.

Example 8 We can immediately obtain the determinant of a **lower triangular** matrix ($a_{ij} = 0$ for $i < j$):

$$\det \begin{bmatrix} a_{11} & & & & \\ * & \cdot & & & \\ \cdot & & \cdot & & \\ \cdot & & & \cdot & \\ * & \cdot & \cdot & * & a_{nn} \end{bmatrix} = a_{11} \cdots a_{nn}$$

since the determinant of the transpose is $a_{11} \cdots a_{nn}$.

3 Expansion by a row or column

It follows from Proposition 9 that column operations affect determinants like row operations.

Proposition 10 *Let A be $n \times n$.*

(i) Let B be obtained from A by interchanging two columns. Then $\det B = -\det A$.

(ii) A matrix A with two identical columns has zero determinant.

(iii) If $A = [\boldsymbol{c}_1 \cdots \boldsymbol{c}_{j-1}\ a\boldsymbol{s} + b\boldsymbol{t}\ \boldsymbol{c}_{j+1} \cdots \boldsymbol{c}_n]$, then

$$\det A = a \det[\boldsymbol{c}_1 \cdots \boldsymbol{c}_{j-1}\ \boldsymbol{s}\ \boldsymbol{c}_{j+1} \cdots \boldsymbol{c}_n] + b \det[\boldsymbol{c}_1 \cdots \boldsymbol{c}_{j-1}\ \boldsymbol{t}\ \boldsymbol{c}_{j+1} \cdots \boldsymbol{c}_n].$$

(iv) Let B be obtained from A by adding c times one column to another column. Then $\det B = \det A$.

Proof. (i) B^t is obtained from A^t by interchanging two rows, so $\det B^t = -\det A^t$ (Proposition 2). Now $\det B = \det B^t = -\det A^t = -\det A$ (Proposition 9).

(ii) $\det A^t = 0$ (Corollary to Proposition 2). So $\det A = \det A^t = 0$ (Proposition 9).

(iii) We have

$$\det A^t = \det[\boldsymbol{c}_1 \cdots \boldsymbol{c}_{j-1}\ a\boldsymbol{s}\ \boldsymbol{c}_{j+1} \cdots \boldsymbol{c}_n]^t + \det[\boldsymbol{c}_1 \cdots \boldsymbol{c}_{j-1}\ b\boldsymbol{t}\ \boldsymbol{c}_{j+1} \cdots \boldsymbol{c}_n]^t.$$

(Proposition 3)

$$= a \det[\boldsymbol{c}_1 \cdots \boldsymbol{c}_{j-1}\ \boldsymbol{s}\ \boldsymbol{c}_{j+1} \cdots \boldsymbol{c}_n]^t + b \det[\boldsymbol{c}_1 \cdots \boldsymbol{c}_{j+1}\ \boldsymbol{t}\ \boldsymbol{c}_{j+1} \cdots \boldsymbol{c}_n]^t$$

(Proposition 4). We now obtain the result by using Proposition 9 for each summand.

(iv) $\det B^t = \det A^t$ (Proposition 5). Now $\det B = \det B^t = \det A^t = \det A$ (Proposition 9).

The matrix obtained by deleting row i and column j of a matrix A will be denoted by A_{ij}. For example, if

$$A = \begin{bmatrix} a_{11} & a_{12} & a_{13} \\ a_{21} & a_{22} & a_{23} \\ a_{31} & a_{32} & a_{33} \end{bmatrix}, \text{ then } A_{32} = \begin{bmatrix} a_{11} & a_{13} \\ a_{21} & a_{23} \end{bmatrix}. \tag{9}$$

The determinant $\det A_{ij}$ is said to be a **minor** of A. The number

$$(-1)^{i+j} \det A_{ij}$$

is said to be the **cofactor** of a_{ij}. For instance, since $2+3$ is odd, the cofactor of a_{32} in (9) is

$$-\det A_{32} = -a_{11}a_{23} + a_{13}a_{21}.$$

Notice that the product of a_{32} and its cofactor is

$$-a_{32} \det A_{32} = -a_{32}a_{11}a_{23} + a_{32}a_{13}a_{21}$$

which consists of two of the terms of $\det A$ (Example 3). Similarly,

$$\begin{aligned}(-1)^{3+1}a_{31} \det A_{31} &= a_{31}a_{12}a_{23} - a_{31}a_{13}a_{22},\\ (-1)^{3+3}a_{33} \det A_{33} &= a_{33}a_{11}a_{22} - a_{33}a_{12}a_{21}\end{aligned}$$

are sums of two terms from $\det A$, and

$$\det A = (-1)^{3+1}a_{31} \det A_{31} + (-1)^{3+2}a_{32} \det A_{32} + (-1)^{3+3}a_{33} \det A_{33}.$$

This is an instance of

Proposition 11 *Let A be $n \times n$. Then*

(i) (expansion by row i) . For each $i, 1 \le i \le n$, we have

$$\det A = \sum_{j=1}^{n} (-1)^{i+j} a_{ij} \det A_{ij}. \tag{10}$$

(ii) (expansion by column j) . For each $j, 1 \le j \le n$, we have

$$\det A = \sum_{i=1}^{n} (-1)^{i+j} a_{ij} \det A_{ij}. \tag{11}$$

In (10), i remains fixed and j varies; in (11), j remains fixed and i varies.
Proof. (i) Fix the integer i. Consider any j. The special products of $\det A_{ij}$ consist of all products of $n-1$ factors a_{uv} in which u takes all values except i, v all values except j. Ignoring signs, the $(n-1)!$ terms of $a_{ij} \det A_{ij}$ consist of all special products in which one factor is a_{ij}. The sum in (10) thus contains

(ignoring signs) all the $n!$ terms of $\det A$, since one factor is $a_{i1}, a_{i2}, \ldots,$ or a_{in} in these terms.

It remains to show that, for a fixed pair i and j, the $(n-1)!$ terms in $(-1)^{i+j} a_{ij} \det A_{ij}$ have the 'right' signs attached. Consider the matrix B obtained by starting with A, interchanging adjacent rows $i-1$ times, then interchanging adjacent columns $j-1$ times, to bring a_{ij} to top left position:

$$b_{11} = a_{ij}.$$

It is clear that $B_{11} = A_{ij}$; the order of rows and columns of A_{ij} is undisturbed by these $i+j-2$ operations. Now

$$\det B = (-1)^{i+j-2} \det A = (-1)^{i+j} \det A; \ \det A = (-1)^{i+j} \det B.$$

The sign of a term $\pm b_{2p(2)} \cdots b_{np(n)}$ of $\det B_{11}$ is evidently the same as the sign of $\pm b_{11} b_{2p(2)} \cdots b_{np(n)}$ in $\det B$. So $\det B$ includes the terms $b_{11} \det B_{11} = a_{ij} \det A_{ij}$. We see that $(n-1)!$ of the terms of $\det A$ are the terms of $(-1)^{i+j} a_{ij} \det A_{ij}$. So $\sum_{j=1}^{n} (-1)^{i+j} a_{ij} \det A_{ij}$ contains all the terms of $\det A$ *with the correct signs.*

To obtain part (ii), we repeat the above proof almost verbatim with the second suffix j fixed instead of the first suffix i.

Example 9

$$\det \begin{bmatrix} a_{11} & a_{12} & 0 & 0 \\ 0 & a_{22} & a_{23} & 0 \\ a_{31} & 0 & a_{33} & a_{34} \\ 0 & a_{42} & 0 & a_{44} \end{bmatrix} = a_{11} \det \begin{bmatrix} a_{22} & a_{23} & 0 \\ 0 & a_{33} & a_{34} \\ a_{42} & 0 & a_{44} \end{bmatrix} + a_{31} \det \begin{bmatrix} a_{12} & 0 & 0 \\ a_{22} & a_{23} & 0 \\ a_{42} & 0 & a_{44} \end{bmatrix}$$

(expand by column 1)

$$= a_{11} \left\{ a_{22} \det \begin{bmatrix} a_{33} & a_{34} \\ 0 & a_{44} \end{bmatrix} - a_{23} \det \begin{bmatrix} 0 & a_{34} \\ a_{42} & a_{44} \end{bmatrix} \right\} + a_{31} a_{12} a_{23} a_{44}$$

(expand by row 1 for first determinant; use Example 8 for second)

$$= a_{11} a_{22} a_{33} a_{44} + a_{11} a_{23} a_{34} a_{42} + a_{31} a_{12} a_{23} a_{44}.$$

The matrix of signs $(-1)^{i+j}$ is a 'checkerboard,' a useful fact when doing calculations. For example,

$$\begin{bmatrix} + & - & + & - \\ - & + & - & + \\ + & - & + & - \\ - & + & - & + \end{bmatrix}.$$

4 The adjoint

Hidden in Proposition 11 is a beautiful formula for A^{-1} when $\det A \neq 0$. The special case $n = 2$ is (15) of Chapter 4.

Definition 3 Let A be $n \times n$. The **adjoint** of A, written adj A, is the transpose of the matrix of cofactors of a_{ij}. That is, the i, j entry of adj A is

$$(-1)^{j+i} \det A_{ji} \qquad (i, j = 1, \dots, n).$$

Example 10 Let $n = 2$. Listing entries and cofactors as pairs, we have $(a_{11}, a_{22}), (a_{12}, -a_{21}), (a_{21}, -a_{12}), (a_{22}, a_{11})$. The adjoint is

$$\text{adj } A = \begin{bmatrix} a_{22} & -a_{21} \\ -a_{12} & a_{11} \end{bmatrix}^t = \begin{bmatrix} a_{22} & -a_{12} \\ -a_{21} & a_{11} \end{bmatrix}.$$

Proposition 12 *Let A be $n \times n$. Then*

$$A \text{ adj } A = (\det A)I. \tag{12}$$

If $\det A \neq 0$,

$$A^{-1} = \frac{\text{adj } A}{\det A}. \tag{13}$$

Proof. The i, j entry of A adj A is

$$\sum_{k=1}^{n} a_{ik}(-1)^{j+k} \det A_{jk}, \tag{14}$$

since the kth entry of column j of adj A is $(-1)^{j+k} \det A_{jk}$.

Suppose first that $i = j$. The sum in (14) is $\det A$ by Proposition 11 (i). So A adj A has entries $\det A$ along the main diagonal.

Now suppose that $i \neq j$. Take A, replace row j by a copy of row i, and write down the expansion by row j of the new matrix. We obtain

$$\sum_{k=1}^{n} a_{ik}(-1)^{j+k} \det A_{jk},$$

since cofactors on row j are the same as those of A. Hence the sum in (14) is the determinant of a matrix with two identical rows, which is 0. Thus A adj A has zeros off the main diagonal. This proves (12).

If $\det A \neq 0$, we may rewrite (13) as

$$A\frac{\text{adj } A}{\det A} = I.$$

Recalling Proposition 11 of Chapter 4, the matrix $\frac{\text{adj } A}{\det A}$ is A^{-1}.

Example 11 Find the inverse of $A = \begin{bmatrix} 1 & 2 & 2 \\ 2 & 1 & 4 \\ -1 & 0 & 1 \end{bmatrix}$ using the adjoint.

Solution. First, $\det A = \det \begin{bmatrix} 1 & 2 & 2 \\ 0 & -3 & 0 \\ 0 & 2 & 3 \end{bmatrix} \begin{matrix} \\ \text{II} - 2\text{I} \\ \text{III} + \text{I} \end{matrix} = -9$. Next,

$$\begin{aligned} \det A_{11} &= 1, & -\det A_{12} &= -6, & \det A_{13} &= 1, \\ -\det A_{21} &= -2 & \det A_{22} &= 3, & -\det A_{23} &= -2, \\ \det A_{31} &= 6, & -\det A_{32} &= 0, & \det A_{33} &= -3, \end{aligned}$$

Thus $A^{-1} = -\frac{1}{9} \begin{bmatrix} 1 & -2 & 6 \\ -6 & 3 & 0 \\ 1 & -2 & -3 \end{bmatrix}$.

We can use Proposition 12 to write down an appealing formula for the solution of a linear system

$$A\boldsymbol{x} = \boldsymbol{b},$$

with A invertible. Obviously

$$\boldsymbol{x} = A^{-1}\boldsymbol{b}$$

on multiplying both sides by A^{-1}. Thus x_j is the inner product of row j of A^{-1} with $(b_1, \ldots, b_n)$,

$$x_j = \frac{1}{\det A} \sum_{i=1}^{n} (-1)^{i+j} b_i \det A_{ij}.$$

The sum here is the expansion by column j of

$$\det[\boldsymbol{a}_1 \cdots \boldsymbol{a}_{j-1}\ \boldsymbol{b}\ \boldsymbol{a}_{j+1} \cdots \boldsymbol{a}_n]$$

where $\boldsymbol{b}$ has been substituted for column j of $A = [\boldsymbol{a}_1 \cdots \boldsymbol{a}_n]$. That is, we have

Proposition 13 (Cramer's Rule). *Let A be an invertible $n \times n$ matrix, $A = [\boldsymbol{a}_1 \cdots \boldsymbol{a}_n]$. The unique solution of the linear system $A\boldsymbol{x} = \boldsymbol{b}$ is*

$$x_j = \frac{\det[\boldsymbol{a}_1 \cdots \boldsymbol{a}_{j-1}\ \boldsymbol{b}\ \boldsymbol{a}_{j+1} \cdots \boldsymbol{a}_n]}{\det A}.$$

For example, if $n = 3$,

$$\boldsymbol{x} = \frac{1}{\det A}\left(\det\begin{bmatrix} b_1 & a_{12} & a_{13} \\ b_2 & a_{22} & a_{23} \\ b_3 & a_{32} & a_{33} \end{bmatrix}, \det\begin{bmatrix} a_{11} & b_1 & a_{13} \\ a_{21} & b_2 & a_{23} \\ a_{31} & b_3 & a_{33} \end{bmatrix}, \det\begin{bmatrix} a_{11} & a_{12} & b_1 \\ a_{21} & a_{22} & b_2 \\ a_{31} & a_{32} & b_3 \end{bmatrix}\right).$$

5 The determinant of a product

The rule here is delightfully simple.

Proposition 14 *Let A and B be $n \times n$. Then*

$$\det AB = \det A \det B.$$

Proof. Let $A = [a_{ij}], B = [b_{ij}]$. Further, write $A = [\boldsymbol{a}_1 \cdots \boldsymbol{a}_n]$ and

$$AB = [\boldsymbol{c}_1 \cdots \boldsymbol{c}_n].$$

Now

$$\boldsymbol{c}_k = b_{1k}\boldsymbol{a}_1 + \cdots + b_{nk}\boldsymbol{a}_n$$

and

$$\det AB = \det[b_{11}\boldsymbol{a}_1 + \cdots + b_{n1}\boldsymbol{a}_n \cdots b_{1n}\boldsymbol{a}_1 + \cdots + b_{nn}\boldsymbol{a}_n].$$

We use Proposition 10 (iii) to 'split up via column 1' :

$$\det AB = \sum_{j=1}^{n} b_{j1} \det[\boldsymbol{a}_j \; b_{12}\boldsymbol{a}_1 + \cdots + b_{n2}\boldsymbol{a}_n \cdots b_{1n}\boldsymbol{a}_1 + \cdots + b_{nn}\boldsymbol{a}_n].$$

We repeat the process for column 2, then column 3, and so on. Finally,

$$\det AB = \sum_{j_1=1}^{n} b_{j_1 1} \sum_{j_2=1}^{n} b_{j_2 2} \cdots \sum_{j_n=1}^{n} b_{j_n n} \det[\boldsymbol{a}_{j_1} \cdots \boldsymbol{a}_{j_n}].$$

All the determinants in which $j_1, \ldots, j_n$ contains a repeated integer are zero (Proposition 10 (ii)). So

$$\det AB = \sum_{p} b_{p(1)1} \cdots b_{p(n)n} \det[\boldsymbol{a}_{p(1)} \cdots \boldsymbol{a}_{p(n)}]$$

where the sum is over all permutations p of $1, \ldots, n$. We now bring $[\boldsymbol{a}_{p(1)} \cdots \boldsymbol{a}_{p(n)}]$ to the form $[\boldsymbol{a}_1 \cdots \boldsymbol{a}_n]$ by interchanging adjacent pairs of columns repeatedly to eliminate disordered pairs. The number of interchanges is odd if p is odd, even if p is even. So

$$\det[\boldsymbol{a}_{p(1)} \cdots \boldsymbol{a}_{p(n)}] = \pm \det[\boldsymbol{a}_1 \cdots \boldsymbol{a}_n]$$

($+$ for even p, $-$ for odd); and

$$\det AB = \sum_{p} \pm b_{p(1)1} \cdots b_{p(n)n} \det[\boldsymbol{a}_1 \cdots \boldsymbol{a}_n]$$

(with signs as above)

$$= \det B^t \det A = \det B \det A$$

by Proposition 9.

Corollary *For P invertible,* $\det P^{-1} = (\det P)^{-1}$.

To see this, $1 = \det I = \det PP^{-1} = \det P \det P^{-1}$.

Example 12 Suppose that

$$P = [\boldsymbol{p}_1 \cdots \boldsymbol{p}_n]$$

is an orthogonal matrix. Show that

$$\det P = \pm 1.$$

Solution. We have $P^t = P^{-1}$ and so $\det P^t = (\det P)^{-1}$. Since $\det P^t = \det P$,

$$1 = (\det P)^2 \quad ; \quad \det P = \pm 1.$$

Example 13 Suppose that any two columns of A are orthogonal:

$$\boldsymbol{a}_i \cdot \boldsymbol{a}_j = 0 \quad (i \neq j).$$

Show that

$$\det A = \pm|\boldsymbol{a}_1| \cdots |\boldsymbol{a}_n|.$$

Solution. This is obvious if some $\boldsymbol{a}_j$ is $\mathbf{0}$. Suppose no $\boldsymbol{a}_j$ is $\mathbf{0}$; then

$$\det A = |\boldsymbol{a}_1| \cdots |\boldsymbol{a}_n| \det \left[\frac{\boldsymbol{a}_1}{|\boldsymbol{a}_1|} \cdots \frac{\boldsymbol{a}_n}{|\boldsymbol{a}_n|} \right]$$

(Proposition 10 (iii))

$$= \pm|\boldsymbol{a}_1| \cdots |\boldsymbol{a}_n|$$

since $\left[\frac{\boldsymbol{a}_1}{|\boldsymbol{a}_1|} \cdots \frac{\boldsymbol{a}_n}{|\boldsymbol{a}_n|} \right]$ is an orthogonal matrix.

Example 14 Let $A = [\boldsymbol{a}_1 \ \boldsymbol{a}_2]$ be 2×2. Show that the parallelogram with vertices $\mathbf{0}, \boldsymbol{a}_1, \boldsymbol{a}_2, \boldsymbol{a}_1 + \boldsymbol{a}_2$ has area $\pm \det A$.

Solution. We have

$$\det A = \det \left[\boldsymbol{a}_1 \quad \boldsymbol{a}_2 - \frac{\boldsymbol{a}_2 \cdot \boldsymbol{a}_1}{\boldsymbol{a}_1 \cdot \boldsymbol{a}_1} \boldsymbol{a}_1 \right]$$

(Proposition 10 (iv)). The vector $\boldsymbol{a}_2 - \frac{\boldsymbol{a}_2 \cdot \boldsymbol{a}_1}{\boldsymbol{a}_1 \cdot \boldsymbol{a}_1} \boldsymbol{a}_1$ is perpendicular to $\boldsymbol{a}_1$ and is the 'altitude' of the parallelogram if $\boldsymbol{a}_1$ is the base. Thus the area of the parallelogram is

$$|\boldsymbol{a}_1| \left| \boldsymbol{a}_2 - \frac{\boldsymbol{a}_2 \cdot \boldsymbol{a}_1}{\boldsymbol{a}_1 \cdot \boldsymbol{a}_1} \boldsymbol{a}_1 \right|.$$

This is $\pm \det A$ by Example 13.

Example 15 (Cauchy) Let $A = [\boldsymbol{a}_1 \ \boldsymbol{a}_2 \ \boldsymbol{a}_3]$ be 3×3. Show that the parallelepiped with vertex $\mathbf{0}$ and vertices $\boldsymbol{a}_1, \boldsymbol{a}_2, \boldsymbol{a}_3$ adjacent to $\mathbf{0}$ has volume $\pm \det A$ (Figure 1).

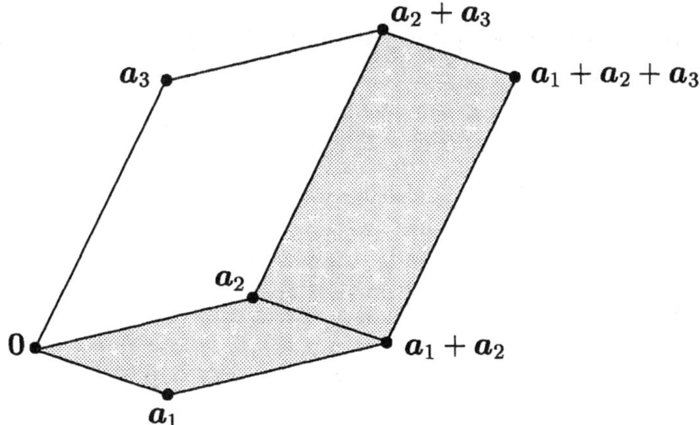

Figure 1. The parallellepiped with volume $\pm\det[\boldsymbol{a}_1, \boldsymbol{a}_2, \boldsymbol{a}_3]$.

Solution. Let $\boldsymbol{v}$ be a normal to the plane P through $\boldsymbol{0}, \boldsymbol{a}_1, \boldsymbol{a}_2$. Let t be a real number such that $\boldsymbol{a}_3 + t\boldsymbol{v}$ lies in this plane,

$$\boldsymbol{a}_3 + t\boldsymbol{v} = x_1\boldsymbol{a}_1 + x_2\boldsymbol{a}_2. \tag{15}$$

The area of the base of the parallelepiped in P is

$$|\boldsymbol{a}_1| \left| \boldsymbol{a}_2 - \frac{\boldsymbol{a}_2 \cdot \boldsymbol{a}_1}{\boldsymbol{a}_1 \cdot \boldsymbol{a}_1} \boldsymbol{a}_1 \right|.$$

Now $|t\boldsymbol{v}|$ is the altitude of the parallellepiped, so the volume is

$$V = |\boldsymbol{a}_1| \left| \boldsymbol{a}_2 - \frac{\boldsymbol{a}_2 \cdot \boldsymbol{a}_1}{\boldsymbol{a}_1 \cdot \boldsymbol{a}_1} \boldsymbol{a}_1 \right| |t\boldsymbol{v}|.$$

Equation (15) gives

$$0 = \det\begin{bmatrix}\boldsymbol{a}_1 & \boldsymbol{a}_2 & \boldsymbol{a}_3 + t\boldsymbol{v}\end{bmatrix} = \det\begin{bmatrix}\boldsymbol{a}_1 & \boldsymbol{a}_2 & \boldsymbol{a}_3\end{bmatrix} + \det\begin{bmatrix}\boldsymbol{a}_1 & \boldsymbol{a}_2 & t\boldsymbol{v}\end{bmatrix}, \tag{16}$$

and

$$\begin{aligned}\det\begin{bmatrix}\boldsymbol{a}_1 & \boldsymbol{a}_2 & t\boldsymbol{v}\end{bmatrix} &= \det\begin{bmatrix}\boldsymbol{a}_1 & \boldsymbol{a}_2 - \frac{\boldsymbol{a}_2\cdot\boldsymbol{a}_1}{\boldsymbol{a}_1\cdot\boldsymbol{a}_1}\boldsymbol{a}_1 & t\boldsymbol{v}\end{bmatrix} \\ &= \pm|\boldsymbol{a}_1| \left| \boldsymbol{a}_2 - \frac{\boldsymbol{a}_2 \cdot \boldsymbol{a}_1}{\boldsymbol{a}_1 \cdot \boldsymbol{a}_1} \boldsymbol{a}_1 \right| |t\boldsymbol{v}| = \pm V\end{aligned}$$

(Example 13). Now $\det A = \pm V$ from (16).

Example 16 (Vandermonde determinant). Let $x_1, \dots, x_n$ be real numbers. Show that the matrix

$$A(x_1, \dots, x_n) = \begin{bmatrix} 1 & x_1 \cdots x_1^{n-1} \\ 1 & x_2 \cdots x_2^{n-1} \\ & \cdots \\ 1 & x_n \cdots x_n^{n-1} \end{bmatrix}$$

has determinant

$$\det A(x_1, \dots, x_n) = (x_2 - x_1)(x_3 - x_1) \cdots (x_n - x_{n-1})$$

where the product on the right contains all $x_j - x_i$ with $i < j$.

Solution. Let us warm up with the case $n = 3$.

$$\det A(x_1, x_2, x_3) = \det \begin{bmatrix} 1 & x_1 & x_1^2 \\ 1 & x_2 & x_2^2 \\ 1 & x_3 & x_3^2 \end{bmatrix}$$

$$= \det \begin{bmatrix} 1 & x_1 & x_1^2 - x_1^2 \\ 1 & x_2 & x_2^2 - x_1 x_2 \\ 1 & x_3 & x_3^2 - x_1 x_3 \end{bmatrix}$$

(subtract x_1 times column 2 from column 3)

$$= \det \begin{bmatrix} 1 & x_1 - x_1 & x_1^2 - x_1^2 \\ 1 & x_2 - x_1 & x_2^2 - x_1 x_2 \\ 1 & x_3 - x_1 & x_3^2 - x_1 x_3 \end{bmatrix}$$

(subtract x_1 times column 1 from column 2).

Since the first row is $[1 \quad 0 \quad 0]$, we expand by that row:

$$\det A(x_1, x_2, x_3) = \det \begin{bmatrix} x_2 - x_1 & x_2^2 - x_1 x_2 \\ x_3 - x_1 & x_3^2 - x_1 x_3 \end{bmatrix}$$

$$= (x_2 - x_1)(x_3 - x_1) \det \begin{bmatrix} 1 & x_2 \\ 1 & x_3 \end{bmatrix}$$

(take out factors from rows 1 and 2)

$$= (x_2 - x_1)(x_3 - x_1)(x_3 - x_2).$$

The last step involved the easy calculation of $\det A(x_2, x_3)$.

We now tackle the case $n > 3$ in the same way. We start by noticing that $A(x_1, \ldots, x_n) = [\mathbf{c}_1 \cdots \mathbf{c}_n]$ has the same determinant as

$$\begin{bmatrix}\mathbf{c}_1 & \mathbf{c}_2 - x_1\mathbf{c}_1 & \mathbf{c}_3 - x_1\mathbf{c}_2 & \cdots & \mathbf{c}_n - x_1\mathbf{c}_{n-1}\end{bmatrix}$$

(perform the column operations starting on the right). The ith entry of $\mathbf{c}_j - x_1\mathbf{c}_{j-1}$ is

$$x_i^{j-1} - x_1 x_i^{j-2} = (x_i - x_1)x_i^{j-2}.$$

In particular the first row is $\begin{bmatrix}1 & 0 & \cdots & 0\end{bmatrix}$. Expanding by this row,

$$\det A(x_1, \ldots, x_n) = a_{11} \det A_{11} = \det[(x_i - x_1)x_i^{j-2}]_{i=2,\ldots,n;j=2,\ldots,n}.$$

We take out $x_2 - x_1, \ldots, x_n - x_1$ as factors of the $n - 1$ rows:

$$\det A(x_1, \ldots, x_n) = (x_2 - x_1)\cdots(x_n - x_1)\det A(x_2, \ldots, x_n).$$

If we continue the process we arrive at $(x_2 - x_1)\cdots(x_n - x_1)\cdots(x_n - x_{n-2})$ $\det\begin{bmatrix}1 & x_{n-1}\\ 1 & x_n\end{bmatrix}$, which gives the result.

Example 17 Let A be $n \times n$. Show that $\det(\text{adj } A) = (\det A)^{n-1}$.

Solution. We have

$$A \text{ adj } A = (\det A)I.$$

Proposition 14 gives

$$\det A \det(\text{adj } A) = (\det A)^n,$$

the right side being the product of the diagonal entries. If $\det A \neq 0$, we get $\det(\text{adj } A) = (\det A)^{n-1}$.

If $\det A = 0$, then

$$A \text{ adj } A = 0, \text{ nullity } (A \text{ adj } A) = n.$$

Now the nullity of A adj A is $\leq$ nullity $A+$ nullity (adj A) (Example 13, Chapter 4). Hence nullity (adj A) > 0 unless nullity $A = n$, that is, unless $A = 0$. Since adj $0 = 0$, the nullity of adj A is positive, and so $\det(\text{adj } A) = 0$ (Proposition 6). We find that $\det(\text{adj } A) = (\det A)^{n-1}$ in the case of $\det A = 0$.

Exercises for Chapter 5

Reminder: Attempt Exercise a.b after reading Section a.

1.1 Determine the parity of the following permutations.

(i) 1, 2, 4, 3, 6, 5 (ii) 1, 4, 5, 2, 3

(iii) 7, 6, 5, 4, 3, 2, 1 (iv) 1, 3, 5, 7, 2, 4, 6, 8

1.2 Show that any permutation of $1, \ldots, n$ can be obtained, starting with $1, \ldots, n$, by interchanging adjacent pairs of integers at most $\frac{1}{2}\, n(n-1)$ times.

Hint: How many disordered pairs are there in $n, n-1, \ldots, 2, 1$?

1.3 Let A be 2×2. Show that

$$\det(zI - A) = z^2 - (a_{11} + a_{22})z + \det A.$$

1.4 Let A be 2×2 and suppose that $A^2 = 0$. Show that

$$\det(zI - A) = z^2.$$

1.5 Show by an example that it is not true in general that

$$\det(A + B) \leq \det A + \det B.$$

2.1 Let B be obtained from the 3×3 matrix A by interchanging row 2 and row 3. By writing out $\det B$ in full, verify that $\det B = -\det A$.

2.2 Evaluate the determinants of the following matrices.

$$\text{(i)} \begin{bmatrix} -2 & 0 & 3 \\ 1 & 0 & -4 \\ 0 & 2 & 6 \end{bmatrix} \qquad \text{(ii)} \begin{bmatrix} -1 & -1 & -2 \\ 3 & 4 & 5 \\ 6 & 8 & 9 \end{bmatrix} \qquad \text{(iii)} \begin{bmatrix} 1 & 2 & 1 \\ 2 & 1 & 2 \\ 1 & 1 & 2 \end{bmatrix}.$$

2.3 Given vectors $\boldsymbol{a}_1, \ldots, \boldsymbol{a}_{n-1}$ in $\mathbb{R}^n$, let $T : \mathbb{R}^n \to \mathbb{R}$,

$$T(\boldsymbol{x}) = \det[\boldsymbol{a}_1 \cdots \boldsymbol{a}_{n-1}\ \boldsymbol{x}].$$

(i) Show that T is a linear mapping.

(ii) If $\boldsymbol{a}_1, \dots, \boldsymbol{a}_{n-1}$ are linearly independent, show that T has rank 1 and write down a basis of Ker T.

2.4 Let A be $n \times n$ of rank r. Show that we can delete $n-r$ rows and $n-r$ columns, leaving a matrix with nonzero determinant. Could we get a similar outcome with fewer deleted rows and columns?

Hint: Delete columns to leave a basis of the column space. What is the dimension of the row space of the remaining $n \times r$ matrix?

3.1 Let A be 3×3. Write out in full the expansion of det A by row 2. Verify that the expression you obtain is det A.

3.2 Evaluate det A where

$$A = \begin{bmatrix} 4 & 1 & -2 & -1 \\ -2 & -3 & 0 & -2 \\ 1 & 2 & 0 & 4 \\ 2 & 0 & 1 & 3 \end{bmatrix}$$

(i) by expanding by column 3 and then row 3 of the 3×3 matrices obtained;

(ii) by row reduction.

Which method is quicker?

3.3 Show that the expansion by the first *row* of

$$\det \begin{bmatrix} a_{11} & a_{12} & \cdots & a_{1n} \\ 0 & a_{22} & \cdots & a_{2n} \\ \vdots & & & \vdots \\ 0 & a_{n2} & \cdots & a_{nn} \end{bmatrix}$$

is $a_{11} \det A_{11}$.

4.1 Show that adj $(A^t) = (\text{adj } A)^t$.

4.2 Show that adj $(cA) = c^{n-1}$adj A.

4.3 Compute adj A if $A = [\boldsymbol{a}_1 \quad k\boldsymbol{a}_1]$ is 2×2, with $\boldsymbol{a}_1 \neq \mathbf{0}$. Verify that in this case

$$\text{nullity (adj } A) + \text{nullity } A = 2.$$

4.4 Let $A = \begin{bmatrix} \boldsymbol{a}_1 & \boldsymbol{a}_2 & \boldsymbol{a}_1 + \boldsymbol{a}_2 \end{bmatrix}$ be 3×3, where $\boldsymbol{a}_1, \boldsymbol{a}_2$ are linearly independent. Compute adj A, and show that

$$\text{nullity (adj } A) + \text{nullity } A = 3.$$

4.5 Let B be an $n \times n$ matrix two of whose rows are zero. Show that adj $B = 0$.

4.6 Apply Cramer's rule to solve the system

$$\begin{aligned} 3x_1 + 2x_2 - x_3 &= 2 \\ x_1 + \ x_2 \qquad &= 5 \\ -x_1 + \ x_2 - x_3 &= 0. \end{aligned}$$

5.1 Using the formula

$$\det(AB) = \det A \det B,$$

give another proof that a matrix A with determinant 0 is not invertible.

5.2 Show that

$$\det AB = \det A \det B$$

for 2×2 matrices $A = [a_{ij}], B = [b_{ij}]$, by writing out both sides in terms of a_{ij} and b_{ij}.

5.3 Let $T : \mathbb{R}^n \to \mathbb{R}^n$ be linear and let A, B be the matrices of T for the bases $\boldsymbol{v}_1, \dots, \boldsymbol{v}_n$ and $\boldsymbol{w}_1, \dots, \boldsymbol{w}_n$ respectively. Show that $\det A = \det B$. The number $\det A$ is said to be the **determinant of** T, written $\det T$.

5.4 Find the area of the parallelogram whose vertices are

$$(0,0), (1,6), (2,-5), (3,1).$$

5.5 Find the area of the parallelogram whose vertices are

$$(1,1), (3,4), (4,8), (6,11).$$

5.6 Find a formula for the area of the triangle whose vertices are $\mathbf{0}, \boldsymbol{v}_1$ and $\boldsymbol{v}_2$ in $\mathbb{R}^2$.

5.7 Show that the triangle with vertices $(a_1, a_2), (b_1, b_2), (c_1, c_2)$ has area

$$\pm\frac{1}{2}\det\begin{bmatrix} a_1 & a_2 & 1 \\ b_1 & b_2 & 1 \\ c_1 & c_2 & 1 \end{bmatrix}.$$

Hint: Move (a_1, a_2) to the origin by a translation and use Problem 5.6.

5.8 Show that the equation of the line in $\mathbb{R}^2$ through distinct points (a_1, a_2) and (b_1, b_2) can be written

$$\det\begin{bmatrix} a_1 & a_2 & 1 \\ b_1 & b_2 & 1 \\ x_1 & x_2 & 1 \end{bmatrix} = 0.$$

5.9 Let A be a 3×3 invertible matrix whose determinant is an integer and whose column vectors have length $\leq K$. If $A\boldsymbol{x} = \boldsymbol{b}$, show that $|\boldsymbol{x}| \leq \sqrt{3}K^2|\boldsymbol{b}|$.

5.10 Let A be $n \times n$. Show that $\det(A^t A) \geq 0$.

5.11 Let $T : \mathbb{R}^2 \to \mathbb{R}^2$ be a reflection in the line through $\mathbf{0}$ and $\boldsymbol{v}$. Show that T is linear. Show that $\det T = -1$.

Hint: Choose a basis to make the problem easy.

5.12 Let A and P be invertible matrices. Show that

$$\text{adj }(P^{-1}AP) = P^{-1}(\text{adj } A)P.$$

Is this formula correct if P is invertible and A is not?

5.13 Show that

$$\det\begin{bmatrix} 1 & x & x^3 \\ 1 & y & y^3 \\ 1 & z & z^3 \end{bmatrix} = (y-x)(z-x)(z-y)(x+y+z).$$

Hint: Use two row operations to simplify, and extract factors from rows where possible.

5.14 Produce a factorization similar to Exercise 5.13 for

$$\det \begin{bmatrix} 1 & x^2 & x^3 \\ 1 & y^2 & y^3 \\ 1 & z^2 & z^3 \end{bmatrix}.$$

5.15 Use the Vandermonde determinant to show that, given distinct real $x_1, \ldots, x_n$ and any numbers $c_1, \ldots, c_n$, there is a polynomial $P(x) = a_0 x^{n-1} + a_1 x^{n-2} + \cdots + a_{n-2} x + a_{n-1}$ such that $P(x_1) = c_1, \ldots, P(x_n) = c_n$.

Chapter 6

Eigenvalues and eigenvectors

1 Finding the eigenvalues and eigenvectors

In the applications of linear mappings $T : \mathbb{R}^n \to \mathbb{R}^n$ in Chapters 6 and 7, it is important to identify the real numbers c and the vectors $\boldsymbol{x} \neq \mathbf{0}$ such that

$$(1) \qquad T\boldsymbol{x} = c\boldsymbol{x}.$$

Definition 1 Let $T : \mathbb{R}^n \to \mathbb{R}^n$ be linear. If $T\boldsymbol{x} = c\boldsymbol{x}$ for some real c and $\boldsymbol{x} \neq \mathbf{0}$, c is an **eigenvalue** of T and $\boldsymbol{x}$ is an **eigenvector** of T.

Note that the eigenvectors for a given eigenvalue c are the nonzero vectors in $\mathrm{Ker}(T - cI)$. The subspace $\mathrm{Ker}(T - cI)$ is the **eigenspace** of T for the eigenvalue c.

Let A be $n \times n$. The eigenvectors, eigenvalues and eigenspaces of A are defined to be those of the mapping $T\boldsymbol{x} = A\boldsymbol{x}$.

Example 1 Let $T : \mathbb{R}^2 \to \mathbb{R}^2$ be a rotation through angle a, where $0 < a < \pi$ or $\pi < a < 2\pi$. Show that T has no eigenvalues.

Solution. If $\boldsymbol{x}$ is an eigenvector, then $T\boldsymbol{x} = \pm\boldsymbol{x}$, since $|T\boldsymbol{x}| = |\boldsymbol{x}|$. But $T\boldsymbol{x} = \boldsymbol{x}$ and $T\boldsymbol{x} = -\boldsymbol{x}$ are not possible for rotations through angles other than multiples of π; here we recall $\boldsymbol{x} \neq \mathbf{0}$. So there are no eigenvectors and consequently no eigenvalues.

Example 2 Let $T : \mathbb{R}^3 \to \mathbb{R}^3$ be a rotation about the axis $\boldsymbol{v}$. Any nonzero multiple of $\boldsymbol{v}$ is seen to be an eigenvector, since

$$T\boldsymbol{v} = \boldsymbol{v}.$$

The eigenvalue here is 1.

Example 3 Let $T : \mathbb{R}^2 \to \mathbb{R}^2$ be the projection of $\boldsymbol{x}$ onto $\boldsymbol{b}$ (Example 2 of Chapter 4). A nonzero multiple $\boldsymbol{x}$ of $\boldsymbol{b}$ is an eigenvector for the eigenvalue 1, since $T\boldsymbol{x} = \boldsymbol{x}$. The line through $\mathbf{0}$ and $\boldsymbol{b}$ is the eigenspace for 1 since $T\boldsymbol{x} \neq \boldsymbol{x}$ for $\boldsymbol{x}$ *not* on this line.

Let $\mathbf{c}$ be chosen perpendicular to $\boldsymbol{b}$. Then $T\boldsymbol{x} = \mathbf{0}$ for any nonzero multiple $\boldsymbol{x}$ of $\mathbf{c}$. Thus $\boldsymbol{x}$ is an eigenvector with eigenvalue 0. The line through $\mathbf{0}$ and $\boldsymbol{c}$ is the eigenspace for the eigenvalue 0, since $T\boldsymbol{x} \neq \mathbf{0}$ for $\boldsymbol{x}$ *not* on this line.

In general it is not easy to spot the eigenvalues and eigenvectors of a given T, and we use the following result.

Proposition 1 *Let A be $n \times n$. The expression* $\det(cI - A)$ *is a polynomial in c,*

$$\det(cI - A) = c^n + b_1 c^{n-1} + \cdots + b_{n-1}c + b_n. \tag{2}$$

The real zeros of this polynomial are the eigenvalues of A.

Before we prove this, we recall a little terminology. An expression

$$P(x) = a_0 x^n + \cdots + a_{n-1}x + a_n$$

with given real $a_0, \ldots, a_n$, is a **polynomial** in x. The **leading coefficient** is a_0, provided $a_0 \neq 0$. The **degree** of the polynomial is n, provided $a_0 \neq 0$. Thus $2x^5 + 3$ has degree 5 and the constant polynomial 7 has degree 0. The constant polynomial 0 does not have a degree, by convention. The real **zeros** of the polynomial are the real numbers w such that $P(w) = 0$, that is,

$$a_0 w^n + \cdots + a_{n-1}w + a_n = 0.$$

If w_1 is a zero of $P(x)$ then it is easy to see that $x - w_1$ divides P without remainder,

$$P(x) = (x - w_1)Q(x)$$

where Q has degree $n - 1$. Continue the process if there are further real zeros: we reach the factorization

$$P(x) = (x - w_1) \cdots (x - w_r)R(x)$$

where R has degree $n - r$, and has no further real zeros. Of course $R(x)$ could be a constant. Certainly $r \le n$, so that *the number of real zeros of* $P(x)$ *is at most* n, the degree of P.

In Chapter 8 we consider complex numbers. We can find complex zeros of polynomials such as $x^4 + x^2 + 1$ (which could never be less than 1 for real x). For the moment, we are only concerned with *real* zeros of polynomials.

Proof of Proposition 1. The signed products of $cI - A$ are all polynomials in c. All have degree less than n, except for $(c - a_{11}) \cdots (c - a_{nn})$. The sum of all signed products is thus a polynomial $c^n + b_1 c^{n-1} + \cdots + b_n$. So $\det(cI - A) = c^n + b_1 c^{n-1} + \cdots + b_n$.

The solutions of

$$\det(cI - A) = 0$$

are the numbers c for which $cI - A$ has a nonzero kernel (Proposition 6 of Chapter 5). Suppose that c satisfies $\det(cI - A) = 0$; then $(cI - A)\boldsymbol{x} = \mathbf{0}$ for some $\boldsymbol{x} \neq \mathbf{0}$. We have $A\boldsymbol{x} = c\boldsymbol{x}$, and c is an eigenvalue. On the other hand, if c is an eigenvalue, then $A\boldsymbol{x} = c\boldsymbol{x}$ for some $\boldsymbol{x} \neq \mathbf{0}$. Now $\boldsymbol{x}$ belongs to $\mathrm{Ker}(cI - A)$, and so $\det(cI - A) = 0$.

Definition 2 The polynomial $\det(cI - A)$ is the **characteristic polynomial** of A.

Example 4 Find the eigenvalues and eigenspaces of

$$A = \begin{bmatrix} 1 & -3 & 3 \\ 3 & -5 & 3 \\ 6 & -6 & 4 \end{bmatrix}.$$

Solution. The characteristic polynomial of A is

$$\det(cI - A) = \det \begin{bmatrix} c-1 & 3 & -3 \\ -3 & c+5 & -3 \\ -6 & 6 & c-4 \end{bmatrix}.$$

Let us produce some zero entries by row reduction:

$$\det(cI - A) = \det \begin{bmatrix} c+2 & -2-c & 0 \\ -3 & c+5 & -3 \\ 0 & -2c-4 & c+2 \end{bmatrix} \begin{matrix} \text{I} - \text{II} \\ \\ \text{III} - 2\text{II} \end{matrix}.$$

Always take out a factor when possible. In this case, factor out $c+2$ from rows 1 and 3:

$$\det(cI-A) = (c+2)^2 \det \begin{bmatrix} 1 & -1 & 0 \\ -3 & c+5 & -3 \\ 0 & -2 & 1 \end{bmatrix}$$

$$= (c+2)^2 \det \begin{bmatrix} 1 & -1 & 0 \\ 0 & c+2 & -3 \\ 0 & -2 & 1 \end{bmatrix} \begin{matrix} \text{II+3I} \\ \\ \\ \end{matrix}$$

$$= (c+2)^2(c-4).$$

The eigenvalues are the solutions of $(c+2)^2(c-4)=0$, that is, $c=-2,4$.

Eigenspace for $c=-2$. We solve $(-2I-A)\boldsymbol{x}=\mathbf{0}$. The coefficient matrix is

$$\begin{bmatrix} -3 & 3 & -3 \\ -3 & 3 & -3 \\ -6 & 6 & -6 \end{bmatrix} \sim \begin{bmatrix} 1 & -1 & 1 \\ 0 & 0 & 0 \\ 0 & 0 & 0 \end{bmatrix} \begin{matrix} \text{II} - \text{I} \\ \text{III} - 2\text{I} \\ \text{I} \times -\frac{1}{3} \end{matrix}.$$

The general solution is $x_1 = x_2 - x_3$ with x_2, x_3 free,that is,

$$\boldsymbol{x} = x_2 \begin{bmatrix} 1 \\ 1 \\ 0 \end{bmatrix} + x_3 \begin{bmatrix} -1 \\ 0 \\ 1 \end{bmatrix}.$$

A basis of the eigenspace is $\boldsymbol{v}_1 = (1,1,0), \boldsymbol{v}_2 = (-1,0,1)$. (Check that $A\boldsymbol{v}_1 = -2\boldsymbol{v}_1, A\boldsymbol{v}_2 = -2\boldsymbol{v}_2$.)

Eigenspace for $c=4$. We solve $(4I-A)\boldsymbol{x}=\mathbf{0}$. The coefficient matrix is

$$\begin{bmatrix} 3 & 3 & -3 \\ -3 & 9 & -3 \\ -6 & 6 & 0 \end{bmatrix} \sim \begin{bmatrix} 1 & 1 & -1 \\ 0 & 12 & -6 \\ 0 & 12 & -6 \end{bmatrix} \begin{matrix} \text{II} + \text{I} \\ \text{III} + 2\text{I} \\ \text{I} \times \frac{1}{3} \end{matrix} \sim \begin{bmatrix} 1 & 0 & -\frac{1}{2} \\ 0 & 1 & -\frac{1}{2} \\ 0 & 0 & 0 \end{bmatrix} \begin{matrix} \text{III} - \text{II} \\ \text{II} \times \frac{1}{12} \\ \text{I} - \text{II} \end{matrix}.$$

The general solution is $x_1 = \frac{1}{2}\, x_3, x_2 = \frac{1}{2}\, x_3$. A basis of the eigenspace is $\boldsymbol{v}_3 = \left(\frac{1}{2}, \frac{1}{2}, 1\right)$. (Check that $A\boldsymbol{v}_3 = 4\boldsymbol{v}_3$.)

Note that you will get a row of zeros in the row reduction when computing eigenspaces. If you do *not* get a zero row, go back and find your error!

Because I prefer not to burden you, in a first course on linear algebra, with numerical techniques for finding eigenvalues to (say) five decimal places, all the polynomials of degree 3 in the book have an integer zero! If b is a zero of $x^n + b_1x^{n-1} + \cdots + b_n$, then

$$x^n + b_1x^{n-1} + \cdots + b_n = (x - b)(x^{n-1} + \cdots + kx + \ell)$$

and so $-b\ell = b_n$. If b and ℓ are integers, we see that b is a divisor of b_n. For instance,

$$x^3 - 3x + 2 = 0$$

has possible integer solutions $\pm 1, \pm 2$. We find the solution $x = 1$ by trial. Now

$$x^3 - 3x + 2 = (x - 1)(x^2 + x - 2) = (x - 1)(x - 1)(x + 2).$$

The zeros are 1 and -2.

If we compute a characteristic polynomial by row operations, we may find factors during the process, as in Example 4. This is a good shortcut.

For numerical evaluation of eigenvalues, see Further Reading.

Definition 3 If $(x - c_1)^k$ divides a polynomial $P(x)$, that is,

$$P(x) = (x - c_1)^k Q(x),$$

but $x - c_1$ does *not* divide Q, we say c_1 has **multiplicity** k as a zero of P.

In Example 4, -2 has multiplicity 2, and 4 has multiplicity 1, as zeros of the characteristic polynomial. We also say -2 has **multiplicity** 2 **as an eigenvalue**.

Example 5 Show that the characteristic polynomial of A is

$$c^n - (a_{11} + \cdots + a_{nn})c^{n-1} + \cdots + (-1)^n \det A.$$

Solution. The only special product that contains a term c^{n-1} when multiplied out is $(c - a_{11}) \dots (c - a_{nn})$, since if we use $n - 1$ factors from the main diagonal, we must take the remaining factor from the main diagonal. Now the coefficient of c^{n-1} is $-a_{11} - a_{22} - \cdots - a_{nn}$.

If $P(c) = c^n + b_1c^{n-1} + \cdots + b_n$ is the characteristic polynomial of A, then $P(0) = b_n$. However,

$$P(0) = \det(0I - A) = \det(-A) = (-1)^n \det A.$$

This shows that $b_n = (-1)^n \det A$.

In Example 4, $(c + 2)^2(c - 4) = c^3 - 12c - 16$. The coefficient b_1 is 0, which is equal to $-(1 - 5 + 4) = -(a_{11} + a_{22} + a_{33})$.

2 Diagonalization

In applications of linear mappings, it is a valuable simplification if for a given linear $T : \mathbb{R}^n \to \mathbb{R}^n$ we are able to choose a basis of $\mathbb{R}^n$ consisting of eigenvectors of T. Let us say $\boldsymbol{v}_1, \dots, \boldsymbol{v}_n$ is a basis of eigenvectors. Then

$$\begin{aligned} T(x_1\boldsymbol{v}_1 + \cdots + x_n\boldsymbol{v}_n) &= x_1 T\boldsymbol{v}_1 + \cdots + x_n T\boldsymbol{v}_n \\ &= c_1 x_1 \boldsymbol{v}_1 + \cdots + c_n x_n \boldsymbol{v}_n \end{aligned}$$

where $\boldsymbol{v}_j$ is an eigenvector for the eigenvalue c_j. The matrix of T for the basis $\boldsymbol{v}_1, \dots, \boldsymbol{v}_n$ is the diagonal matrix

$$D = \begin{bmatrix} c_1 & & \\ & \ddots & \\ & & c_n \end{bmatrix}. \tag{3}$$

Proposition 2 *Let $T : \mathbb{R}^n \to \mathbb{R}^n$ be linear. The following statements are equivalent.*

(i) There is a basis of $\mathbb{R}^n$ consisting of eigenvectors of T.

(ii) T has diagonal matrix for some basis.

Proof. We just saw that (i) implies (ii).

Now suppose that T has the matrix (3) for some basis $\boldsymbol{v}_1, \dots, \boldsymbol{v}_n$. Then $T\boldsymbol{v}_j = c_j\boldsymbol{v}_j$ and there is a basis consisting of eigenvectors of T.

If the matrix of a linear mapping T is diagonal for some basis, we say T is **diagonalizable** .

An $n \times n$ matrix is **diagonalizable** if

$$A = PDP^{-1}$$

for some diagonal D and invertible P.

Proposition 3 *Let A be $n \times n$. The following statements are equivalent.*

(i) A is diagonalizable.

(ii) There is a basis of $\mathbb{R}^n$ consisting of eigenvectors of A.

Proof. If (i) holds, then $A = PDP^{-1}$ (D diagonal, $P = [\boldsymbol{p}_1 \cdots \boldsymbol{p}_n]$ invertible). By Proposition 15 of Chapter 4, the mapping $T\boldsymbol{x} = A\boldsymbol{x}$ has matrix $P^{-1}AP$ for the basis $\boldsymbol{p}_1, \ldots, \boldsymbol{p}_n$. Since $P^{-1}AP = D$, Proposition 2 yields (ii).

Now suppose that (ii) holds. Let $P = [\boldsymbol{p}_1 \cdots \boldsymbol{p}_n]$ where the $\boldsymbol{p}_i$ are linearly independent eigenvectors of A. Then P has rank n and P^{-1} exists. Moreover,

$$A\boldsymbol{p}_j = c_j\boldsymbol{p}_j \tag{4}$$

yields

$$A[\boldsymbol{p}_1 \cdots \boldsymbol{p}_n] = [\boldsymbol{p}_1 \cdots \boldsymbol{p}_n]\begin{bmatrix} c_1 & & \\ & \ddots & \\ & & c_n \end{bmatrix}, \tag{5}$$

that is, $AP = PD$ where D is diagonal. So

$$A = PDP^{-1}. \tag{6}$$

Example 6 Let A be as in Example 4, then $\boldsymbol{v}_1, \boldsymbol{v}_2, \boldsymbol{v}_3$ is a basis of $\mathbb{R}^3$. (We cannot have $\boldsymbol{v}_3$ in $\text{Span}\{\boldsymbol{v}_1, \boldsymbol{v}_2\}$, since $\boldsymbol{0}$ is the only vector $\boldsymbol{v}$ with

$$A\boldsymbol{v} = -2\boldsymbol{v} = 4\boldsymbol{v}.)$$

So $\boldsymbol{v}_1, \boldsymbol{v}_2, \boldsymbol{v}_3$ is a basis of eigenvectors and, as in (4)–(6),

$$A = PDP^{-1}$$

where

$$P = [\boldsymbol{v}_1\ \boldsymbol{v}_2\ \boldsymbol{v}_3] = \begin{bmatrix} 1 & -1 & \frac{1}{2} \\ 1 & 0 & \frac{1}{2} \\ 0 & 1 & 1 \end{bmatrix}, \quad D = \begin{bmatrix} -2 & & \\ & -2 & \\ & & 4 \end{bmatrix}.$$

Note that the entries of D have to be in the order corresponding to the order of $\boldsymbol{v}_1, \boldsymbol{v}_2, \boldsymbol{v}_3$. It is also true that

$$A = P_1 D_1 P_1^{-1}$$

where

$$P_1 = \begin{bmatrix} 1 & \frac{1}{2} & -1 \\ 1 & \frac{1}{2} & 0 \\ 0 & 1 & 1 \end{bmatrix} \quad \text{and} \quad D_1 = \begin{bmatrix} -2 & & \\ & 4 & \\ & & -2 \end{bmatrix}.$$

Example 7 If possible, find an invertible P and diagonal D such that $A = PDP^{-1}$, where

$$A = \begin{bmatrix} 2 & 4 & 2 \\ 1 & -1 & -1 \\ 1 & 1 & 3 \end{bmatrix}.$$

Solution.

$$\begin{aligned}
\det(cI - A) &= \det \begin{bmatrix} c-2 & -4 & -2 \\ -1 & c+1 & 1 \\ -1 & -1 & c-3 \end{bmatrix} \\
&= \det \begin{bmatrix} 1 & 1 & 3-c \\ 0 & c+2 & 4-c \\ 0 & -2-c & -2+(c-2)(c-3) \end{bmatrix} \begin{matrix} \text{I} \leftrightarrow \text{III} \\ \text{I} \times -1 \\ \text{II} + \text{I} \\ \text{III} - (c-2)\text{I} \end{matrix} \\
&= \det \begin{bmatrix} c+2 & 4-c \\ -2-c & c^2-5c+4 \end{bmatrix} \\
&= (c+2) \det \begin{bmatrix} 1 & 4-c \\ -1 & c^2-5c+4 \end{bmatrix} \quad \text{(factor } c+2 \text{ from column 1)} \\
&= (c+2)(c^2-6c+8) = (c+2)(c-4)(c-2).
\end{aligned}$$

Eigenvalues are $-2, 4, 2$.

Eigenspace for $c = -2$. $(-2I - A)\boldsymbol{x} = \mathbf{0}$ has coefficient matrix

$$\begin{bmatrix} -4 & -4 & -2 \\ -1 & -1 & 1 \\ -1 & -1 & -5 \end{bmatrix} \sim \begin{bmatrix} 1 & 1 & -1 \\ 0 & 0 & -6 \\ 0 & 0 & -6 \end{bmatrix} \begin{matrix} \text{II} \leftrightarrow \text{I} \\ \text{II} - 4\text{I} \\ \text{III} - \text{I} \\ \text{I} \times -1 \end{matrix} \sim \begin{bmatrix} 1 & 1 & 0 \\ 0 & 0 & 1 \\ 0 & 0 & 0 \end{bmatrix} \begin{matrix} \text{III} - \text{II} \\ \text{II} \times -\frac{1}{6} \\ \text{I} + \text{II} \end{matrix}$$

Clearly $\boldsymbol{v}_1 = (-1, 1, 0)$ is a basis of the eigenspace.

Eigenspace for $c = 4$. $(4I - A)\boldsymbol{x} = \mathbf{0}$ has coefficient matrix

$$\begin{bmatrix} 2 & -4 & -2 \\ -1 & 5 & 1 \\ -1 & -1 & 1 \end{bmatrix} \sim \begin{bmatrix} 1 & 1 & -1 \\ 0 & 6 & 0 \\ 0 & -6 & 0 \end{bmatrix} \begin{matrix} \text{I} \leftrightarrow \text{III} \\ \text{III} + 2\text{I} \\ \text{II} - \text{I} \\ \text{I} \times -1 \end{matrix} \sim \begin{bmatrix} 1 & 0 & -1 \\ 0 & 1 & 0 \\ 0 & 0 & 0 \end{bmatrix} \begin{matrix} \text{III} + \text{II} \\ \text{II} \times \frac{1}{6} \\ \text{I} - \text{II} \end{matrix}$$

Clearly $\boldsymbol{v}_2 = (1, 0, 1)$ is a basis of the eigenspace.

Eigenspace for $c = 2$. $(2I - A)\boldsymbol{x} = \mathbf{0}$ has coefficient matrix

$$\begin{bmatrix} 0 & -4 & -2 \\ -1 & 3 & 1 \\ -1 & -1 & -1 \end{bmatrix} \sim \begin{bmatrix} 1 & 1 & 1 \\ 0 & 4 & 2 \\ 0 & 0 & 0 \end{bmatrix} \begin{matrix} \mathrm{I} \leftrightarrow \mathrm{III} \\ \mathrm{II} - \mathrm{I} \\ \mathrm{III} + \mathrm{II} \\ \mathrm{I} \times -1 \end{matrix} \sim \begin{bmatrix} 1 & 0 & \frac{1}{2} \\ 0 & 1 & \frac{1}{2} \\ 0 & 0 & 0 \end{bmatrix} \begin{matrix} \mathrm{II} \times \frac{1}{4} \\ \mathrm{I} - \mathrm{II} \end{matrix} .$$

Clearly $\boldsymbol{v}_3 = \left(-\frac{1}{2}, -\frac{1}{2}, 1\right)$ is a basis of the eigenspace.

After checking that $P = [\boldsymbol{v}_1\ \boldsymbol{v}_2\ \boldsymbol{v}_3]$ has rank 3, we may assert that $A = PDP^{-1}$, where

$$P = \begin{bmatrix} -1 & 1 & -\frac{1}{2} \\ 1 & 0 & -\frac{1}{2} \\ 0 & 1 & 1 \end{bmatrix}, \quad D = \begin{bmatrix} -2 & & \\ & 4 & \\ & & 2 \end{bmatrix}.$$

3 Further results on eigenvectors and eigenvalues

With the help of the following result, we can see without calculation that $\boldsymbol{v}_1, \boldsymbol{v}_2, \boldsymbol{v}_3$ in Example 7 must be independent.

Proposition 4 *Let* $T : \mathbb{R}^n \to \mathbb{R}^n$ *be linear. Let* $\boldsymbol{v}_1, \ldots, \boldsymbol{v}_m$ *be eigenvectors for the distinct eigenvalues* $c_1, \ldots, c_m$ *respectively. Then* $\boldsymbol{v}_1, \ldots, \boldsymbol{v}_m$ *is a linearly independent set.*

Proof Suppose the set is *not* linearly independent. For some $j \geq 2$, which we take as small as possible,

$$x_1\boldsymbol{v}_1 + \cdots + x_j\boldsymbol{v}_j = \mathbf{0} \tag{7}$$

with not all x_i equal to 0. Clearly at least two x_i must be nonzero. Apply T to both sides of (7):

$$x_1c_1\boldsymbol{v}_1 + \cdots + x_jc_j\boldsymbol{v}_j = T\mathbf{0} = \mathbf{0}. \tag{8}$$

Subtract c_j times (7) from (8):

$$x_1(c_1 - c_j)\boldsymbol{v}_1 + \cdots + x_{j-1}(c_{j-1} - c_j)\boldsymbol{v}_{j-1} = \mathbf{0}.$$

At least one x_i in this equation is nonzero, and since $c_i - c_j \neq 0$ we obtain an equation similar to (7) with $j-1$ in place of j. This is not possible, so $\boldsymbol{v}_1, \ldots, \boldsymbol{v}_m$ are linearly independent.

Corollary *If the characteristic polynomial of A factorises as $(c-c_1)\ldots(c-c_n)$ with distinct real c_i, A is diagonalizable.*

Proof. Choose $\boldsymbol{v}_j \neq \mathbf{0}$ such that $A\boldsymbol{v}_j = c_j\boldsymbol{v}_j$ (Proposition 1). Now $\boldsymbol{v}_1, \ldots, \boldsymbol{v}_n$ are independent (Proposition 4), hence they are a basis of $\mathbb{R}^n$ (Proposition 7 of Chapter 3). The result now follows from Proposition 3.

There is a nice general rule that links *dimension* of the eigenspace for eigenvalue c with *multiplicity* of c.

Proposition 5 *Let c_1 be an eigenvalue of A, with multiplicity s. Then the eigenspace for c_1 has dimension $\leq s$.*

Proof. We first note that, for invertible P,

$$\begin{aligned}\det(cI - P^{-1}AP) &= \det(P^{-1}(cI-A)P) \\ &= (\det P)^{-1}\det(cI-A)\det P = \det(cI-A).\end{aligned}$$

That is, $P^{-1}AP$ has the same characteristic polynomial as A.

Now take a basis $\boldsymbol{v}_1, \ldots, \boldsymbol{v}_r$ of the eigenspace for c_1 and add $\boldsymbol{v}_{r+1}, \ldots, \boldsymbol{v}_n$ to get a basis $\boldsymbol{v}_1, \ldots, \boldsymbol{v}_n$ of $\mathbb{R}^n$ (Proposition 6 of Chapter 3). Let

$$P = [\boldsymbol{v}_1 \cdots \boldsymbol{v}_n],$$

then P is invertible. Now P can be used to 'partially diagonalize A' as follows:

$$AP = [c_1\boldsymbol{v}_1 \cdots c_1\boldsymbol{v}_r \; A\boldsymbol{v}_{r+1} \cdots A\boldsymbol{v}_n].$$

So

$$P^{-1}AP = P^{-1}[c_1\boldsymbol{v}_1 \cdots c_1\boldsymbol{v}_r \; A\boldsymbol{v}_{r+1} \cdots A\boldsymbol{v}_n]. \tag{9}$$

We know that $P\boldsymbol{e}_j = \boldsymbol{v}_j$ for $j \leq r$ and therefore $P^{-1}\boldsymbol{v}_j = \boldsymbol{e}_j$ for $j \leq r$. Inserting this in (9),

$$P^{-1}AP = [c_1\boldsymbol{e}_1 \cdots c_1\boldsymbol{e}_r \; P^{-1}A\boldsymbol{v}_{r+1} \cdots P^{-1}A\boldsymbol{v}_n].$$

Finally,

$$\begin{aligned} cI - P^{-1}AP &= [(c-c_1)\boldsymbol{e}_1 \cdots (c-c_1)\boldsymbol{e}_r \ \ * \cdots *], \\ \det(cI - P^{-1}AP) &= (c-c_1)^r \det[\boldsymbol{e}_1 \cdots \boldsymbol{e}_r \ \ * \cdots *] = (c-c_1)^r Q(c), \text{ say.} \end{aligned}$$

So c_1 has multiplicity at least r, that is, $s \geq r$.

In the examples we have done so far, we have $s = 1$ except for the eigenvalue $c_1 = -2$ in Example 4, where $s = 2 =$ dimension of eigenspace for c_1.

Example 8 Since the rotation in Example 1 is without eigenvalues, the characteristic polynomial cannot have a zero on the real line. Explicitly,

$$\det(cI - T) = \begin{bmatrix} c - \cos a & \sin a \\ -\sin a & c - \cos a \end{bmatrix} = (c - \cos a)^2 + \sin^2 a.$$

This is positive for all c, since $(c - \cos a)^2 \geq 0$ and $\sin^2 a > 0$ for $0 < a < 2\pi, a \neq \pi$.

Example 9 Show that

$$A = \begin{bmatrix} 2 & 0 & 0 \\ 4 & 2 & 0 \\ * & * & 3 \end{bmatrix}$$

is not diagonalizable.

Solution. Clearly $cI - A$ is lower triangular, and

$$\det(cI - A) = (c-2)^2(c-3).$$

Consider the eigenspace for $c = 2$. The system $(2I - A)\boldsymbol{x} = \boldsymbol{0}$ has coefficient matrix

$$\begin{bmatrix} 0 & 0 & 0 \\ -4 & 0 & 0 \\ * & * & -1 \end{bmatrix}$$

of rank 2. The eigenspace has dimension 1. The eigenspace for $c = 3$ also has dimension 1 (Proposition 5), so we cannot find more than two independent eigenvectors. Proposition 3 tells us that A is not diagonalizable. You should try to construct larger examples along similar lines.

Think of the diagonal matrix D in Proposition 3 as a block matrix

$$D = \begin{bmatrix} c_1 I_1 & & \\ & \ddots & \\ & & c_n I_n \end{bmatrix}$$

where I_j is an identity matrix. Modify some or all of the $c_j I_j$ in D by adding 1's just above the main diagonal, to get

$$\begin{bmatrix} c_j & 1 & & & \\ & c_j & 1 & & \\ & & \ddots & & \\ & & & c_j & 1 \\ & & & & c_j \end{bmatrix},$$

and write J for the matrix this produces in place of D. For example,

$$J = \begin{bmatrix} 7 & 1 & 0 & 0 & 0 \\ 0 & 7 & 0 & 0 & 0 \\ 0 & 0 & -1 & 0 & 0 \\ 0 & 0 & 0 & 2 & 0 \\ 0 & 0 & 0 & 0 & 2 \end{bmatrix} \text{ or } J = \begin{bmatrix} -5 & 1 & 0 & 0 & 0 \\ 0 & -5 & 1 & 0 & 0 \\ 0 & 0 & -5 & 0 & 0 \\ 0 & 0 & 0 & 3 & 1 \\ 0 & 0 & 0 & 0 & 3 \end{bmatrix}.$$

The following result acts as a substitute for diagonalization where it is not available. For *any* matrix A whose characteristic polynomial has n zeros counted with multiplicities, there is a matrix J of this type with $A = PJP^{-1}$ (P invertible); J is the **Jordan form** of A. We omit the proof; see Further Reading. It is not hard to compute Jordan form for a small matrix, as we now show. See also the application to differential equations in Chapter 8.

Example 10 Find invertible P such that

$$\begin{bmatrix} 2 & 0 & 0 \\ 4 & 2 & 0 \\ 1 & -1 & 3 \end{bmatrix} = P \begin{bmatrix} 2 & 1 & 0 \\ 0 & 2 & 0 \\ 0 & 0 & 3 \end{bmatrix} P^{-1}.$$

How did I 'guess' the Jordan form

$$\begin{bmatrix} 2 & 1 & 0 \\ 0 & 2 & 0 \\ 0 & 0 & 3 \end{bmatrix}?$$

I will explain at the end of the solution.

Solution. The eigenvalues of the matrix

$$A = \begin{bmatrix} 2 & 0 & 0 \\ 4 & 2 & 0 \\ 1 & -1 & 3 \end{bmatrix}$$

are 2 (with multiplicity 2) and 3, as we saw in Example 9.

Eigenspace for $c = 2$. The coefficient matrix of $(2I - A)\boldsymbol{x} = \mathbf{0}$ is

$$\begin{bmatrix} 0 & 0 & 0 \\ -4 & 0 & 0 \\ -1 & 1 & -1 \end{bmatrix}.$$

The general solution is clearly $x_3(0,1,1)$. Let $\boldsymbol{v}_1 = (0,1,1)$.

Eigenspace for $c = 3$. The coefficient matrix of $(3I - A)\boldsymbol{x} = \mathbf{0}$ is

$$\begin{bmatrix} -1 & 0 & 0 \\ -4 & -1 & 0 \\ -1 & 1 & 0 \end{bmatrix}.$$

It is easy to see that $\boldsymbol{v}_3 = (0,0,1)$ is a basis of the eigenspace.

Now let $P = [\boldsymbol{v}_1\ \boldsymbol{v}_2\ \boldsymbol{v}_3]$, $\boldsymbol{v}_2 = (x,y,z)$. Now $P^{-1}AP$ is the matrix of the mapping $T\boldsymbol{x} = A\boldsymbol{x}$ for the basis $\boldsymbol{v}_1, \boldsymbol{v}_2, \boldsymbol{v}_3$. We would like to choose $\boldsymbol{v}_2$ so that

$$P^{-1}AP = \begin{bmatrix} 2 & 1 & 0 \\ 0 & 2 & 0 \\ 0 & 0 & 3 \end{bmatrix}, \text{ that is, } T\boldsymbol{v}_2 = \boldsymbol{v}_1 + 2\boldsymbol{v}_2.$$

Equivalently,

$$(2I - A)\boldsymbol{v}_2 = \begin{bmatrix} 0 \\ -1 \\ -1 \end{bmatrix},$$

or

$$\begin{bmatrix} 0 & 0 & 0 \\ -4 & 0 & 0 \\ -1 & 1 & -1 \end{bmatrix} \begin{bmatrix} x \\ y \\ z \end{bmatrix} = \begin{bmatrix} 0 \\ -1 \\ -1 \end{bmatrix}.$$

We select $(x,y,z) = (1/4, -3/4, 0)$, giving

$$P = \begin{bmatrix} 0 & \frac{1}{4} & 0 \\ 1 & -\frac{3}{4} & 0 \\ 1 & 0 & 1 \end{bmatrix}.$$

Now P is invertible (expand det P by column 3). You should check that

$$AP = P \begin{bmatrix} 2 & 1 & 0 \\ 0 & 2 & 0 \\ 0 & 0 & 3 \end{bmatrix}.$$

I should explain that since A is not diagonalizable, and since the 'block' $3I$ in $P^{-1}AP$ is guaranteed by the existence of an eigenvector for eigenvalue 3, the only possibility for Jordan form is $\begin{bmatrix} 2 & 1 & 0 \\ 0 & 2 & 0 \\ 0 & 0 & 3 \end{bmatrix}$.

4 The Cayley-Hamilton theorem

It is quite remarkable that any matrix is a 'zero of its characteristic polynomial'.

Proposition 6 (Cayley-Hamilton) *Let A be $n \times n$ with characteristic polynomial $c^n + b_1c^{n-1} + \cdots + b_{n-1}c + b_n$. Then*

$$A^n + b_1A^{n-1} + \cdots + b_{n-1}A + b_nI = 0.$$

It is convenient to prove a preparatory lemma. We shall use the notation $B(c), C(c), \ldots$ for matrices whose entries are polynomials in c.

Lemma *If $C = B(c)(cI - A)$ is a matrix of constants, then $C = 0$.*

Proof. If $B(c) \neq 0$, we can write

$$B(c) = c^rB_r + c^{r-1}B_{r-1} + \cdots + B_0$$

where B_j is a matrix of constants obtained by taking terms in c^j from $B(c)$, and $B_r \neq 0$. Now

$$B(c)(cI - A) = c^{r+1}B_r + \text{ terms in } c^r, c^{r-1}, \ldots, 1.$$

This is *not* a matrix of constants. So we must have $B(c) = 0$, and $C = 0$.

Proof of Proposition 6. We write $b_0 = 1$ for convenience. Let

$$f(c) = b_0c^n + b_1c^{n-1} + \cdots + b_{n-1}c + b_n,$$

so that $f(A)$ denotes $b_0A^n + b_1A^{n-1} + \cdots + b_{n-1}A + b_nI$. Now

$$f(A) - f(c)I = \sum_{j=1}^{n} b_{n-j}(A^j - c^jI).$$

It is easy to check that

$$A^j - c^jI = (A^{j-1} + A^{j-2}c + \cdots + Ac^{j-2} + c^{j-1}I)(A - cI)$$

(multiply out the right side and cancel all but two terms). Hence

$$\begin{aligned}(10)\quad f(A) - f(c)I &= \sum_{j=1}^{n} b_{n-j}(A^{j-1} + A^{j-2}c + \cdots + Ac^{j-2} + c^{j-1}I)(A - cI)\\ &= G(c)(A - cI), \text{ let us say.}\end{aligned}$$

The remaining ingredient of the proof is

$$(11)\qquad \text{adj } (A - cI)(A - cI) = \det(A - cI)I = f(c)I$$

(Proposition 12 of Chapter 5). Combining (10) and (11),

$$f(A) = (G(c) + \text{adj } (A - cI))(A - cI).$$

The left side is a matrix of constants, and must be 0 by the lemma.

Example 11 Let A be as in Example 4. Then

$$\begin{aligned}f(A) = (A + 2I)^2(A - 4I) &= \begin{bmatrix} 3 & -3 & 3 \\ 3 & -3 & 3 \\ 6 & -6 & 6 \end{bmatrix} \begin{bmatrix} 3 & -3 & 3 \\ 3 & -3 & 3 \\ 6 & -6 & 6 \end{bmatrix} \begin{bmatrix} -3 & -3 & 3 \\ 3 & -9 & 3 \\ 6 & -6 & 0 \end{bmatrix}\\ &= \begin{bmatrix} 18 & -18 & 18 \\ 18 & -18 & 18 \\ 36 & -36 & 36 \end{bmatrix} \begin{bmatrix} -3 & -3 & 3 \\ 3 & -9 & 3 \\ 6 & -6 & 0 \end{bmatrix} = 0.\end{aligned}$$

5 Markov chains

In the mathematical model known as a **Markov chain** , we are given a family of objects F each of which is in one of n possible 'states' $1^\circ, 2^\circ, \ldots, n^\circ$. The objects can move from one state to another, and we record this at fixed time intervals. For simplicity, make the time interval one year.

Example 12 F is the set of arable fields in Lincoln county. The three 'states' correspond to the crops growing in the fields: 1°, corn; 2°, barley; 3°, alfalfa.

Example 13 F is the set of inhabitants of Washington county. The four 'states': living in 1°, Capital City; 2°, one of the other urban areas; 3°, a suburban area; 4°, a rural area.

Example 14 F is the set of voters in Eutopia. The three 'states': supporter of 1°, Conservative party; 2°, Liberal party; 3°, Democratic party.

The other ingredient of a Markov chain is a definite set of probabilities of moving between states. These probabilities do not vary with time. Write p_{ij} for the probability of a member of F moving from state j to state i after a year. The matrix $A = [p_{ij}]$ $(i, j = 1, \dots, n)$ is the **transition matrix** of the Markov chain. Suppose in Example 12 the transition matrix is

$$\begin{bmatrix} \frac{1}{4} & 0 & \frac{1}{3} \\ \frac{1}{2} & \frac{1}{3} & \frac{1}{3} \\ \frac{1}{4} & \frac{2}{3} & \frac{1}{3} \end{bmatrix}.$$

The first column reveals that a corn field remains under corn with probability $\frac{1}{4}$; changes to barley with probability $\frac{1}{2}$; and changes to alfalfa with probability $\frac{1}{4}$. Of course *each column of a transition matrix sums to* 1, that is,

$$\sum_{i=1}^{n} p_{ij} = 1 \qquad \text{for } j = 1, \dots, n,$$

since for fixed j, the state after one year must change to one of the states $1°, \dots, n°$. A **probability vector** is a column vector whose entries are nonnegative with sum 1.

We need an **initial probability vector** $\boldsymbol{q}_0$ to tell us the distribution of F into different states in year 0.

Example 15 If 1/5 of the voters of Eutopia support the Conservatives, 2/5 are Liberals, and 2/5 are Democrats, in year 0, the initial probability vector is

$$\boldsymbol{q}_0 = \begin{bmatrix} \frac{1}{5} \\ \frac{2}{5} \\ \frac{2}{5} \end{bmatrix}.$$

In general, we write $\boldsymbol{q}_i$ for the probability vector in year i. If $\boldsymbol{q}_i = (q_{i1}, \dots, q_{in})$, this tells us that the proportion of members of F that are in state j^0, in year i, is q_{ij}. Obviously $q_{ij} \geq 0, \sum_{j=1}^{n} q_{ij} = 1$.

Proposition 7 *Let A be the transition matrix of a Markov chain with probability vector $\boldsymbol{q}_i$ in year i. Then $\boldsymbol{q}_i = A^i \boldsymbol{q}_0$ $(i = 1, 2, \dots)$.*

Proof. The probability of a member of F being in state j in year $i+1$ is the sum of

(probability of $k°$ in year i) × (probability of moving from $k°$ to $j°$)

taken over $k = 1, \dots, n$. That is,

$$q_{i+1,j} = \sum_{k=1}^{n} q_{ik} p_{jk}.$$

The right side is the j-th entry of $A\boldsymbol{q}_i$; so $\boldsymbol{q}_{i+1} = A\boldsymbol{q}_i$. Clearly $\boldsymbol{q}_1 = A\boldsymbol{q}_0, \boldsymbol{q}_2 = A(A\boldsymbol{q}_0) = A^2\boldsymbol{q}_0, \boldsymbol{q}_3 = A(A^2\boldsymbol{q}_0) = A^3\boldsymbol{q}_0$, and we continue until we reach $\boldsymbol{q}_i = A^i\boldsymbol{q}_0$.

Example 16 Suppose that the transition matrix in Example 14 is

$$A = \begin{bmatrix} \frac{1}{4} & \frac{7}{10} & \frac{1}{3} \\ \frac{1}{6} & \frac{1}{10} & \frac{1}{3} \\ \frac{7}{12} & \frac{1}{5} & \frac{1}{3} \end{bmatrix}.$$

Then the proportion of Liberals in Eutopia after 7 years is the second entry of $A^7 \begin{bmatrix} \frac{1}{5} \\ \frac{2}{5} \\ \frac{2}{5} \end{bmatrix}$, assuming the initial probability vector of Example 15. The use of a numerical computer program such as MATLAB has obvious merits here. Your institution almost certainly has a computer lab where you can use MATLAB, or a similar computer program, to compute high powers of a given matrix. These programs will do almost anything in linear algebra (the Gram-Schmidt process from Chapter 7; computing eigenvalues; and so on). They always come equipped with instructions for inputting a matrix and operating on it, so there is no need to take up space here on this topic. For a course in numerical linear algebra, see Further Reading.

6 The steady state vector

What I find most interesting in the theory of Markov chains is the *long-term behaviour* of the family F. We try to establish that the entries of $\boldsymbol{q}_j$ each tend to a limit as j tends to infinity. Recall the following facts about limits of sequences. If $u_1, u_2, \ldots$ and $v_1, v_2, \ldots$ are infinite sequences, and $\lim_{j\to\infty} u_j$, $\lim_{j\to\infty} v_j$ exist, then

$$\lim_{j\to\infty}(u_j + v_j) = \lim_{j\to\infty} u_j + \lim_{j\to\infty} v_j, \tag{12}$$

$$\lim_{j\to\infty} u_j v_j = \lim_{j\to\infty} u_j \cdot \lim_{j\to\infty} v_j. \tag{13}$$

Another crucial fact is that

$$\lim_{j\to\infty} k^j = 0 \tag{14}$$

if $|k| < 1$, while of course 1^j has limit 1. The sequence $(-1)^j$ does not have a limit.

If $A_1, A_2, \ldots$ is a sequence of $m \times n$ matrices, then $\lim_{j\to\infty} A_j = A$ means that the limit of any given entry of A_j is the corresponding entry of A. For example,

$$\lim_{j\to\infty}\begin{bmatrix} 1 & \frac{1}{2} - 3^{-j} \\ 1 - \frac{1}{j} & 4 + 2^{-j} + \frac{1}{j^2} \end{bmatrix} = \begin{bmatrix} 1 & \frac{1}{2} \\ 1 & 4 \end{bmatrix}.$$

It is easy to deduce from (12), (13) that if $\lim_{j\to\infty} A_j$, $\lim_{j\to\infty} B_j$ exist (where A_j is $m \times n$ and B_j is $n \times p$) then

$$\lim_{j\to\infty} A_j B_j = \lim_{j\to\infty} A_j \cdot \lim_{j\to\infty} B_j.$$

Proposition 8 *Let A be diagonalizable,*

$$A = P\begin{bmatrix} c_1 & & \\ & \ddots & \\ & & c_n \end{bmatrix} P^{-1},$$

where $|c_i| < 1$ $(i = 1, \ldots, n)$ except that some of the c_i are 1. Then

$$\lim_{j\to\infty} A^j = PD_1P^{-1}$$

where D_1 is a diagonal matrix with ith diagonal entry 1 if $c_i = 1$ and 0 if $c_i \neq 1$.

Proof. Note that, D being diagonal with $d_{ii} = c_i$,

$$\begin{aligned} A^j &= PDP^{-1} \cdot PDP^{-1} \dots PDP^{-1} \\ &= PD^j P^{-1}. \end{aligned}$$

Of course

$$\begin{bmatrix} a_1 & & \\ & \ddots & \\ & & a_n \end{bmatrix} \begin{bmatrix} b_1 & & \\ & \ddots & \\ & & b_n \end{bmatrix} = \begin{bmatrix} a_1 b_1 & & \\ & \ddots & \\ & & a_n b_n \end{bmatrix}$$

can be used repeatedly to get

$$D^j = \begin{bmatrix} c_1^j & & \\ & \ddots & \\ & & c_n^j \end{bmatrix}.$$

We have, using (14),

$$\lim_{j \to \infty} D^j = \begin{bmatrix} d_1 & & \\ & \ddots & \\ & & d_n \end{bmatrix} = D_1$$

where $d_i = 0$ if $|c_i| < 1, d_i = 1$ if $c_i = 1$. Now

$$\lim_{j \to \infty} A^j = P \lim_{j \to \infty} D^j \cdot P^{-1} = PD_1 P^{-1}.$$

Proposition 8 does give some hope for Markov chains in view of the following result.

Proposition 9 *Let c be an eigenvalue of the transition matrix A. Then $|c| \leq 1$. Moreover, 1 is an eigenvalue of A.*

Proof. The matrices A and A^t have the same characteristic polynomial, since

$$\det(cI - A) = \det(cI - A)^t = \det(cI - A^t)$$

(Proposition 9 of Chapter 5). So A and A^t have the same characteristic polynomial, and c is an eigenvalue of A^t. Choose $\boldsymbol{x} \neq \mathbf{0}$ such that

$$A^t \boldsymbol{x} = c\boldsymbol{x}.$$

Let x_j have the largest absolute value among the coordinates of $\boldsymbol{x}$. Then

$$p_{1j}x_1 + \cdots + p_{nj}x_n = cx_j.$$

However,

$$\begin{aligned} |p_{1j}x_1 + \cdots + p_{nj}x_n| &\le p_{1j}|x_1| + \cdots + p_{nj}|x_n| \\ &\le (p_{1j} + \cdots + p_{nj})|x_j| = |x_j|. \end{aligned}$$

Therefore $|cx_j| \le |x_j|$. Divide by $|x_j|$ to get $|c| \le 1$.

To see that 1 is an eigenvalue of A we show that 1 is an eigenvalue of A^t. Now

$$A^t \begin{bmatrix} 1 \\ 1 \\ \vdots \\ 1 \end{bmatrix} = \begin{bmatrix} 1 \\ 1 \\ \vdots \\ 1 \end{bmatrix}$$

because each column of A sums to 1. This proves that 1 is an eigenvalue of A^t.

Example 17 Show that the eigenvalues of the transition matrix $A = \begin{bmatrix} 0 & 1 \\ 1 & 0 \end{bmatrix}$ are 1 and -1 but that no other 2×2 transition matrix has the eigenvalue -1.

Solution. For this A,

$$\det(cI - A) = \det \begin{bmatrix} c & -1 \\ -1 & c \end{bmatrix} = c^2 - 1 = (c-1)(c+1).$$

The eigenvalues are 1 and -1.

Now let

$$A = \begin{bmatrix} a & b \\ 1-a & 1-b \end{bmatrix} \quad (0 \le a \le 1,\ 0 \le b \le 1).$$

Then

$$-I - A = \begin{bmatrix} -1-a & -b \\ a-1 & -2+b \end{bmatrix}.$$

If -1 is an eigenvalue,

$$\begin{aligned} 0 = \det(-I - A) &= (-1-a)(-2+b) + ab - b \\ &= 2 + 2a - 2b. \end{aligned}$$

Hence $2+2a=2b\le 2$. So $a\le 0$, and in fact $a=0$. We deduce that $b=1$ and $A=\begin{bmatrix}0 & 1\\ 1 & 0\end{bmatrix}$.

Proposition 10 *Let A be the transition matrix of a Markov chain. Suppose that (i) A is diagonalizable; (ii) 1 is the only eigenvalue of modulus 1. Then for any initial probability vector $\boldsymbol{q}_0$, the probability vector $\boldsymbol{q}_j$ tends to a limit $\boldsymbol{q}$ (the* **steady-state** *vector) . Moreover, $\boldsymbol{q}$ is a probability vector and is an eigenvector for the eigenvalue 1.*

This proposition says that, given some input about the transition matrix, there are long-term proportions of the family F in each of the possible states.

Proof. By Proposition 9, we can apply Proposition 8 to get

$$\lim_{j\to\infty} A^j = PD_1P^{-1}$$

where D_1 is diagonal with entries 1 (corresponding to eigenvalues 1 of A) and 0 (for other eigenvalues). So

$$\lim_{j\to\infty}\boldsymbol{q}_j = \lim_{j\to\infty} A^j\boldsymbol{q}_0 = PD_1P^{-1}\boldsymbol{q}_0 = \boldsymbol{q}, \text{ say.}$$

The coordinates of $\boldsymbol{q}_j$ are nonnegative, so their limits (the coordinates of $\boldsymbol{q}$) are nonnegative. Similarly, the coordinates of $\boldsymbol{q}$ sum to 1. To see that $\boldsymbol{q}$ is an eigenvector for the eigenvalue 1, we notice that

$$\boldsymbol{q} = \lim_{j\to\infty} A^{j+1}\boldsymbol{q}_0 = A\lim_{j\to\infty} A^j\boldsymbol{q}_0 = A\boldsymbol{q}.$$

Example 18 Let $A=\begin{bmatrix}a & b\\ 1-a & 1-b\end{bmatrix}$ where $0<a<1, 0\le b\le 1$. Then

$$\begin{aligned}\det(cI-A) &= (c-a)(c-1+b)-b+ab\\ &= c^2+(b-a-1)c+a-b\\ &= (c-1)(c+b-a).\end{aligned}$$

The eigenvalues are 1 and $a-b$, which satisfies $-1<a-b<1$. So A is diagonalizable (Corollary to Proposition 4).

Eigenspace for $c = 1$. $(I - A)\boldsymbol{x} = \mathbf{0}$ has coefficient matrix

$$\begin{bmatrix} 1-a & -b \\ -1+a & b \end{bmatrix}.$$

Obviously $x \begin{bmatrix} b \\ 1-a \end{bmatrix}$ is the general solution. The steady state vector is $\boldsymbol{q} = (xb, x(1-a))$ for some x. Since $\boldsymbol{q}$ is a probability vector,

$$xb + x(1-a) = 1,$$
$$x = \frac{1}{1+b-a}, \boldsymbol{q} = \left(\frac{b}{1+b-a}, \frac{1-a}{1+b-a}\right).$$

For instance, if the probability of going from state i to state i is $1/4$ for $i = 1$ and $1/2$ for $i = 2$, then

$$\boldsymbol{q} = \left(\frac{2}{5}, \frac{3}{5}\right).$$

Perhaps it is surprising that the initial probability vector is irrelevant to the value of $\boldsymbol{q}$. We can get a general result of this kind.

Proposition 11 *Make the same assumptions as in Proposition 10. Suppose in addition that the eigenvalue 1 has multiplicity 1. Then* $\boldsymbol{q}$ *is the same vector regardless of the value of* $\boldsymbol{q}_0$.

Proof. The eigenspace for the eigenvalue 1 has dimension 1 (Proposition 5). Let $\boldsymbol{v}$ be an eigenvector from this eigenspace. If

$$\boldsymbol{v} = \begin{bmatrix} v_1 \\ \vdots \\ v_n \end{bmatrix},$$

we must have

$$\boldsymbol{q} = \begin{bmatrix} \dfrac{v_1}{v_1 + \cdots + v_n} \\ \vdots \\ \dfrac{v_n}{v_1 + \cdots + v_n} \end{bmatrix}$$

since this is the only multiple of $\boldsymbol{v}$ whose coordinates add up to 1. (The equation $v_1 + \cdots + v_n = 0$ is impossible, because Proposition 10 guarantees that there is an eigenvector for the eigenvalue 1 whose coordinates sum to 1.)

Example 19 Let F be the set of car owners in the U.S. The four states are ownership of: $1°$, a sports utility vehicle; $2°$, a supercompact car; $3°$, a family sedan; $4°$, a car not of types $1°$ to $3°$. Once you own an S.U.V. you cannot bear to give it up; likewise for a supercompact. This degree of loyalty does not apply to $3°, 4°$, but family sedan owners remain in states $1° - 3°$. The transition matrix is

$$A = \begin{bmatrix} 1 & 0 & * & * \\ 0 & 1 & * & * \\ 0 & 0 & a & * \\ 0 & 0 & 0 & b \end{bmatrix}$$

where $0 < a,\ b < 1$. Now

$$\det(cI - A) = (c-1)^2(c-a)(c-b).$$

Thus the eigenvalues are either 1 or have absolute value less than 1. Let us suppose that $b \neq a$.

Eigenspace for $c = 1$. The coefficient matrix of $(I - A)\boldsymbol{x} = \mathbf{0}$ is

$$\begin{bmatrix} 0 & 0 & * & * \\ 0 & 0 & * & * \\ 0 & 0 & 1-a & * \\ 0 & 0 & 0 & 1-b \end{bmatrix}.$$

Evidently x_1, x_2 are free while $x_3 = x_4 = 0$. A basis of the eigenspace is $\boldsymbol{e}_1, \boldsymbol{e}_2$. Pick any eigenvectors from the eigenspaces for a, b, say $\boldsymbol{u}_1, \boldsymbol{u}_2$. Then $x_1\boldsymbol{e}_1 + x_2\boldsymbol{e}_2, \boldsymbol{u}_1, \boldsymbol{u}_2$ is a linearly independent set (any nonzero (x_1, x_2)) by Proposition 4. Consequently $\boldsymbol{e}_1, \boldsymbol{e}_2, \boldsymbol{u}_1, \boldsymbol{u}_2$ are independent and A is diagonalizable. Proposition 10 is applicable; the possible steady state vectors $\boldsymbol{q}$ are $(q_1, 1 - q_1, 0, 0)$ where $0 \leq q_1 \leq 1$.

Suppose for example

$$A = \begin{bmatrix} 1 & 0 & \frac{1}{3} & \frac{1}{5} \\ 0 & 1 & \frac{1}{3} & \frac{1}{5} \\ 0 & 0 & \frac{1}{3} & 0 \\ 0 & 0 & 0 & \frac{3}{5} \end{bmatrix}$$

and the initial probability vector is $\left(\frac{1}{8}, \frac{3}{8}, \frac{1}{4}, \frac{1}{4}\right)$. To find the steady state vector we proceed as follows.

Eigenspace for $c = 1/3$. The coefficient matrix of $\left(\frac{1}{3}\, I - A\right) \boldsymbol{x} = \mathbf{0}$ is

$$\begin{bmatrix} -\frac{2}{3} & 0 & -\frac{1}{3} & -\frac{1}{5} \\ 0 & -\frac{2}{3} & -\frac{1}{3} & -\frac{1}{5} \\ 0 & 0 & 0 & 0 \\ 0 & 0 & 0 & -\frac{4}{15} \end{bmatrix} \sim \begin{bmatrix} 1 & 0 & \frac{1}{2} & \frac{3}{10} \\ 0 & 1 & \frac{1}{2} & \frac{3}{10} \\ 0 & 0 & 0 & 1 \\ 0 & 0 & 0 & 0 \end{bmatrix} \begin{matrix} \text{I} \times -\frac{3}{2} \\ \text{II} \times -\frac{3}{2} \\ \text{III} \leftrightarrow \text{IV} \\ \text{III} \times -\frac{15}{4} \end{matrix}$$

$$\sim \begin{bmatrix} 1 & 0 & \frac{1}{2} & 0 \\ 0 & 1 & \frac{1}{2} & 0 \\ 0 & 0 & 0 & 1 \\ 0 & 0 & 0 & 0 \end{bmatrix} \begin{matrix} \text{I} - \text{III} \times \frac{3}{10} \\ \text{II} - \text{III} \times \frac{3}{10} \end{matrix} .$$

A basis of the eigenspace is $(-1, -1, 2, 0)$.

Eigenspace for $c = 3/5$. The coefficient matrix of $\left(\frac{3}{5}\, I - A\right) \boldsymbol{x} = \mathbf{0}$ is

$$\begin{bmatrix} -\frac{2}{5} & 0 & -\frac{1}{3} & -\frac{1}{5} \\ 0 & -\frac{2}{5} & -\frac{1}{3} & -\frac{1}{5} \\ 0 & 0 & \frac{4}{15} & 0 \\ 0 & 0 & 0 & 0 \end{bmatrix} \sim \begin{bmatrix} 1 & 0 & \frac{5}{6} & \frac{1}{2} \\ 0 & 1 & \frac{5}{6} & \frac{1}{2} \\ 0 & 0 & 1 & 0 \\ 0 & 0 & 0 & 0 \end{bmatrix} \begin{matrix} \text{I} \times -\frac{5}{2} \\ \text{II} \times -\frac{5}{2} \\ \text{III} \times \frac{15}{4} \end{matrix}$$

$$\sim \begin{bmatrix} 1 & 0 & 0 & \frac{1}{2} \\ 0 & 1 & 0 & \frac{1}{2} \\ 0 & 0 & 1 & 0 \\ 0 & 0 & 0 & 0 \end{bmatrix} \begin{matrix} \text{II} - \frac{5}{6}\, \text{II} \\ \text{I} - \frac{5}{6}\, \text{III} \end{matrix} .$$

A basis of the eigenspace is $(-1, -1, 0, 2)$.

Take

$$P = \begin{bmatrix} 1 & 0 & -1 & -1 \\ 0 & 1 & -1 & -1 \\ 0 & 0 & 2 & 0 \\ 0 & 0 & 0 & 2 \end{bmatrix} ;$$

then

$$[P|I] \sim \begin{bmatrix} 1 & 0 & -1 & -1 & 1 & 0 & 0 & 0 \\ 0 & 1 & -1 & -1 & 0 & 1 & 0 & 0 \\ 0 & 0 & 1 & 0 & 0 & 0 & \frac{1}{2} & 0 \\ 0 & 0 & 0 & 1 & 0 & 0 & 0 & \frac{1}{2} \end{bmatrix} \begin{matrix} \text{III} \times \frac{1}{2} \\ \text{IV} \times \frac{1}{2} \\ \\ \\ \end{matrix} \sim \begin{bmatrix} 1 & 0 & 0 & 0 & 1 & 0 & \frac{1}{2} & \frac{1}{2} \\ 0 & 1 & 0 & 0 & 0 & 1 & \frac{1}{2} & \frac{1}{2} \\ 0 & 0 & 1 & 0 & 0 & 0 & \frac{1}{2} & 0 \\ 0 & 0 & 0 & 1 & 0 & 0 & 0 & \frac{1}{2} \end{bmatrix} \begin{matrix} \text{I} + \text{IV} \\ \text{II} + \text{IV} \\ \text{I} + \text{III} \\ \text{II} + \text{III} \end{matrix}.$$

So

$$P^{-1} = \begin{bmatrix} 1 & 0 & \frac{1}{2} & \frac{1}{2} \\ 0 & 1 & \frac{1}{2} & \frac{1}{2} \\ 0 & 0 & \frac{1}{2} & 0 \\ 0 & 0 & 0 & \frac{1}{2} \end{bmatrix}$$

and, as in the proof of Proposition 10, the steady state vector is

$$\boldsymbol{q} = \begin{bmatrix} 1 & 0 & -1 & -1 \\ 0 & 1 & -1 & -1 \\ 0 & 0 & 2 & 0 \\ 0 & 0 & 0 & 2 \end{bmatrix} \begin{bmatrix} 1 & 0 & 0 & 0 \\ 0 & 1 & 0 & 0 \\ 0 & 0 & 0 & 0 \\ 0 & 0 & 0 & 0 \end{bmatrix} \begin{bmatrix} 1 & 0 & \frac{1}{2} & \frac{1}{2} \\ 0 & 1 & \frac{1}{2} & \frac{1}{2} \\ 0 & 0 & \frac{1}{2} & 0 \\ 0 & 0 & 0 & \frac{1}{2} \end{bmatrix} \begin{bmatrix} \frac{1}{8} \\ \frac{3}{8} \\ \frac{1}{4} \\ \frac{1}{4} \end{bmatrix}$$

$$= \begin{bmatrix} 1 & 0 & 0 & 0 \\ 0 & 1 & 0 & 0 \\ 0 & 0 & 0 & 0 \\ 0 & 0 & 0 & 0 \end{bmatrix} \begin{bmatrix} 1 & 0 & \frac{1}{2} & \frac{1}{2} \\ 0 & 1 & \frac{1}{2} & \frac{1}{2} \\ 0 & 0 & \frac{1}{2} & 0 \\ 0 & 0 & 0 & \frac{1}{2} \end{bmatrix} \begin{bmatrix} \frac{1}{8} \\ \frac{3}{8} \\ \frac{1}{4} \\ \frac{1}{4} \end{bmatrix} = \begin{bmatrix} 1 & 0 & \frac{1}{2} & \frac{1}{2} \\ 0 & 1 & \frac{1}{2} & \frac{1}{2} \\ 0 & 0 & 0 & 0 \\ 0 & 0 & 0 & 0 \end{bmatrix} \begin{bmatrix} \frac{1}{8} \\ \frac{3}{8} \\ \frac{1}{4} \\ \frac{1}{4} \end{bmatrix} = \begin{bmatrix} \frac{3}{8} \\ \frac{5}{8} \\ 0 \\ 0 \end{bmatrix}.$$

Exercises for Chapter 6

Reminder: attempt exercise a.b after reading Section a.

1.1 Find the eigenvalues and eigenspaces of the projection in Example 3 by means of Proposition 1.

1.2 Let $A = \begin{bmatrix} a & b \\ c & d \end{bmatrix}$. If $(a-d)^2 < -4bc$, show that A has no eigenvalues.

1.3 Show that if $T : \mathbb{R}^n \to \mathbb{R}^n$ is invertible, then 0 is not an eigenvalue of T.

1.4 Let $T : \mathbb{R}^n \to \mathbb{R}^n$ be invertible. Show that if c is an eigenvalue of T, then c^{-1} is an eigenvalue of T^{-1}.

1.5 In Exercise 1.4, show that the eigenspace of T for eigenvalue c has the same dimension as the eigenspace of T^{-1} for eigenvalue c^{-1}.

Hint: $\text{Ker}(c^{-1}I - T^{-1}) = T^{-1}(\text{Ker}(cI - T))$.

1.6 Find the coefficient of the linear term in the characteristic polynomial of $[a_{ij}]_{1 \le i,j \le 3}$.

1.7 If $AB = BA$, $\boldsymbol{v}$ is an eigenvector of A, and $B\boldsymbol{v} \neq \mathbf{0}$, show that $B\boldsymbol{v}$ is an eigenvector of A.

In questions 2.1–2.5, test A for diagonalizability. If A is diagonalizable, find P such that $P^{-1}AP$ is diagonal.

2.1 $A = \begin{bmatrix} 1 & 3 \\ 0 & 2 \end{bmatrix}$

2.2 $A = \begin{bmatrix} 3 & 7 \\ 2 & 5 \end{bmatrix}$

2.3 $A = \begin{bmatrix} 4 & -2 & 0 \\ 3 & -1 & 0 \\ 5 & 1 & -1 \end{bmatrix}$

2.4 $A = \begin{bmatrix} 1 & 2 & 2 \\ 1 & 1 & 3 \\ 2 & 1 & 2 \end{bmatrix}$

2.5 $A = \begin{bmatrix} 1 & 1 & 1 & 1 \\ 0 & 1 & 0 & 0 \\ 0 & 0 & 3 & 3 \\ 0 & 0 & 0 & 4 \end{bmatrix}$

2.6 Let $T : \mathbb{R}^n \to \mathbb{R}^n$ be invertible. If T is diagonalizable, show that T^{-1} is diagonalizable.

2.7 Recall that the **trace** of an $n \times n$ matrix $A = [a_{ij}]$ is $\text{tr}A = \sum_{i=1}^{n} a_{ii}$. Prove that $\text{tr}AB = \text{tr}BA$.

2.8 Prove that the trace of a matrix whose characteristic polynomial factors as $(c - c_1)^{h_1} \cdots (c - c_k)^{h_k}$ ($c_1, \ldots, c_k$ real) is $\sum_{i=1}^{k} h_i c_i$.

2.9 The matrix A has characteristic polynomial $(c - c_1)^2(c - c_2)^3(c - c_3)^4$ where $c_1 < c_2 < c_3$. If A is diagonalizable, what are the dimensions of the eigenspaces for c_1, c_2, c_3?

2.10 If A is diagonalizable and Q is invertible, show that $Q^{-1}AQ$ is diagonalizable.

3.1 Let $A = \begin{bmatrix} 5 & -2 & 0 \\ 2 & 1 & 0 \\ 1 & 1 & 5 \end{bmatrix}$. Show that A is not diagonalizable. Find an invertible matrix P such that

$$P^{-1}AP = \begin{bmatrix} a & 1 & 0 \\ 0 & a & 0 \\ 0 & 0 & b \end{bmatrix}$$

where a and b are real numbers (which you should find first).

4.1 Let A be an $n \times n$ matrix. Show that any positive power A^m can be written as a linear combination of $I, A, A^2, \ldots, A^{n-1}$:

$$A^m = c_0 I + c_1 A + \cdots + c_{n-1} A^{n-1}.$$

Hint: Try $m = n$, then $m = n + 1, n + 2, \ldots$.

4.2 Verify the Cayley-Hamilton theorem for

$$A = \begin{bmatrix} 1 & 4 & 1 \\ 0 & 1 & 1 \\ 1 & 0 & 1 \end{bmatrix}.$$

4.3 Give a shorter proof of the Cayley-Hamilton theorem for a diagonalizable matrix A.

4.4 Use the Cayley-Hamilton theorem to find the inverse of the matrix $\begin{bmatrix} a & b \\ c & d \end{bmatrix}$, if it exists.

Hint: $A(A + b_1 I) = -b_2 I$.

5.1 Let A be a transition matrix with eigenvalue $c \neq 1$ and an eigenvector $\boldsymbol{x}$ for c. Show that $x_1 + \cdots + x_n = 0$.

Hint: $A\boldsymbol{x} = c\boldsymbol{x}$. Add up the coordinates on both sides.

6.1 Let A be an $n \times n$ transition matrix whose rows are a rearrangement of $\boldsymbol{e}_1, \ldots, \boldsymbol{e}_n$, $A \neq I$. If $\boldsymbol{q}_0$ has n distinct coordinates, show that there is no steady state vector.

For each of the following transition matrices, show that $\lim_{m \to \infty} A^m$ exists.

6.2 $A = \begin{bmatrix} 0.1 & 0.3 & 0.5 \\ 0.2 & 0.3 & 0.5 \\ 0.7 & 0.4 & 0 \end{bmatrix}$

6.3 $A = \begin{bmatrix} 0.2 & 0.5 & 0.3 \\ 0.3 & 0.1 & 0.3 \\ 0.5 & 0.4 & 0.4 \end{bmatrix}$

6.4 $A = \begin{bmatrix} 0 & 0.5 & 0 \\ 1 & 0 & 0 \\ 0 & 0.5 & 1 \end{bmatrix}$

6.5 With A as in Exercise 6.2 and initial probability vector $\left(\frac{1}{3}, \frac{1}{3}, \frac{1}{3}\right)$, find the steady state vector.

6.6 With A as in Exercise 6.3 and initial probability vector $\left(\frac{1}{4}, \frac{1}{8}, \frac{5}{8}\right)$, find the steady state vector.

6.7 Let $A = \begin{bmatrix} 1 & 0 & \frac{1}{4} & \frac{1}{8} \\ 0 & 1 & \frac{1}{4} & \frac{1}{8} \\ 0 & 0 & \frac{1}{2} & \frac{1}{8} \\ 0 & 0 & 0 & \frac{5}{8} \end{bmatrix}$.

If the initial probability vector is $\left(\frac{1}{5}, \frac{2}{5}, \frac{1}{5}, \frac{1}{5}\right)$, find the steady state vector.

Chapter 7

Symmetric matrices and quadratic forms

1 Quadratic forms

An $n \times n$ matrix A is **symmetric** if

$$A^t = A.$$

A 2×2 symmetric matrix takes the form

$$A = \begin{bmatrix} f & g \\ g & h \end{bmatrix}.$$

Associated with this matrix is the quadratic form in two variables

$$(1) \qquad Q(\boldsymbol{x}) = fx_1^2 + 2gx_1x_2 + hx_2^2.$$

We can write Q in terms of A as follows:

$$(2) \quad Q(\boldsymbol{x}) = [fx_1 + gx_2 \quad gx_1 + hx_2] \begin{bmatrix} x_1 \\ x_2 \end{bmatrix} = [x_1 \; x_2] \begin{bmatrix} f & g \\ g & h \end{bmatrix} \begin{bmatrix} x_1 \\ x_2 \end{bmatrix} = \boldsymbol{x}^t A \boldsymbol{x}.$$

We are interpreting a 1×1 matrix as being a real number here.

Quadratic forms in two or more variables come up frequently in applications. For example, a quadratic form can be used to approximate how a function of several variables behaves at a stationary point. See Further Reading. In this chapter we will use symmetric matrices to learn about quadratic forms and vice versa.

Definition 1 A **quadratic form** in n variables is a function $Q : \mathbb{R}^n \to \mathbb{R}$ of the type

$$Q(\boldsymbol{x}) = Q(x_1, \ldots, x_n) = \sum_{i=1}^{n} \sum_{j=1}^{n} a_{ij} x_i x_j$$

where the coefficients a_{ij} are real numbers and $A = [a_{ij}]_{1 \le i,j \le n}$ is symmetric. The matrix A is the **matrix of** Q.

Example 1 $n = 2$. Here

$$\begin{aligned} Q(\boldsymbol{x}) &= a_{11}x_1^2 + a_{12}x_1x_2 + a_{21}x_2x_1 + a_{22}x_2^2 \\ &= a_{11}x_1^2 + 2a_{12}x_1x_2 + a_{22}x_2^2, \end{aligned}$$

since $a_{12} = a_{21}$. This is, of course, the same as (1) with different labeling of coefficients.

Example 2 Find the matrix of the quadratic form

$$Q(x_1, x_2, x_3) = -x_1^2 + 7x_2^2 + 3x_1x_3 - 8x_2x_3.$$

Solution. The diagonal terms are $-1, 7, 0$, the coefficients of x_1^2, x_2^2, x_3^2 respectively. We obtain a_{ij} for $i \neq j$ by *halving* the coefficient of x_ix_j, since we have not written separate terms for x_ix_j and x_jx_i. So the matrix of Q is

$$A = \begin{bmatrix} -1 & 0 & \frac{3}{2} \\ 0 & 7 & -4 \\ \frac{3}{2} & -4 & 0 \end{bmatrix}.$$

You should check that $\boldsymbol{x}^t A \boldsymbol{x} = Q(x_1, x_2, x_3)$, as in (2).

In general, if A is $n \times n$,

(3)
$$\begin{aligned} \boldsymbol{x}^t A \boldsymbol{x} &= [x_1 \cdots x_n] \begin{bmatrix} a_{11}x_1 + \cdots + a_{1n}x_n \\ \vdots \\ a_{n1}x_1 + \cdots + a_{nn}x_n \end{bmatrix} \\ &= \sum_{i=1}^{n} x_i \sum_{j=1}^{n} a_{ij}x_j = \sum_{i=1}^{n} \sum_{j=1}^{n} a_{ij}x_ix_j. \end{aligned}$$

This gives us a generalization of (2). We record this in

Proposition 1 *Let $A = [a_{ij}]$ be $n \times n$. Then*

(i) $\boldsymbol{x}^t A\boldsymbol{y} = \boldsymbol{x} \cdot (A\boldsymbol{y}) = \boldsymbol{y}^t A^t \boldsymbol{x} = \boldsymbol{y} \cdot (A^t \boldsymbol{x})$.

(ii) If A is symmetric, then

$$\boldsymbol{x} \cdot A\boldsymbol{y} = \boldsymbol{y} \cdot A\boldsymbol{x}.$$

(iii) If A is symmetric and Q is the quadratic form with matrix A, then

$$Q(\boldsymbol{x}) = \boldsymbol{x}^t A\boldsymbol{x}.$$

Proof. We start by noting that

$$\boldsymbol{x}^t \boldsymbol{w} = \boldsymbol{x} \cdot \boldsymbol{w}; \tag{4}$$

both sides are $x_1 w_1 + \cdots + x_n w_n$. In particular,

$$\boldsymbol{x}^t A\boldsymbol{y} = \boldsymbol{x} \cdot A\boldsymbol{y}.$$

Since $\boldsymbol{x} \cdot \boldsymbol{w} = \boldsymbol{w} \cdot \boldsymbol{x}$ always holds,

$$\boldsymbol{x} \cdot A\boldsymbol{y} = (A\boldsymbol{y}) \cdot \boldsymbol{x} = (A\boldsymbol{y})^t \boldsymbol{x} = \boldsymbol{y}^t A^t \boldsymbol{x} = \boldsymbol{y} \cdot (A^t \boldsymbol{x}).$$

Here we used (4) twice. We establish (i) by combining these equations.

If A is symmetric, we use $A = A^t$ and (i) to get (ii):

$$\boldsymbol{x} \cdot A\boldsymbol{y} = \boldsymbol{y} \cdot A\boldsymbol{x}.$$

We have already proved (iii); see (3).

Example 3 Write out the quadratic form whose matrix is

$$\begin{bmatrix} 1 & 0 & 3 & 0 \\ 0 & 2 & 2 & 1 \\ 3 & 2 & -1 & 4 \\ 0 & 1 & 4 & 0 \end{bmatrix}.$$

Solution. $Q(\boldsymbol{x}) = x_1^2 + 2x_2^2 - x_3^2 + 6x_1x_3 + 4x_2x_3 + 2x_2x_4 + 8x_3x_4$. The quickest way to get this is to read off the terms in x_i^2 from the main diagonal, and then *double* the a_{ij} above the main diagonal to get the coefficients of $x_i x_j$ with $i < j$.

We now make a digression to study orthogonality a little further. The reason for this will become clear in Section 4.

2 Orthogonal sets

A set of vectors $\boldsymbol{v}_1, \ldots, \boldsymbol{v}_k$ in $\mathbb{R}^n$ is **orthogonal** if no $\boldsymbol{v}_i$ is $\boldsymbol{0}$ and

$$\boldsymbol{v}_i \cdot \boldsymbol{v}_j = 0 \quad \text{for } i \neq j.$$

We say that $\boldsymbol{v}_1, \ldots, \boldsymbol{v}_k$ is an **orthonormal** set if it is orthogonal and every $\boldsymbol{v}_i$ has length 1. Alternatively, we can write

$$\boldsymbol{v}_i \cdot \boldsymbol{v}_j = \begin{cases} 0 & \text{for } i \neq j \\ 1 & \text{for } i = j. \end{cases}$$

We already came across orthonormal sets when discussing orthogonal *matrices* in Chapter 4. The set of column vectors of an $n \times n$ orthogonal matrix P is an orthonormal set $\boldsymbol{v}_1, \ldots, \boldsymbol{v}_n$ in $\mathbb{R}^n$. The set of row vectors is also an orthonormal set. Here we use $P^tP = I$ and $PP^t = I$.

Example 4 Let $\boldsymbol{v}_1 = (1, 0, 1), \boldsymbol{v}_2 = (0, 1, 0), \boldsymbol{v}_3 = (1, 0, -1)$. Then $\boldsymbol{v}_1, \boldsymbol{v}_2, \boldsymbol{v}_3$ is an orthogonal set. If we divide each $\boldsymbol{v}_i$ by its length and use $\left|\frac{\boldsymbol{v}_i}{|\boldsymbol{v}_i|}\right| = \frac{1}{|\boldsymbol{v}_i|}\,|\boldsymbol{v}_i| = 1$, we get an orthonormal set $\boldsymbol{u}_1 = \left(\frac{1}{\sqrt{2}}, 0, \frac{1}{\sqrt{2}}\right), \boldsymbol{u}_2 = (0, 1, 0)$, $\boldsymbol{u}_3 = \left(\frac{1}{\sqrt{2}}, 0, -\frac{1}{\sqrt{2}}\right)$. This process is called **normalizing** . Now

$$P = [\boldsymbol{u}_1 \quad \boldsymbol{u}_2 \quad \boldsymbol{u}_3] = \begin{bmatrix} \frac{1}{\sqrt{2}} & 0 & \frac{1}{\sqrt{2}} \\ 0 & 1 & 0 \\ \frac{1}{\sqrt{2}} & 0 & -\frac{1}{\sqrt{2}} \end{bmatrix}$$

is an orthogonal matrix.

In Chapter 1 we used the simple fact that, given nonzero vectors $\boldsymbol{u}$ and $\boldsymbol{v}$,

$$\boldsymbol{v} - \frac{\boldsymbol{v} \cdot \boldsymbol{u}}{\boldsymbol{u} \cdot \boldsymbol{u}}\,\boldsymbol{u}$$

is orthogonal to $\boldsymbol{u}$:

$$\text{(5)} \qquad \left(\boldsymbol{v} - \frac{\boldsymbol{v} \cdot \boldsymbol{u}}{\boldsymbol{u} \cdot \boldsymbol{u}}\right) \cdot \boldsymbol{u} = \boldsymbol{v} \cdot \boldsymbol{u} - \frac{\boldsymbol{v} \cdot \boldsymbol{u}}{\boldsymbol{u} \cdot \boldsymbol{u}}\,\boldsymbol{u} \cdot \boldsymbol{u} = 0.$$

This observation enables us to construct an orthonormal basis of any given subspace of $\mathbb{R}^n$. Before we do this there is a fundamental fact that we record.

Proposition 2 *An orthogonal set is linearly independent.*

Proof. Let $\boldsymbol{v}_1, \dots, \boldsymbol{v}_k$ be an orthogonal set. Suppose that

$$x_1\boldsymbol{v}_1 + \cdots + x_k\boldsymbol{v}_k = \mathbf{0}. \tag{6}$$

Take the inner product of both sides with $\boldsymbol{v}_i$:

$$(x_1\boldsymbol{v}_1 + \cdots + x_k\boldsymbol{v}_k) \cdot \boldsymbol{v}_i = \mathbf{0} \cdot \boldsymbol{v}_i = 0. \tag{7}$$

The left side of (7) is

$$x_i\boldsymbol{v}_i \cdot \boldsymbol{v}_i = x_i|\boldsymbol{v}_i|^2$$

since the terms $x_j\boldsymbol{v}_j \cdot \boldsymbol{v}_i$ with $j \neq i$ are zero. So

$$x_i|\boldsymbol{v}_i|^2 = 0.$$

Since $|\boldsymbol{v}_i| > 0$, we have $x_i = 0$. As i is any of $1, \dots, k$, the only solution of (6) is the zero solution.

Proposition 3 *Let $\boldsymbol{w}_1, \dots \boldsymbol{w}_m$ be a linearly independent set in $\mathbb{R}^n$. Let*

$$\begin{aligned} \boldsymbol{v}_1 &= \boldsymbol{w}_1, \\ \boldsymbol{v}_2 &= \boldsymbol{w}_2 - \frac{\boldsymbol{w}_2 \cdot \boldsymbol{v}_1}{\boldsymbol{v}_1 \cdot \boldsymbol{v}_1}\,\boldsymbol{v}_1 \end{aligned}$$

and generally, once $\boldsymbol{v}_1, \dots, \boldsymbol{v}_{j-1}$ have been defined, let

$$\boldsymbol{v}_j = \boldsymbol{w}_j - \sum_{i=1}^{j-1} \frac{\boldsymbol{w}_j \cdot \boldsymbol{v}_i}{\boldsymbol{v}_i \cdot \boldsymbol{v}_i}\,\boldsymbol{v}_i. \tag{8}$$

Then $\boldsymbol{v}_1, \dots, \boldsymbol{v}_m$ is an orthogonal set. Also

$$\operatorname{Span}\{\boldsymbol{v}_1, \dots, \boldsymbol{v}_m\} = \operatorname{Span}\{\boldsymbol{w}_1, \dots, \boldsymbol{w}_m\}. \tag{9}$$

Proof. Firstly, $\boldsymbol{v}_j$ is not zero, since if $\boldsymbol{v}_j = \mathbf{0}$, then $\boldsymbol{w}_j$ is a combination of $\boldsymbol{v}_1, \dots, \boldsymbol{v}_{j-1}$, and indeed a combination of $\boldsymbol{w}_1, \dots, \boldsymbol{w}_{j-1}$, which is not the case. Note that we need $\boldsymbol{v}_1 \neq \mathbf{0}$ in order to define $\boldsymbol{v}_2$, then $\boldsymbol{v}_2 \neq \mathbf{0}$ in order to define $\boldsymbol{v}_3$, and so on.

Secondly, we can prove in succession that $\boldsymbol{v}_2 \cdot \boldsymbol{v}_1 = 0$, $\boldsymbol{v}_3 \cdot \boldsymbol{v}_k = 0$ for $k < 3$, $\boldsymbol{v}_4 \cdot \boldsymbol{v}_k = 0$ for $k < 4$, and so on. We already saw in (5) that $\boldsymbol{v}_2 \cdot \boldsymbol{v}_1 = 0$.

Once we have got $\boldsymbol{v}_j \cdot \boldsymbol{v}_k = 0$ for a certain r and all $k < j \leq r$, we look at $\boldsymbol{v}_{r+1}$. We have, for $k < r + 1$,

$$\begin{aligned}\boldsymbol{v}_{r+1} \cdot \boldsymbol{v}_k &= \boldsymbol{w}_{r+1} \cdot \boldsymbol{v}_k - \sum_{i=1}^{r} \frac{\boldsymbol{w}_{r+1} \cdot \boldsymbol{v}_i}{\boldsymbol{v}_i \cdot \boldsymbol{v}_i} \boldsymbol{v}_i \cdot \boldsymbol{v}_k \\ &= \boldsymbol{w}_{r+1} \cdot \boldsymbol{v}_k - \frac{\boldsymbol{w}_{r+1} \cdot \boldsymbol{v}_k}{\boldsymbol{v}_k \cdot \boldsymbol{v}_k} \boldsymbol{v}_k \cdot \boldsymbol{v}_k = 0.\end{aligned}$$

Eventually we obtain $\boldsymbol{v}_j \cdot \boldsymbol{v}_k = 0$ for all $j \leq m$ and all $k < j$. So $\boldsymbol{v}_1, \ldots, \boldsymbol{v}_m$ are orthogonal. Hence $\boldsymbol{v}_1, \ldots, \boldsymbol{v}_m$ is an independent set (Proposition 2). Now

$$W = \text{Span}\{\boldsymbol{w}_1, \ldots, \boldsymbol{w}_m\}$$

is an m-dimensional subspace of $\mathbb{R}^n$. Thus $\boldsymbol{v}_1, \ldots, \boldsymbol{v}_m$ is a basis of W (Proposition 7 of Chapter 3). This proves (9).

The formula (8) for an orthogonal basis of W is called the **Gram-Schmidt process** .

Example 5 Find an orthogonal basis of the column space of

$$A = \begin{bmatrix} 1 & 2 & 3 \\ -1 & 0 & 1 \\ 0 & 1 & 2 \\ 1 & 3 & 0 \end{bmatrix}.$$

Solution. Let $A = [\boldsymbol{w}_1\ \boldsymbol{w}_2\ \boldsymbol{w}_3]$,

$$\begin{aligned}\boldsymbol{v}_1 &= \boldsymbol{w}_1, \\ \boldsymbol{v}_2 &= \boldsymbol{w}_2 - \frac{\boldsymbol{w}_2 \cdot \boldsymbol{v}_1}{\boldsymbol{v}_1 \cdot \boldsymbol{v}_1} \boldsymbol{v}_1 = (2,0,1,3) - \frac{5}{3}(1,-1,0,1) \\ 3\boldsymbol{v}_2 &= (6,0,3,9) - (5,-5,0,5) = (1,5,3,4).\end{aligned}$$

You should check that $\boldsymbol{v}_2 \cdot \boldsymbol{v}_1 = 0$. Let us replace $\boldsymbol{v}_2$ by $3\boldsymbol{v}_2$, which we rename $\boldsymbol{v}_2$. This simplification will not spoil our construction of an orthogonal basis. Let

$$\begin{aligned}\boldsymbol{v}_3 &= \boldsymbol{w}_3 - \frac{\boldsymbol{w}_3 \cdot \boldsymbol{v}_1}{\boldsymbol{v}_1 \cdot \boldsymbol{v}_1} \boldsymbol{v}_1 - \frac{\boldsymbol{w}_3 \cdot \boldsymbol{v}_2}{\boldsymbol{v}_2 \cdot \boldsymbol{v}_2} \boldsymbol{v}_2 \\ &= (3,1,2,0) - \frac{2}{3}(1,-1,0,1) - \frac{14}{51}(1,5,3,4), \\ 51\boldsymbol{v}_3 &= (153,51,102,0) - (34,-34,0,34) - (14,70,42,56) \\ &= (105,15,60,-90) = 15(7,1,4,-6).\end{aligned}$$

At this stage, you should check that $\boldsymbol{v}_3 \cdot \boldsymbol{v}_1 = \boldsymbol{v}_3 \cdot \boldsymbol{v}_2 = 0$. Replace $\boldsymbol{v}_3$ by $\frac{51}{15}\,\boldsymbol{v}_3$, which we rename $\boldsymbol{v}_3$. Our orthogonal basis of W is

$$\boldsymbol{v}_1 = (1, -1, 0, 1), \boldsymbol{v}_2 = (1, 5, 3, 4), \boldsymbol{v}_3 = (7, 1, 4, -6).$$

Example 6 Let $\boldsymbol{w}_1, \boldsymbol{w}_2$ be a basis of a plane S in $\mathbb{R}^2$. Then $\boldsymbol{v}_1 = \boldsymbol{w}_1$, $\boldsymbol{v}_2 = \boldsymbol{w}_2 - \frac{\boldsymbol{w}_2 \cdot \boldsymbol{v}_1}{\boldsymbol{v}_1 \cdot \boldsymbol{v}_1}\,\boldsymbol{v}_1$ is an orthogonal basis of S. Given $\boldsymbol{w}_3$ in $\mathbb{R}^3$, $\boldsymbol{w}_3$ not in S, the vector

$$\boldsymbol{w}_3 - \frac{\boldsymbol{w}_3 \cdot \boldsymbol{v}_1}{\boldsymbol{v}_1 \cdot \boldsymbol{v}_1} - \frac{\boldsymbol{w}_3 \cdot \boldsymbol{v}_2}{\boldsymbol{v}_2 \cdot \boldsymbol{v}_2}\,\boldsymbol{v}_2$$

is orthogonal to S, which means that

$$\frac{\boldsymbol{w}_3 \cdot \boldsymbol{v}_1}{\boldsymbol{v}_1 \cdot \boldsymbol{v}_1}\,\boldsymbol{v}_1 + \frac{\boldsymbol{w}_3 \cdot \boldsymbol{v}_2}{\boldsymbol{v}_2 \cdot \boldsymbol{v}_2}\,\boldsymbol{v}_2$$

is the projection of $\boldsymbol{w}_3$ into the plane S (Figure 1).

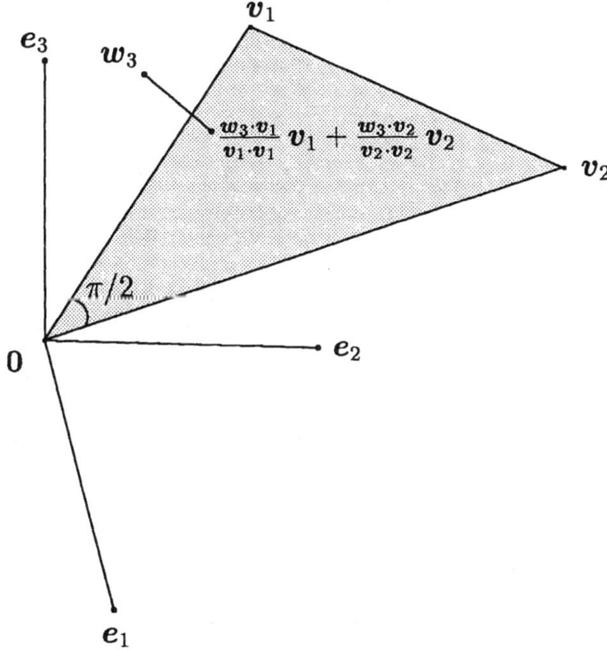

Figure 1. Segment joining $\boldsymbol{w}_3$ to its projection into $S = \operatorname{Span}\{\boldsymbol{v}_1, \boldsymbol{v}_2\}$ is perpendicular to S.

We can carry over this *projection* idea to arbitrary dimensions.

Definition 2 Let $\boldsymbol{v}_1, \ldots, \boldsymbol{v}_m$ be an orthogonal set in $\mathbb{R}^n$, with $m < n$. Given $\boldsymbol{v}$ in $\mathbb{R}^n$, the vector

$$\boldsymbol{v}' = \frac{\boldsymbol{v} \cdot \boldsymbol{v}_1}{\boldsymbol{v}_1 \cdot \boldsymbol{v}_1} \boldsymbol{v}_1 + \cdots + \frac{\boldsymbol{v} \cdot \boldsymbol{v}_m}{\boldsymbol{v}_m \cdot \boldsymbol{v}_m} \boldsymbol{v}_m$$

is the **projection** or **orthogonal projection** of $\boldsymbol{v}$ into $V = \text{Span}\{\boldsymbol{v}_1, \ldots, \boldsymbol{v}_m\}$.

Proposition 4 *Let $\boldsymbol{v}_1, \ldots, \boldsymbol{v}_m$ be an orthogonal set in $\mathbb{R}^n$. Suppose $\boldsymbol{v}$ is not in $V = \text{Span}\{\boldsymbol{v}_1, \ldots, \boldsymbol{v}_m\}$. The projection $\boldsymbol{v}'$ of $\boldsymbol{v}$ into V has the following properties.*

(i) $\boldsymbol{v} - \boldsymbol{v}'$ is orthogonal to every point of V.

(ii) $\boldsymbol{v}'$ is closer to $\boldsymbol{v}$ than any other point of V.

Proof. Part (i) follows from the Gram-Schmidt process applied to $\boldsymbol{v}_1, \ldots, \boldsymbol{v}_m, \boldsymbol{v}$. The first m steps just regurgitate $\boldsymbol{v}_1, \ldots, \boldsymbol{v}_m$ (see (8)). The final step gives $\boldsymbol{v} - \boldsymbol{v}'$ on the right hand side of (8). Proposition 3 tells us that $\boldsymbol{v} - \boldsymbol{v}'$ is orthogonal to $\boldsymbol{v}_1, \ldots, \boldsymbol{v}_m$. So $\boldsymbol{v} - \boldsymbol{v}'$ is orthogonal to any combination of $\boldsymbol{v}_1, \ldots, \boldsymbol{v}_m$. This proves (i).

Part (ii) is now essentially Pythagoras's theorem. If $\boldsymbol{w}$ is a point of V different from $\boldsymbol{v}'$,

$$\begin{aligned} |\boldsymbol{v} - \boldsymbol{w}|^2 &= |\boldsymbol{v} - \boldsymbol{v}' + \boldsymbol{v}' - \boldsymbol{w}|^2 \\ &= |\boldsymbol{v} - \boldsymbol{v}'|^2 + |\boldsymbol{v}' - \boldsymbol{w}|^2 \end{aligned}$$

(since $\boldsymbol{v} - \boldsymbol{v}'$ is orthogonal to $\boldsymbol{v}' - \boldsymbol{w}$). So $|\boldsymbol{v} - \boldsymbol{w}|^2 > |\boldsymbol{v} - \boldsymbol{v}'|^2$ and $\boldsymbol{v}'$ is the closest point of V to $\boldsymbol{v}$. This is the reasoning that we used in computing distance from a point to a hyperplane at the end of Chapter 1.

Example 7 Find the closest point to $(0, 1, 1, 5)$ of the column space of the matrix A in Example 5.

Solution. Let $\boldsymbol{v}_1, \boldsymbol{v}_2, \boldsymbol{v}_3$ be the basis in Example 5. According to Proposition 4, the closest point to $\boldsymbol{v} = (0, 1, 1, 5)$ in the column space is

$$\begin{aligned} \boldsymbol{v}' &= \frac{\boldsymbol{v} \cdot \boldsymbol{v}_1}{\boldsymbol{v}_1 \cdot \boldsymbol{v}_1} \boldsymbol{v}_1 + \frac{\boldsymbol{v} \cdot \boldsymbol{v}_2}{\boldsymbol{v}_2 \cdot \boldsymbol{v}_2} \boldsymbol{v}_2 + \frac{\boldsymbol{v} \cdot \boldsymbol{v}_3}{\boldsymbol{v}_3 \cdot \boldsymbol{v}_3} \boldsymbol{v}_3 \\ &= \frac{4}{3} (1, -1, 0, 1) + \frac{28}{51} (1, 5, 3, 4) - \frac{25}{102} (7, 1, 4, -6) \\ &= \frac{1}{102} (17, 119, 68, 510) = \frac{1}{6} (1, 7, 4, 30). \end{aligned}$$

Reminder: The above calculation of 'closest point' requires $\boldsymbol{v}_1, \boldsymbol{v}_2, \boldsymbol{v}_3$ **orthogonal**. My students occasionally forget this!

One of the merits of an orthogonal basis is the ease of calculating coordinates for the basis.

Proposition 5 *Let $\boldsymbol{v}_1, \ldots, \boldsymbol{v}_m$ be an orthogonal basis of a subspace V of $\mathbb{R}^n$. Let $\boldsymbol{v}$ be in V. Then*

$$\boldsymbol{v} = \frac{\boldsymbol{v} \cdot \boldsymbol{v}_1}{\boldsymbol{v}_1 \cdot \boldsymbol{v}_1}\, \boldsymbol{v}_1 + \cdots + \frac{\boldsymbol{v} \cdot \boldsymbol{v}_m}{\boldsymbol{v}_m \cdot \boldsymbol{v}_m}\, \boldsymbol{v}_m. \tag{10}$$

If $\boldsymbol{v}_1, \ldots, \boldsymbol{v}_m$ is an orthonormal set, then

$$\boldsymbol{v} = (\boldsymbol{v} \cdot \boldsymbol{v}_1)\boldsymbol{v}_1 + \cdots + (\boldsymbol{v} \cdot \boldsymbol{v}_m)\boldsymbol{v}_m.$$

Proof. Just as in the proof of Proposition 3, we see that $\boldsymbol{w} = \boldsymbol{v} - \frac{\boldsymbol{v} \cdot \boldsymbol{v}_1}{\boldsymbol{v}_1 \cdot \boldsymbol{v}_1}\, \boldsymbol{v}_1 - \cdots - \frac{\boldsymbol{v} \cdot \boldsymbol{v}_m}{\boldsymbol{v}_m \cdot \boldsymbol{v}_m}\, \boldsymbol{v}_m$ is orthogonal to each $\boldsymbol{v}_i$. Hence $\boldsymbol{w}$ is orthogonal to any vector in $\mathrm{Span}\{\boldsymbol{v}_1, \ldots, \boldsymbol{v}_m\}$. In particular, $\boldsymbol{w} \cdot \boldsymbol{w} = 0$, so $\boldsymbol{w}$ is $\mathbf{0}$.

The second part of the proposition follows since $\boldsymbol{v}_i \cdot \boldsymbol{v}_i = 1$ in this case.

Example 8 Find an orthogonal basis of the hyperplane

$$x_1 - 2x_2 + x_3 + 3x_4 = 0$$

in $\mathbb{R}^4$.

Solution. We noted in Chapter 3 that a basis of the hyperplane is $\boldsymbol{w}_1 = (2, 1, 0, 0), \boldsymbol{w}_2 = (-1, 0, 1, 0), \boldsymbol{w}_3 = (-3, 0, 0, 1)$. Let $\boldsymbol{v}_1 = \boldsymbol{w}_1$,

$$\begin{aligned} \boldsymbol{v}_2 &= \boldsymbol{w}_2 - \frac{\boldsymbol{w}_2 \cdot \boldsymbol{v}_1}{\boldsymbol{v}_1 \cdot \boldsymbol{v}_1}\, \boldsymbol{v}_1 = (-1, 0, 1, 0) + \frac{2}{5}\, (2, 1, 0, 0), \\ 5\boldsymbol{v}_2 &= (-1, 2, 5, 0). \end{aligned}$$

Write, instead, $\boldsymbol{v}_2 = (-1, 2, 5, 0)$. Let

$$\begin{aligned} \boldsymbol{v}_3 &= \boldsymbol{w}_3 - \frac{\boldsymbol{w}_3 \cdot \boldsymbol{v}_1}{\boldsymbol{v}_1 \cdot \boldsymbol{v}_1}\, \boldsymbol{v}_1 - \frac{\boldsymbol{w}_3 \cdot \boldsymbol{v}_2}{\boldsymbol{v}_2 \cdot \boldsymbol{v}_2}\, \boldsymbol{v}_2 \\ &= (-3, 0, 0, 1) + \frac{6}{5}\, (2, 1, 0, 0) - \frac{3}{30}\, (-1, 2, 5, 0) \\ 10\boldsymbol{v}_3 &= (-5, 10, -5, 10). \end{aligned}$$

Write, instead, $\boldsymbol{v}_3 = (-1, 2, -1, 2)$. The required orthogonal basis is

$$(2, 1, 0, 0), (-1, 2, 5, 0), (-1, 2, -1, 2).$$

3 Orthogonal complement

Definition 3 Let V be a subspace of $\mathbb{R}^n$. We write $V^\perp$ for the set of vectors $\boldsymbol{w}$ such that $\boldsymbol{w} \cdot \boldsymbol{v} = 0$ for every $\boldsymbol{v}$ in V. Then $V^\perp$ is the **orthogonal complement** of V.

The symbol $\perp$ is pronounced 'perp'.

Example 9 Let V be a plane through $\mathbf{0}$ in $\mathbb{R}^3$. Clearly $V^\perp$ is the normal to V passing through $\mathbf{0}$. Notice that $(V^\perp)^\perp = V$ since the set of points perpendicular to the normal is just V. We also observe that every vector $\boldsymbol{u}$ in $\mathbb{R}^3$ can be written as the sum $\boldsymbol{v} + \boldsymbol{w}$ ($\boldsymbol{v}$ in V, $\boldsymbol{w}$ in $V^\perp$); just take $\boldsymbol{v}$ to be the projection of $\boldsymbol{u}$ into V.

For any subspace V, we already have information about $V^\perp$ from Chapter 3. Let $\boldsymbol{v}_1, \ldots, \boldsymbol{v}_m$ be a basis of V. Obviously if $\boldsymbol{w}$ is in $V^\perp$, then

$$\boldsymbol{w} \cdot \boldsymbol{v}_i = 0 \quad (i = 1, \ldots, m). \tag{11}$$

On the other hand, if (11) holds for a vector $\boldsymbol{w}$, then $\boldsymbol{w}$ is orthogonal to any combination of $\boldsymbol{v}_1, \ldots, \boldsymbol{v}_m$, and so $\boldsymbol{w}$ is in $V^\perp$. Summarizing, $V^\perp$ is the set of solutions of (11), or

$$V^\perp = \text{Nul } A \tag{12}$$

where A is the $m \times n$ matrix whose rows are the $\boldsymbol{v}_i$.

Proposition 6 *Let V be a subspace of $\mathbb{R}^n$.*

(i) $V^\perp$ is a subspace of $\mathbb{R}^n$ and $\dim V + \dim V^\perp = n$.

(ii) $(V^\perp)^\perp = V$.

(iii) Every point $\boldsymbol{v}$ in $\mathbb{R}^n$ can be written in exactly one way as

$$\boldsymbol{v} = \boldsymbol{v}' + \boldsymbol{v}'' \quad (\boldsymbol{v}' \textit{ in } V, \boldsymbol{v}'' \textit{ in } V^\perp).$$

Proof. Let A be a matrix whose rows are an orthogonal basis $\boldsymbol{v}_1, \ldots, \boldsymbol{v}_m$ of V. We get (i) from (12), because Nul A is a subspace whose dimension is $n -$ rank $A = n - m$. Here we use the independence of the rows of A.

To get (ii), we first notice that $(V^\perp)^\perp$ includes any vector $\boldsymbol{v}$ in V, since $\boldsymbol{v} \cdot \boldsymbol{w} = 0$ for every $\boldsymbol{w}$ in $V^\perp$. Moreover,

$$(\dim V^\perp)^\perp + \dim V^\perp = n$$

from (i), so $(V^\perp)^\perp$ has the same dimension as V. Now $(V^\perp)^\perp$ must equal V (Proposition 9 of Chapter 3).

Given $\boldsymbol{v}$, define $\boldsymbol{v}'$ by formula (8); then $\boldsymbol{v}'$ is in V, and we recall that $\boldsymbol{v}'' = \boldsymbol{v} - \boldsymbol{v}'$ is orthogonal to every point of V. There could be no other split of $\boldsymbol{v}$ as $\boldsymbol{u} + \boldsymbol{w}$ ($\boldsymbol{u}$ in V, $\boldsymbol{w}$ in $V^\perp$). For if

$$\boldsymbol{v} = \boldsymbol{u} + \boldsymbol{w} = \boldsymbol{v}' + \boldsymbol{v}'' \quad (\boldsymbol{u} \text{ in } V, \boldsymbol{w} \text{ in } V^\perp), \tag{13}$$

then

$$\boldsymbol{v}' - \boldsymbol{u} = \boldsymbol{w} - \boldsymbol{v}''.$$

Since $\boldsymbol{w} - \boldsymbol{v}''$ is a vector in $V^\perp$, it is orthogonal to $\boldsymbol{v}' - \boldsymbol{u}$. So $\boldsymbol{w} - \boldsymbol{v}''$ is orthogonal to itself, and has length 0. In other words, $\boldsymbol{w} = \boldsymbol{v}''$. Now (13) yields $\boldsymbol{u} = \boldsymbol{v}'$.

Example 10 Let V be the linear span of $\boldsymbol{y}_1 = (1, 3, 0, 1)$ and $\boldsymbol{y}_2 = (-1, 0, 2, 1)$. Find an orthogonal basis of $V^\perp$.

Solution. The solution space of $\boldsymbol{y}_1 \cdot \boldsymbol{x} = \boldsymbol{y}_2 \cdot \boldsymbol{x} = 0$ is found using

$$\begin{bmatrix} 1 & 3 & 0 & 1 \\ -1 & 0 & 2 & 1 \end{bmatrix} \sim \begin{bmatrix} 1 & 3 & 0 & 1 \\ 0 & 3 & 2 & 2 \end{bmatrix}^{\mathrm{II}+\mathrm{I}} \sim \begin{bmatrix} 1 & 0 & -2 & -1 \\ 0 & 1 & \frac{2}{3} & \frac{2}{3} \end{bmatrix}\begin{matrix} \mathrm{I}-\mathrm{II} \\ \mathrm{II}\times\frac{1}{3}. \end{matrix}$$

The general solution is $\left(2x_3 + x_4, \frac{-2x_3}{3} - \frac{2x_4}{3}, x_3, x_4\right)$. So a basis of $V^\perp$ is $\boldsymbol{w}_1 = \left(2, -\frac{2}{3}, 1, 0\right), \boldsymbol{w}_2 = \left(1, -\frac{2}{3}, 0, 1\right)$. It is easier to use $\boldsymbol{u}_1 = (6, -2, 3, 0)$, $\boldsymbol{u}_2 = (3, -2, 0, 3)$ in place of $\boldsymbol{w}_1, \boldsymbol{w}_2$. Let $\boldsymbol{v}_1 = \boldsymbol{u}_1$,

$$\boldsymbol{v}_2 = \boldsymbol{u}_2 - \frac{\boldsymbol{u}_2 \cdot \boldsymbol{u}_1}{\boldsymbol{u}_1 \cdot \boldsymbol{u}_1}\,\boldsymbol{u}_1 = (3, -2, 0, 3) - \frac{22}{49}\,(6, -2, 3, 0),$$
$$49\boldsymbol{v}_2 = (15, -54, -66, 147) = 3(5, -18, -22, 49).$$

An orthogonal basis of $V^\perp$ is $(6, -2, 3, 0), (5, -18, -22, 49)$.

4 Orthogonal diagonalization

Our excursion into orthogonality was motivated by the following result.

Proposition 7 *Let* $\boldsymbol{v}$ *and* $\boldsymbol{w}$ *be vectors from distinct eigenspaces of the symmetric matrix* A. *Then*

$$\boldsymbol{v} \cdot \boldsymbol{w} = 0.$$

Proof. We have $A\boldsymbol{v} = c_1\boldsymbol{v}, A\boldsymbol{w} = c_2\boldsymbol{w}$, and $c_1 \neq c_2$. Now

$$\boldsymbol{w} \cdot (A\boldsymbol{v}) = \boldsymbol{v} \cdot (A\boldsymbol{w})$$

from Proposition 1 (ii). The left side is $\boldsymbol{w} \cdot c_1\boldsymbol{v} = c_1\boldsymbol{w} \cdot \boldsymbol{v} = c_1\boldsymbol{v} \cdot \boldsymbol{w}$ and the right side is $\boldsymbol{v} \cdot c_2\boldsymbol{w} = c_2\boldsymbol{v} \cdot \boldsymbol{w}$. So

$$(c_1 - c_2)\boldsymbol{v} \cdot \boldsymbol{w} = 0.$$

Since $c_1 - c_2 \neq 0$, we must have $\boldsymbol{v} \cdot \boldsymbol{w} = 0$.

Example 11 Verify the result of Proposition 7 for

$$A = \begin{bmatrix} f & g \\ g & h \end{bmatrix}$$

where $g \neq 0$.

Solution. The characteristic equation is

$$c^2 - (f + h)c + fh - g^2 = 0.$$

The solutions of the quadratic equation are

$$c_1 = \frac{f + h + \sqrt{(f + h)^2 - 4fh + 4g^2}}{2}, c_2 = \frac{f + h - \sqrt{(f + h)^2 - 4fh + 4g^2}}{2}$$

which we rewrite as

$$c_1 = \frac{f + h + \sqrt{(f - h)^2 + 4g^2}}{2}, c_2 = \frac{f + h - \sqrt{(f - h)^2 + 4g^2}}{2}.$$

Since the number D under the square root sign is positive, $c_1 \neq c_2$.

Eigenspace for c_1. The first equation of the system $(c_1 I - A)\boldsymbol{x} = \mathbf{0}$ is

$$\left(\frac{h-f}{2} + \frac{\sqrt{D}}{2}\right) x_1 - g x_2 = 0.$$

A basis of the eigenspace is $\boldsymbol{v}_1 = \left(g, \frac{h-f}{2} + \frac{\sqrt{D}}{2}\right)$.

Eigenspace for c_2. Repeating the above calculation with $\sqrt{D}$ replaced by $-\sqrt{D}$, a basis of the eigenspace is $\boldsymbol{v}_2 = \left(g, \frac{h-f}{2} - \frac{\sqrt{D}}{2}\right)$. Now

$$\boldsymbol{v}_1 \cdot \boldsymbol{v}_2 = g^2 + \frac{(h-f)^2}{4} - \frac{D^2}{4} = 0.$$

Definition 4 Let A be $n \times n$ and let V be a subspace of $\mathbb{R}^n$. We write $A : V \to V$ if $A\boldsymbol{v}$ is in V whenever $\boldsymbol{v}$ is in V.

Example 12 Let A be the matrix of a rotation of $\mathbb{R}^3$ about the axis L. Then $A : L \to L$ and $A : L^\perp \to L^\perp$.

We now come to an interesting result with a geometrical flavor.

Proposition 8 *Let V be a subspace of $\mathbb{R}^n$. Write S for the set of $\boldsymbol{v}$ in V with $|\boldsymbol{v}| = 1$. Let A be a symmetric matrix, $A : V \to V$. The vector $\boldsymbol{v}_1$ in S where $\boldsymbol{v}^t A \boldsymbol{v}$ is largest is an eigenvector of A.*

Proof. The result is obvious if $\dim V = 1$; $A\boldsymbol{v}_1$ is in V and must be a multiple of $\boldsymbol{v}_1$. So we may suppose that $\dim V \geq 2$.

Firstly, there *is* a point $\boldsymbol{v}_1$ in S where $\boldsymbol{v}^t A \boldsymbol{v}$ is largest. This follows from a standard result in calculus about a continuous function on a closed bounded set. Pick any $\boldsymbol{w}$ in S with $\boldsymbol{w} \cdot \boldsymbol{v}_1 = 0$.

Let

$$\boldsymbol{z} = \boldsymbol{v}_1 \cos y + \boldsymbol{w} \sin y,$$

where y is an arbitrary angle. We have $|\boldsymbol{z}|^2 = |\boldsymbol{v}_1|^2 \cos^2 y + |\boldsymbol{w}|^2 \sin^2 y = \cos^2 y + \sin^2 y = 1$, so $\boldsymbol{z}$ is also in S. Consequently $\boldsymbol{z}^t A \boldsymbol{z}$ has a maximum value when $y = 0$ and $\boldsymbol{z} = \boldsymbol{v}_1$. Now

$$\begin{aligned} \boldsymbol{z}^t A \boldsymbol{z} &= (\boldsymbol{v}_1 \cos y + \boldsymbol{w} \sin y)^t A (\boldsymbol{v}_1 \cos y + \boldsymbol{w} \sin y) \\ &= \boldsymbol{v}_1^t A \boldsymbol{v}_1 \cos^2 y + 2\boldsymbol{v}_1^t A \boldsymbol{w} \sin y \cos y + \boldsymbol{w}^t A \boldsymbol{w} \sin^2 y \end{aligned}$$

using Proposition 1 (ii). The derivative of the right side is

$$-2\boldsymbol{v}_1^t A\boldsymbol{v}_1 \sin y \cos y + 2\boldsymbol{v}_1^t A\boldsymbol{w}(\cos^2 y - \sin^2 y) + 2\boldsymbol{w}^t A\boldsymbol{w} \sin y \cos y.$$

The derivative at $y = 0$ is $2\boldsymbol{v}_1^t A\boldsymbol{w}$. Since this is where the maximum occurs,

$$\boldsymbol{v}_1^t A\boldsymbol{w} = 0 = \boldsymbol{w} \cdot A\boldsymbol{v}_1.$$

Summarizing, *any $\boldsymbol{w}$ in S that is orthogonal to $\boldsymbol{v}_1$ is orthogonal to $A\boldsymbol{v}_1$*. Let $\boldsymbol{v}_1, \boldsymbol{v}_2, \ldots, \boldsymbol{v}_j$ be an orthonormal basis of V. Then $A\boldsymbol{v}_1$ is a vector in V that is orthogonal to $\boldsymbol{v}_2, \ldots, \boldsymbol{v}_r$. So

$$A\boldsymbol{v}_1 = (A\boldsymbol{v}_1 \cdot \boldsymbol{v}_1)\boldsymbol{v}_1 + (A\boldsymbol{v}_1 \cdot \boldsymbol{v}_2)\boldsymbol{v}_2 + \cdots + (A\boldsymbol{v}_1 \cdot \boldsymbol{v}_r)\boldsymbol{v}_r$$

(Proposition 5)

$$= (A\boldsymbol{v}_1 \cdot \boldsymbol{v}_1)\boldsymbol{v}_1.$$

This proves that $\boldsymbol{v}_1$ is an eigenvector of A.

Example 13 Let

$$A = \begin{bmatrix} 8 & & & \\ & 5 & & \\ & & 7 & \\ & & & -3 \end{bmatrix}$$

and $V = \text{Span}\{\boldsymbol{e}_2, \boldsymbol{e}_3, \boldsymbol{e}_4\}$. Then $A : V \to V$. Now

$$\boldsymbol{v}^t A\boldsymbol{v} = 5v_2^2 + 7v_3^2 - 3v_4^2$$

for $\boldsymbol{v} = (0, v_2, v_3, v_4)$ in V. If S is the set of $(0, v_2, v_3, v_4)$ with $v_2^2 + v_3^2 + v_4^2 = 1$, then

$$\boldsymbol{v}^t A\boldsymbol{v} = 5v_2^2 + 7v_3^2 - 3v_4^2 \le 7(v_2^2 + v_3^2 + v_4^2) = 7$$

with equality at $(0, 0, \pm 1, 0)$. The proposition tells us that the vectors $(0, 0, \pm 1, 0)$ are eigenvalues of A, which is easy to verify.

Proposition 8 is a stepping stone to the following important result.

Proposition 9 *Let A be $n \times n$ symmetric. There is an orthonormal basis of $\mathbb{R}^n$ consisting of eigenvectors of A.*

Proof. We apply Proposition 8 repeatedly. In the first step, $V = \mathbb{R}^n$. Obviously, $A : V \to V$, so Proposition 8 supplies an eigenvector $\boldsymbol{v}_1$. In the next step, $V = (\text{Span}\{\boldsymbol{v}_1\})^\perp$; Proposition 8 supplies an eigenvector $\boldsymbol{v}_2$ in V, as we see in the next paragraph. In the jth step, $V = (\text{Span}\{\boldsymbol{v}_1, \ldots, \boldsymbol{v}_{j-1}\})^\perp$; where the unit vectors $\boldsymbol{v}_1, \ldots, \boldsymbol{v}_{j-1}$ have been chosen already and satisfy $A\boldsymbol{v}_i = c_i\boldsymbol{v}_i$.

We need to be sure that Proposition 8 is applicable to $V = (\text{Span}\{\boldsymbol{v}_1, \ldots, \boldsymbol{v}_{j-1}\})^\perp$, for each $j = 2, \ldots, n$. If $\boldsymbol{v}$ is in V, then Proposition 1 (ii) gives

$$\boldsymbol{v}_i \cdot A\boldsymbol{v} = \boldsymbol{v} \cdot A\boldsymbol{v}_i = \boldsymbol{v} \cdot c_i\boldsymbol{v}_i = 0 \quad (i = 1, \ldots, j-1).$$

So $A\boldsymbol{v}$ is orthogonal to $\boldsymbol{v}_1, \ldots, \boldsymbol{v}_{j-1}$, and is a vector in V. This gives the required hypothesis $A : V \to V$. Now Proposition 8 supplies a unit eigenvector $\boldsymbol{v}_j$ in V. By definition of V, $\boldsymbol{v}_j \cdot \boldsymbol{v}_i = 0$ for $i < j$. The process concludes when $j = n$.

Example 14 Let $A = \begin{bmatrix} 4 & 2 & 2 \\ 2 & 4 & 2 \\ 2 & 2 & 4 \end{bmatrix}$. Find an orthonormal basis of $\mathbb{R}^3$ consisting of eigenvectors of A.

Solution. The characteristic polynomial of A is

$$\det \begin{bmatrix} c-4 & -2 & -2 \\ -2 & c-4 & -2 \\ -2 & -2 & c-4 \end{bmatrix} = \det \begin{bmatrix} c-2 & 2-c & 0 \\ -2 & c-4 & -2 \\ 0 & 2-c & c-2 \end{bmatrix} \begin{matrix} \text{I} - \text{II} \\ \text{III} - \text{II} \\ \\ \end{matrix}$$

$$= (c-2)^2 \det \begin{bmatrix} 1 & -1 & 0 \\ -2 & c-4 & -2 \\ 0 & -1 & 1 \end{bmatrix} \text{ (factors from rows 1, 3)}$$

$$= (c-2)^2 \det \begin{bmatrix} 1 & -1 & 0 \\ 0 & c-6 & -2 \\ 0 & -1 & 1 \end{bmatrix} \begin{matrix} \text{II} + 2\text{I} \\ \\ \\ \end{matrix} = (c-2)^2(c-8).$$

Eigenspace for $c = 2$. The coefficient matrix of $(2I - A)\boldsymbol{x} = \boldsymbol{0}$ is

$$\begin{bmatrix} -2 & -2 & -2 \\ -2 & -2 & -2 \\ -2 & -2 & -2 \end{bmatrix} \sim \begin{bmatrix} 1 & 1 & 1 \\ 0 & 0 & 0 \\ 0 & 0 & 0 \end{bmatrix} \begin{matrix} \text{II} - \text{I} \\ \text{III} - \text{I} \\ \text{I} \times -\frac{1}{2}. \end{matrix}$$

A basis of the eigenspace is $\boldsymbol{w}_1 = (-1, 1, 0), \boldsymbol{w}_2 = (-1, 0, 1)$. Let

$$\begin{aligned}\boldsymbol{v}_1 = \boldsymbol{w}_1, \boldsymbol{v}_2 = \boldsymbol{w}_2 - \frac{\boldsymbol{w}_2 \cdot \boldsymbol{v}_1}{\boldsymbol{v}_1 \cdot \boldsymbol{v}_1}\boldsymbol{v}_1 \\ = (-1,0,1) - \frac{1}{2}(-1,1,0) = \left(-\frac{1}{2}, -\frac{1}{2}, 1\right).\end{aligned}$$

This gives an orthogonal basis $(-1,1,0), (-1,-1,2)$ of the eigenspace. Normalize to obtain an orthonormal basis $\boldsymbol{u}_1 = \left(-\frac{1}{\sqrt{2}}, \frac{1}{\sqrt{2}}, 0\right), \boldsymbol{u}_2 = \left(-\frac{1}{\sqrt{6}}, -\frac{1}{\sqrt{6}}, \frac{2}{\sqrt{6}}\right)$.

Eigenspace for $c = 8$. The coefficient matrix of $(8I - A)\boldsymbol{x} = \boldsymbol{0}$ is

$$\begin{bmatrix} 4 & -2 & -2 \\ -2 & 4 & -2 \\ -2 & -2 & 4 \end{bmatrix} \sim \begin{bmatrix} 1 & 1 & -2 \\ 0 & 6 & -6 \\ 0 & -6 & 6 \end{bmatrix} \begin{matrix} \text{I} \leftrightarrow \text{III} \\ \text{II} - \text{I} \\ \text{III} + 2\text{I} \\ \text{I} \times -\frac{1}{2} \end{matrix} \sim \begin{bmatrix} 1 & 0 & -1 \\ 0 & 1 & -1 \\ 0 & 0 & 0 \end{bmatrix} \begin{matrix} \text{III} + \text{II} \\ \text{II} \times \frac{1}{6} \\ \text{I} - \text{II} \end{matrix}.$$

A basis of the eigenspace is $\boldsymbol{u}_3 = \left(\frac{1}{\sqrt{3}}, \frac{1}{\sqrt{3}}, \frac{1}{\sqrt{3}}\right)$, which is orthogonal to $\boldsymbol{u}_1, \boldsymbol{u}_2$, as predicted in Proposition 7.

Proposition 9 tells us that when A is symmetric, there is a diagonalization

$$A = PDP^{-1}$$

where the columns of P are an orthonormal basis of eigenvectors and D is diagonal. Of course P is an orthogonal matrix, so $P^{-1} = P^t$, and

$$A = PDP^t.$$

This equation (with D diagonal, P orthogonal) is an **orthogonal diagonalization** of A. In Example 14, we obtain

$$A = \begin{bmatrix} -\frac{1}{\sqrt{2}} & -\frac{1}{\sqrt{6}} & \frac{1}{\sqrt{3}} \\ \frac{1}{\sqrt{2}} & -\frac{1}{\sqrt{6}} & \frac{1}{\sqrt{3}} \\ 0 & \frac{2}{\sqrt{6}} & \frac{1}{\sqrt{3}} \end{bmatrix} \begin{bmatrix} 2 & & \\ & 2 & \\ & & 8 \end{bmatrix} \begin{bmatrix} -\frac{1}{\sqrt{2}} & \frac{1}{\sqrt{2}} & 0 \\ -\frac{1}{\sqrt{6}} & -\frac{1}{\sqrt{6}} & \frac{1}{\sqrt{6}} \\ \frac{1}{\sqrt{3}} & \frac{1}{\sqrt{3}} & \frac{1}{\sqrt{3}} \end{bmatrix}.$$

Are there any nonsymmetric matrices that have an orthogonal diagonalization? The answer is a resounding no. If $A = PDP^t$ (P orthogonal, D diagonal), then

$$A^t = (PDP^t)^t = (P^t)^t D^t P^t = PDP^t = A,$$

and A is symmetric!

It is interesting that Proposition 9 contains the statement that the characteristic polynomial of a symmetric matrix has n real zeros, counted with multiplicity. If there were less than n real zeros, we would not be able to pick n linearly independent eigenvectors, because of Proposition 5 of Chapter 6. In the next chapter we examine this again, from the point of view of complex numbers.

5 Diagonalizing a quadratic form

Let $Q(\boldsymbol{y})$ be a quadratic form in $\boldsymbol{y} = (y_1, \dots, y_n)$ and let $\boldsymbol{v}_1, \dots, \boldsymbol{v}_n$ be any basis of $\mathbb{R}^n$. Let

$$\boldsymbol{y} = S\boldsymbol{x}$$

be the change of coordinates corresponding to

$$y_1\boldsymbol{e}_1 + \cdots + y_n\boldsymbol{e}_n = x_1\boldsymbol{v}_1 + \cdots + x_n\boldsymbol{v}_n. \tag{14}$$

If we write Q as a function of $\boldsymbol{x}$,

$$Q(\boldsymbol{y}) = (S\boldsymbol{x})^t AS\boldsymbol{x} = \boldsymbol{x}^t(S^tAS)\boldsymbol{x} = Q_1(\boldsymbol{x}), \text{ say.}$$

Of course S^tAS is symmetric, so the change of coordinates gives a quadratic form Q_1 with matrix S^tAS. Note that $Q_1(\boldsymbol{x}) = Q(S\boldsymbol{x})$.

Example 15 Let $\boldsymbol{v}_1 = (1,3), \boldsymbol{v}_2 = (2,5)$,

$$Q(y_1, y_2) = y_1^2 - 3y_1y_2 - y_2^2.$$

Then $S = \begin{bmatrix} 1 & 2 \\ 3 & 5 \end{bmatrix}, A = \begin{bmatrix} 1 & -3/2 \\ -3/2 & -1 \end{bmatrix}$. Thus $Q(\boldsymbol{y}) = Q_1(\boldsymbol{x})$ where Q_1 has matrix

$$\begin{bmatrix} 1 & 3 \\ 2 & 5 \end{bmatrix} \begin{bmatrix} 1 & -3/2 \\ -3/2 & 1 \end{bmatrix} \begin{bmatrix} 1 & 2 \\ 3 & 5 \end{bmatrix} = \begin{bmatrix} -17 & -59/2 \\ -59/2 & -51 \end{bmatrix}.$$

In other words,

$$Q_1(x_1, x_2) = -17x_1^2 - 59x_1x_2 - 51x_2^2.$$

You should check that this expression matches

$$Q(S\boldsymbol{x}) = (x_1 + 2x_2)^2 - 3(x_1 + 2x_2)(3x_1 + 5x_2) - (3x_1 + 5x_2)^2.$$

Naturally it is desirable to *simplify* $Q(\boldsymbol{x})$ by changing coordinates. If $\boldsymbol{y} = S\boldsymbol{x}$ is a change of coordinates, and

$$Q(S\boldsymbol{x}) = c_1x_1^2 + c_2x_2^2 + \cdots + c_nx_n^2 \tag{15}$$

for some real $c_1, \ldots, c_n$, we refer to (15) as a **diagonalization** of Q. The form $\sum_{i=1}^{n} c_ix_i^2$ is a **diagonal quadratic form** . If, in addition, S is an orthogonal matrix, (15) is an **orthogonal diagonalization** of Q.

In matrix terms, (15) is equivalent to

$$S^tAS = D,$$

where A is the matrix of Q and $D = \begin{bmatrix} c_1 & & \\ & \ddots & \\ & & c_n \end{bmatrix}$. So Proposition 9 tells us that any quadratic form has an orthogonal diagonalization. (Note that if $A = PDP^t$ with P orthogonal, then $D = P^tAP$.)

Example 16 Orthogonally diagonalize the quadratic form

$$Q(\boldsymbol{y}) = 2y_1^2 + 2y_2^2 + 5y_3^2 - 2y_1y_2 + 4y_1y_3 - 4y_2y_3.$$

Solution. The matrix of Q is

$$A = \begin{bmatrix} 2 & -1 & 2 \\ -1 & 2 & -2 \\ 2 & -2 & 5 \end{bmatrix}$$

with characteristic polynomial

$$f(c) = \det \begin{bmatrix} c-2 & 1 & -2 \\ 1 & c-2 & 2 \\ -2 & 2 & c-5 \end{bmatrix} = (c-2)(c^2 - 7c + 6) - (c-1)$$
$$- 2(2c - 2)$$

(expanding by row 1)

$$= (c-2)(c-1)(c-6) - 5(c-1)$$
$$= (c-1)(c^2 - 8c + 7) = (c-1)^2(c-7)$$

Eigenspace for $c = 1$. The coefficient matrix of $(I - A)\boldsymbol{x} = \mathbf{0}$ is

$$\begin{bmatrix} -1 & 1 & -2 \\ 1 & -1 & 2 \\ -2 & 2 & -4 \end{bmatrix} \sim \begin{bmatrix} 1 & -1 & 2 \\ 0 & 0 & 0 \\ 0 & 0 & 0 \end{bmatrix} \begin{matrix} \text{II} + \text{I} \\ \text{III} - 2\text{I} \\ \text{I} \times -1 \end{matrix}.$$

A basis of the eigenspace is $\boldsymbol{w}_1 = (1, 1, 0), \boldsymbol{w}_2 = (-2, 0, 1)$. Let $\boldsymbol{v}_1 = \boldsymbol{w}_1, \boldsymbol{v}_2 = \boldsymbol{w}_2 - \frac{\boldsymbol{w}_2 \cdot \boldsymbol{v}_1}{\boldsymbol{v}_1 \cdot \boldsymbol{v}_1} \boldsymbol{v}_1 = (-2, 0, 1) + (1, 1, 0) = (-1, 1, 1)$. An orthonormal basis of the eigenspace is

$$\boldsymbol{u}_1 = \left(\frac{1}{\sqrt{2}}, \frac{1}{\sqrt{2}}, 0 \right), \quad \boldsymbol{u}_2 = \left(-\frac{1}{\sqrt{3}}, \frac{1}{\sqrt{3}}, \frac{1}{\sqrt{3}} \right).$$

Eigenspace for $c = 7$. The coefficient matrix of $(7I - A)\boldsymbol{x} = \mathbf{0}$ is

$$\begin{bmatrix} 5 & 1 & -2 \\ 1 & 5 & 2 \\ -2 & 2 & 2 \end{bmatrix} \sim \begin{bmatrix} 1 & 5 & 2 \\ 0 & -24 & -12 \\ 0 & 12 & 6 \end{bmatrix} \begin{matrix} \text{I} \leftrightarrow \text{II} \\ \text{II} - 5\text{I} \\ \text{III} + 2\text{I} \end{matrix} \sim \begin{bmatrix} 1 & 0 & -\frac{1}{2} \\ 0 & 1 & \frac{1}{2} \\ 0 & 0 & 0 \end{bmatrix} \begin{matrix} \text{III} + \frac{1}{2}\text{II} \\ \text{II} \times -\frac{1}{24} \\ \text{I} - 5\text{II} \end{matrix}.$$

A basis of the eigenspace is $\boldsymbol{u}_3 = \left(\frac{1}{\sqrt{6}}, -\frac{1}{\sqrt{6}}, \frac{2}{\sqrt{6}} \right)$. We conclude that

$$A = PDP^t, \text{ or equivalently } D = P^t AP,$$

where

$$P = \begin{bmatrix} \frac{1}{\sqrt{2}} & -\frac{1}{\sqrt{3}} & \frac{1}{\sqrt{6}} \\ \frac{1}{\sqrt{2}} & \frac{1}{\sqrt{3}} & -\frac{1}{\sqrt{6}} \\ 0 & \frac{1}{\sqrt{3}} & \frac{2}{\sqrt{6}} \end{bmatrix}, \quad D = \begin{bmatrix} 1 & & \\ & 1 & \\ & & 7 \end{bmatrix}.$$

This gives

$$Q(\boldsymbol{y}) = x_1^2 + x_2^2 + 7x_3^2$$

where

$$\boldsymbol{x} = P^{-1}\boldsymbol{y} = P^t\boldsymbol{y} = \left(\frac{y_1 + y_2}{\sqrt{2}}, \frac{-y_1 + y_2 + y_3}{\sqrt{3}}, \frac{y_1 - y_2 + 2y_3}{\sqrt{6}} \right).$$

You should check that $x_1^2 + x_2^2 + 7x_3^2 = 2y_1^2 + 2y_2^2 + 5y_3^2 - 2y_1y_2 + 4y_1y_3 - 4y_2y_3$.

Example 17 Orthogonally diagonalize $Q(y_1, y_2) = 2y_1^2 + 4y_1y_2 + y_2^2$.

Solution. The matrix of Q is

$$A = \begin{bmatrix} -2 & 2 \\ 2 & 1 \end{bmatrix}.$$

We apply Example 11 with $f = -2, h = 1, g = 2, (f-h)^2 + 4g^2 = 25$. The eigenvalues are $c_1 = (-1+\sqrt{25})/2 = 2, c_2 = (-1-\sqrt{25})/2 = -3$, and

$$\boldsymbol{v}_1 = \left(2, \frac{3}{2} + \frac{\sqrt{25}}{2}\right) = (2,4), \; v_2 = \left(2, \frac{3}{2} - \frac{\sqrt{25}}{2}\right) = (2,-1)$$

is an orthogonal basis of eigenvectors. Normalizing yields

$$D = P^t A P$$

with $P = \begin{bmatrix} \frac{1}{\sqrt{5}} & \frac{2}{\sqrt{5}} \\ \frac{2}{\sqrt{5}} & -\frac{1}{\sqrt{5}} \end{bmatrix}$ and $D = \begin{bmatrix} 2 & \\ & -3 \end{bmatrix}$. In terms of quadratic forms,

$$Q(P\boldsymbol{x}) = 2x_1^2 - 3x_2^2.$$

6 Sylvester's law of inertia

This tells us that, no matter how we diagonalize a quadratic form, the number of positive c_i in (15), the number of negative c_i, and the number of zero c_i, will come out the same.

Example 18 In Example 16, let us diagonalize by 'completing the square':

$$\begin{aligned} Q(\boldsymbol{x}) &= 2(y_1^2 - y_1y_2 + 2y_1y_3) + 2y_2^2 + 5y_3^2 - 4y_2y_3 \\ &= 2\left(y_1 - \frac{1}{2}\,y_2 + y_3\right)^2 + \frac{3}{2}\,y_2^2 + 3y_3^2 - 4y_2y_3 \\ &= 2\left(y_1 - \frac{1}{2}\,y_2 + y_3\right)^2 + \frac{3}{2}\left(y_2^2 - \frac{8}{3}\,y_2y_3\right) + 3y_3^2 \\ &= 2\left(y_1 - \frac{1}{2}\,y_2 + y_3\right)^2 + \frac{3}{2}\left(y_2 - \frac{4}{3}\,y_3\right)^2 + \frac{1}{3}\,y_3^2 \\ &= 2z_1^2 + \frac{3}{2}\,z_2^2 + \frac{1}{3}\,z_3^2 \end{aligned}$$

where $z = Uy, U = \begin{bmatrix} 1 & -\frac{1}{2} & 1 \\ 0 & 1 & -\frac{4}{3} \\ 0 & 0 & 1 \end{bmatrix}$ being invertible. If $c_1 = 2, c_2 = \frac{3}{2}, c_3 = \frac{1}{3}$, the number of positive c_i is 3, the number of negative c_i is 0, the number of zero c_i is 0, just as in our previous diagonalization $Q(y) = x_1^2 + x_2^2 + 7x_3^2$.

Let us establish a preliminary result (it appears as an exercise in Chapter 4).

Proposition 10 *Let A, P_1, P_2 by $n \times n$ with P_1, P_2 invertible. Then AP_1 and P_2A have the same rank as A.*

Proof. The equation

$$P_2 A\boldsymbol{x} = \mathbf{0}$$

is equivalent to $A\boldsymbol{x} = \mathbf{0}$ since $T\boldsymbol{y} = P_2\boldsymbol{y}$ is a one-to-one mapping. Hence P_2A has the same nullity as A. By Proposition 10 of Chapter 3, P_2A has the same rank as A.

By Proposition 12 of Chapter 3, rank B = rank B^t for any matrix B. Hence rank AP_1 = rank $P_1^tA^t$ = rank A^t (by what we already proved) = rank A, as required.

Proposition 11 (Sylvester's law of inertia). *Suppose Q is a quadratic form in n variables and*

$$\begin{aligned} &(16) \qquad Q(P_1\boldsymbol{x}) = a_1x_1^2 + \cdots + a_rx_r^2 - (a_{r+1}x_{r+1}^2 - \cdots - a_{r+s}x_{r+s}^2) \\ &\qquad\qquad Q(P_2\boldsymbol{z}) = A_1z_1^2 + \cdots + A_Rz_R^2 - (A_{R+1}z_{R+1}^2 - \cdots - A_{R+S}z_{R+S}^2) \end{aligned}$$

where P_1, P_2 are invertible and the a_i and A_i are positive. Then $r + s$ is the rank of the matrix A of Q, and

$$r = R, \qquad s = S.$$

Proof. First of all,

$$D = \begin{bmatrix} a_1 & & & & & \\ & \ddots & & & & \\ & & a_r & & & \\ & & & -a_{r+1} & & \\ & & & & \ddots & \\ & & & & & -a_{r+s} \\ & & & & & & \\ \end{bmatrix}$$

has the same rank as A by Proposition 10 ($D = P_1^t A P_1$). So

$$r + s = \text{rank } A$$

and similarly $R + S = \text{rank } A$. Now

$$r + s = R + S. \tag{17}$$

Suppose that $R < r$. We can choose $\boldsymbol{y} \neq \boldsymbol{0}$ in $\mathbb{R}^n$ such that, when $\boldsymbol{x}$ and $\boldsymbol{z}$ are defined via $\boldsymbol{y} = P_1\boldsymbol{x} = P_2\boldsymbol{z}$, we have $x_{r+1} = \cdots = x_{r+s} = 0$ and $z_1 = \cdots = z_R = z_{R+S+1} = \cdots = z_n = 0$. For the total number of equations is $n - S + s = n - (r + s - R) + s = n + R - r$, which is less than n, and of course these are linear equations in $y_1, \ldots, y_n$. Now $Q(\boldsymbol{y}) = a_1x_1^2 + \cdots + a_rx_r^2 > 0$ ($x_1, \ldots, x_r$ cannot all be zero) and similarly $Q(\boldsymbol{y}) = -(A_{R+1}z_{R+1}^2 + \cdots + A_{R+S}z_{R+S}^2) < 0$. This is not possible, so $R \geq r$. Similarly, $r \geq R$. This gives $r = R$, and now $s = S$ from (17).

Forms $Q(\boldsymbol{y})$ which have $r = n$ in (16) are **positive definite** quadratic forms. Since this requires in particular an orthogonal diagonalization

$$Q(P\boldsymbol{x}) = c_1x_1^2 + \cdots + c_nx_n^2 \tag{18}$$

with all $c_i > 0$, we find that the positive definite forms are those whose matrices have all eigenvalues positive. If, on the other hand, $s = n$ in (16), Q is **negative definite** and this is equivalent to all eigenvalues of the matrix of Q being negative. Forms which are neither positive definite nor negative definite are **indefinite** . We say that a symmetric matrix A is **positive definite** if all eigenvalues of A are positive. **Negative definite** and **indefinite** matrices are defined similarly.

Another way of stating that Q is positive definite is:

$$Q(\boldsymbol{y}) > 0 \quad \text{for } \boldsymbol{y} \neq \mathbf{0}.$$

This is obviously a consequence of (18) with positive $c_1, \ldots, c_n$, but could *not* be true for a set of $c_1, \ldots, c_n$ that includes a nonpositive number.

7 Quadric curves

Let $\boldsymbol{y} = P\boldsymbol{x}$ be a change of coordinates in $\mathbb{R}^2$ with $P = [\boldsymbol{u}_1 \ \boldsymbol{u}_2]$ orthogonal. We write $\boldsymbol{u}_2 > \boldsymbol{u}_1$ if $\boldsymbol{u}_2$ is obtained from $\boldsymbol{u}_1$ by anticlockwise rotation through $\pi/2$. If $\boldsymbol{u}_2 > \boldsymbol{u}_1$ does not hold, we replace $\boldsymbol{u}_2$ by $-\boldsymbol{u}_2$ and relabel $-\boldsymbol{u}_2$ as $\boldsymbol{u}_2$; this gives a new pair $\boldsymbol{u}_1, \boldsymbol{u}_2$ with $\boldsymbol{u}_2 > \boldsymbol{u}_1$. In the equation

$$y_1\boldsymbol{e}_1 + y_2\boldsymbol{e}_2 = x_1\boldsymbol{u}_1 + x_2\boldsymbol{u}_2,$$

$\boldsymbol{x}$ gives the coordinates with respect to axes $\boldsymbol{u}_1, \boldsymbol{u}_2$ obtained by *rotating* the pair $\boldsymbol{e}_1, \boldsymbol{e}_2$.

Example 19 Let $\boldsymbol{u}_1 = \left(-\frac{1}{\sqrt{10}}, -\frac{3}{\sqrt{10}}\right), \boldsymbol{u}_2 = \left(-\frac{3}{\sqrt{10}}, \frac{1}{\sqrt{10}}\right)$. Since $\boldsymbol{u}_1, \boldsymbol{u}_2$ are in the third and second quadrants respectively, we replace $\boldsymbol{u}_2$ by $-\boldsymbol{u}_2$. Thus if we now write $\boldsymbol{u}_1 = \left(-\frac{1}{\sqrt{10}}, -\frac{3}{\sqrt{10}}\right), \boldsymbol{u}_2 = \left(\frac{3}{\sqrt{10}}, -\frac{1}{\sqrt{10}}\right)$ we have $\boldsymbol{u}_2 > \boldsymbol{u}_1$.

A **quadric curve** is a curve in $\mathbb{R}^2$ (not the empty set) with equation

$$Q(\boldsymbol{y}) = 1, \tag{19}$$

where Q is a quadratic form of rank 2. We may orthogonally diagonalize Q:

$$Q(\boldsymbol{y}) = c_1x_1^2 + c_2x_2^2\ , \ y_1\boldsymbol{e}_1 + y_2\boldsymbol{e}_2 = x_1\boldsymbol{u}_1 + x_2\boldsymbol{u}_2. \tag{20}$$

Here $c_1c_2 \neq 0$ and at least one of c_1, c_2 is positive. If c_1, c_2 are both positive the curve (19) is an **ellipse** (Figure 3; see also Exercise 1.2 of Chapter 1). If $c_1c_2 < 0$, the curve (19) is a **hyperbola** (Figure 2). The most obvious difference between these quadric curves is that only the hyperbola contains points tending to infinity.

To sketch a quadric curve, find the orthogonal diagonalization (20), with $\boldsymbol{u}_2 > \boldsymbol{u}_1$. Draw in the lines through $\mathbf{0}$ and $\boldsymbol{u}_1, \boldsymbol{u}_2$. Our sketch is now done **with respect to the x_1, x_2 axes**, about which the curve is symmetric, so we only need insert some points in quadrant 1. In the case of an ellipse, these include $\left(\frac{1}{\sqrt{c_1}}, 0\right)$ and $\left(0, \frac{1}{\sqrt{c_2}}\right)$ (the **semiaxes**) .

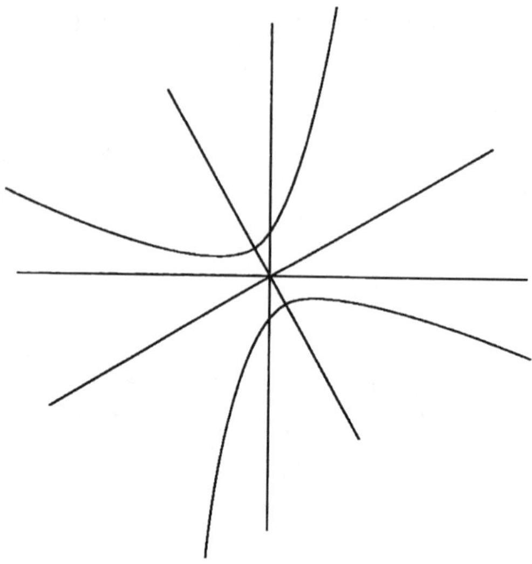

Figure 2. The hyperbola $-2y_1^2 + 4y_1y_2 + y_2^2 = 1$.

Example 20 To sketch the curve

$$-2y_1^2 + 4y_1y_2 + y_2^2 = 1$$

we refer to Example 17. We found $\boldsymbol{u}_1 = \left(\frac{1}{\sqrt{5}}, \frac{2}{\sqrt{5}}\right)$ and $\boldsymbol{u}_2 = \left(-\frac{2}{\sqrt{5}}, \frac{1}{\sqrt{5}}\right)$ as an orthonormal pair of eigenvectors with $\boldsymbol{u}_2 > \boldsymbol{u}_1$. The curve is

$$2x_1^2 - 3x_2^2 = 1$$

which is a hyperbola (Figure 2). It can readily be sketched after graphing a few values of (x_1, x_2) such as $\left(\frac{1}{\sqrt{2}}, 0\right), (\sqrt{2}, 1), (\sqrt{13/2}, 2)$. For large positive x_1, x_2, the curve is very close to the line

$$\sqrt{2}x_1 = \sqrt{3}x_2$$

since

$$\sqrt{2}x_1 - \sqrt{3}x_2 = \frac{1}{\sqrt{2}x_1 + \sqrt{3}x_2}.$$

This straight line is an **asymptote** of the hyperbola (so is the line $\sqrt{2}x_1 = -\sqrt{3}x_2$). As a final touch for a correct sketch, we note that the curve cuts the x_1 axis at right angles; this can be checked via a simple calculation with derivatives.

Example 21 To sketch the curve

$$y_1^2 + 2y_1y_2 + 3y_2^2 = 1$$

we apply Example 11 with $f = g = 1, h = 3, (f - h)^2 + 4g^2 = 8$. The eigenvalues are $c_1 = \frac{4+2\sqrt{2}}{2} = 2 + \sqrt{2}, c_2 = 2 - \sqrt{2}$. We obtain eigenvectors $\boldsymbol{v}_1 = \left(1, \frac{2}{2} + \frac{2\sqrt{2}}{2}\right) = (1, 1 + \sqrt{2})$ and $\boldsymbol{v}_2 = (1, 1 - \sqrt{2})$. Let $\boldsymbol{u}_1 = \frac{\boldsymbol{v}_1}{|\boldsymbol{v}_1|}$ and $\boldsymbol{u}_2 = -\frac{\boldsymbol{v}_2}{|\boldsymbol{v}_2|}$; then $\boldsymbol{u}_2 > \boldsymbol{u}_1$. (It is reasonable to use approximations to two decimal places in drawing $\boldsymbol{u}_1, \boldsymbol{u}_2$.) Since the eigenvalues are positive, the curve is an ellipse; it can be sketched by working similarly to Example 20 (Figure 3).

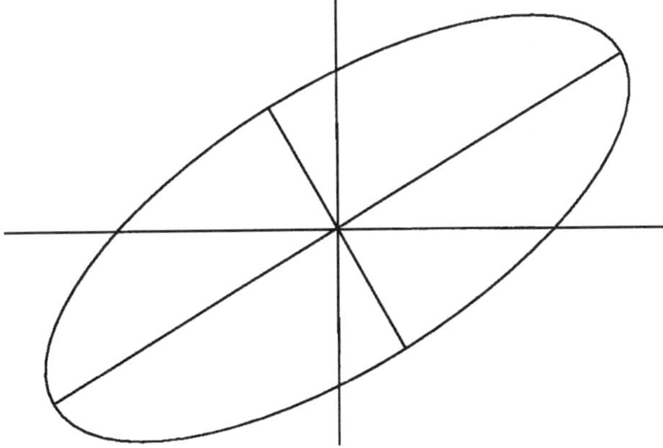

Figure 3. The ellipse $y_1^2 + 2y_1y_2 + 3y_2^2 = 1$.

Exercises for Chapter 7

Reminder: attempt Exercise a.b after reading Section a.

For each of the following matrices, write down the quadratic form whose matrix is A.

1.1 $A = \begin{bmatrix} 2 & -9 \\ -9 & 7 \end{bmatrix}$

1.2 $A = \begin{bmatrix} 3 & 5 & -2 \\ 5 & -4 & -3 \\ -2 & -3 & -1 \end{bmatrix}$

1.3 $A = \begin{bmatrix} 2 & 3 & 0 & 1 \\ 3 & 4 & -2 & 1 \\ 0 & -2 & 6 & 0 \\ 1 & 1 & 0 & -5 \end{bmatrix}$

For each of the following quadratic forms Q, write down the matrix of Q.

1.4 $Q(\boldsymbol{x}) = x_1^2 + 17x_1x_2 - 3x_2^2$

1.5 $Q(\boldsymbol{x}) = x_1x_2 + 2x_1x_3 + 5x_2x_3$

1.6 $Q(\boldsymbol{x}) = x_1^2 + x_2^2 + 2x_3^2 - x_4^2 + x_1x_2 + 3x_1x_3 + 2x_1x_4 + 4x_2x_4 + 6x_3x_4$

1.7 If $B = AP$, where A is symmetric and P is orthogonal, show that $A^2 = BB^t$

For each of the following linearly independent sets, find an orthogonal basis of the linear span V of the set. Aim for the simplest looking basis.

2.1 $(1, 2, 3), (-1, 2, 3)$

2.2 $(0, -1, 2, 1), (1, 3, -1, 1), (1, 0, 2, -2)$

2.3 $(1, 0, 1, -1, 0), (0, 1, 2, 0, 1), (0, 0, -1, -1, 1)$

2.4 Find the orthogonal projection of $(1, 4, -1, 0)$ onto the subspace Span$\{(1, 0, 1, 1), (1, 1, 1, 1)\}$.

2.5 Find the orthogonal projection e_1' of e_1 onto the subspace V in Exercise 2.3. You should check that $e_1 - e_1'$ is orthogonal to a basis of V.

3.1 Let Tv denote the projection of v into a given subspace V of $\mathbb{R}^n$, $\dim V = k$. Show that T is a linear mapping and find Im T. Verify that rank $T = k$, nullity $T = n - k$.

3.2 Find the orthogonal projection of z into the hyperplane

$$a_1x_1 + a_2x_2 + \cdots + a_nx_n = 0.$$

Hint: Let V be the hyperplane. Find an orthogonal basis of $V^\perp$, rather than V. The answer is a combination of z, a.

For each of the following symmetric matrices A, find an orthogonal matrix P such that P^tAP is diagonal.

4.1 $A = \begin{bmatrix} 4 & 1 \\ 1 & 4 \end{bmatrix}$

4.2 $A = \begin{bmatrix} 12 & 5 \\ 5 & -12 \end{bmatrix}$

4.3 $A = \begin{bmatrix} 3 & 0 & 0 \\ 0 & 0 & \frac{5}{2} \\ 0 & \frac{5}{2} & 12 \end{bmatrix}$

4.4 $A - \begin{bmatrix} 0 & 3 & 3 \\ 3 & 0 & 17 \\ 3 & 17 & 0 \end{bmatrix}$

4.5 Let V be an eigenspace of A. Show that $A : V \to V$.

4.6 Let $A : V \to V$ and let $B = PAP^{-1}$ (P invertible). Show that $B : W \to W$ where W is the set of Pv, v in V.

4.7 Let

$$A = \left[\begin{array}{c|c} B & 0 \\ \hline 0 & D \end{array}\right],$$

where B is $m \times m$ and D is $n \times n$. Show that there are subspaces V_1 and V_2, $V_1 \cap V_2 = \mathbf{0}$, such that $\mathbb{R}^{m+n} = V_1 + V_2$ and $A : V_1 \to V_1, A : V_2 \to V_2$.

For each of the following quadratic forms Q, give a change of coordinates $\boldsymbol{x} = P^t\boldsymbol{y}$ (P orthogonal) that diagonalizes Q. Write Q in the form $Q = \sum_{i=1}^{n} c_i x_i^2$ and verify this equation when $\boldsymbol{x} = P^t\boldsymbol{y}$.

5.1 $Q(\boldsymbol{x}) = 7y_1^2 + 7y_2^2 + 4y_3^2 + 4y_1y_3$

5.2 $Q(\boldsymbol{x}) = 2y_1^2 + y_2^2 + 2y_3^2 + 4y_1y_2 + 2y_1y_3 + 4y_2y_3$

For each of the following quadratic forms decide whether the form is positive definite, negative definite, or indefinite.

6.1 $Q(\boldsymbol{x}) = 2x_1^2 + 2x_2^2 + 2x_3^2 - 2x_1x_2 - 2x_1x_3 - 2x_2x_3$

6.2 $Q(\boldsymbol{x}) = 4x_1^2 + 3x_2^2 + 5x_3^2 + 4x_1x_2 + 4x_1x_3$

6.3 Let A be a positive definite $n \times n$ matrix. Let A_m be the matrix formed from the first m columns of the first m rows of A. Show that $\det A_m > 0$, for $m = 1, \ldots, n$.

6.4 Let $Q(\boldsymbol{x}) = ax_1^2 + 2bx_1x_2 + cx_2^2$. If $a > 0$ and $ac - b^2 > 0$, show that Q is positive definite. (Hint: complete the square.) How does this relate to Exercise 6.3?

For each of the following quadratic forms $Q(\boldsymbol{y})$, give a rotation of coordinates that reduces the curve $Q(\boldsymbol{y}) = 1$ to the form

$$c_1x_1^2 + c_2x_2^2 = 1.$$

Determine whether the curve is an ellipse or a hyperbola and draw it with reasonable accuracy.

7.1 $Q(\boldsymbol{y}) = 6y_1^2 + 3y_2^2 + 4y_1y_2$

7.2 $Q(\boldsymbol{y}) = 9y_1^2 + 2y_2^2 + 24y_1y_2$

Chapter 8

Vector Spaces

1 Complex numbers

We can enlarge the real number system $\mathbb{R}$ in order to be able to solve any polynomial equation

$$x^n + b_1 x^{n-1} + \cdots + b_{n-1}x + b_n = 0. \tag{1}$$

The simplest equation that cannot be solved in $\mathbb{R}$ is

$$x^2 + 1 = 0. \tag{2}$$

If we introduce a new number i which is a solution of (2), and define $\mathbb{C}$, the **complex numbers**, to consist of all

$$a + bi$$

with a and b real, the outcome is extremely satisfactory. Not only does $\mathbb{C}$ obey all the algebraic laws we are accustomed to use in $\mathbb{R}$, but an arbitrary polynomial can be factored:

$$x^n + b_1 x^{n-1} + \cdots + b_{n-1}x + b_n = (x - c_1)(x - c_2)\ldots(x - c_n)$$

where $c_1, \ldots, c_n$ are complex numbers. Here the b_j can be taken to be any **complex** numbers. (In a sense, we have accomplished more than the original purpose of enlarging $\mathbb{R}$.) This is such an enormous advantage that complex numbers are used in every branch of mathematics.

In this section I give an introduction to complex numbers. Define $\mathbb{C}$, the set of complex numbers, to be the Euclidean plane $\mathbb{R}^2$ with an additional operation of multiplication of any $\boldsymbol{a}$ and $\boldsymbol{b}$,

$$\boldsymbol{ab} = (a_1b_1 - a_2b_2, a_1b_2 + a_2b_1). \tag{3}$$

The product of $\boldsymbol{a}$ and $\boldsymbol{b}$ is another vector in $\mathbb{R}^2$. Obviously we cannot write the product as $\boldsymbol{a} \cdot \boldsymbol{b}$, which has been reserved for inner product.

Complex numbers can be written more neatly by defining i as

$$i = (0, 1).$$

We also agree to identify the real number y with the vector $(y, 0)$. Now (a_1, a_2) can be written $a_1 + ia_2$, and (3) reads

$$(a_1 + ia_2)(b_1 + ib_2) = a_1b_1 - a_2b_2 + i(a_1b_2 + a_2b_1).$$

In particular, i satisfies (2):

$$i^2 = 0 - 1 + i(0 + 0) = -1.$$

Another useful fact to note is that $0(b_1 + ib_2) = 0$. We may write a for $a_1 + ia_2, b$ for $b_1 + ib_2$, and so on.

Proposition 1 *For any a, b, c, in $\mathbb{C}$,*

1° *$a + b$ and ab are in $\mathbb{C}$;*

2° *$a + b = b + a$, and $ab = ba$;*

3° *$a + (b + c) = (a + b) + c$ and $a(bc) = (ab)c$;*

4° *$a(b + c) = ab + ac$;*

5° *There is a number 0 in $\mathbb{C}$ such that $a + 0 = a$ for all a;*

6° *There is a number 1 in $\mathbb{C}$ such that $a1 = a$ for all a;*

7° *For each a in $\mathbb{C}$ there is $-a$ in $\mathbb{C}$ such that $a + (-a) = 0$;*

8° *For each a in $\mathbb{C}, a \neq 0$, there is a^{-1} in $\mathbb{C}$ such that $aa^{-1} = 1$.*

Notice that $1° - 8°$ are exactly the same as the algebraic rules of $\mathbb{R}$. We have, however, no ordering of $\mathbb{C}$ that resembles the relation $a < b$ in $\mathbb{R}$.

Proposition 1 gives the following consequence. If $ab = 0$, then $a = 0$ or $b = 0$. For if $a \neq 0$, then $0 = a^{-1}0 = a^{-1}ab = 1b = b$. Here we used $8°, 2°, 3°, 6°$.

Proof of Proposition 1. Rules which do not involve multiplication, such as 7°, have already been covered (Chapter 1).

1° It is clear from (3) that ab is in $\mathbb{C}$ whenever a and b are in $\mathbb{C}$.

2° $$\begin{aligned} ba &= b_1a_1 - b_2a_2 + i(b_1a_2 + b_2a_1) \\ &= a_1b_1 - a_2b_2 + i(a_2b_1 + a_1b_2) = ab. \end{aligned}$$

3° $$\begin{aligned} a(bc) &= (a_1 + ia_2)(b_1c_1 - b_2c_2 + i(b_1c_2 + b_2c_1)) \\ &= a_1b_1c_1 - a_1b_2c_2 - a_2b_1c_2 - a_2b_2c_1 + i(a_1b_1c_2 + a_1b_2c_1 + a_2b_1c_1 - a_2b_2c_2); \\ (ab)c &= (a_1b_1 - a_2b_2 + i(a_1b_2 + a_2b_1))(c_1 + ic_2) \\ &= a_1b_1c_1 - a_2b_2c_1 - a_1b_2c_2 - a_2b_1c_2 + i(a_1b_1c_2 - a_2b_2c_2 + a_1b_2c_1 + a_2b_1c_1) \\ &= a(bc). \end{aligned}$$

4° $$\begin{aligned} a(b+c) &= (a_1 + ia_2)(b_1 + c_1 + i(b_2 + c_2)) \\ &= a_1(b_1 + c_1) - a_2(b_2 + c_2) + i(a_1(b_2 + c_2) + a_2(b_1 + c_1)) \\ &= a_1b_1 - a_2b_2 + i(a_1b_2 + a_2b_1) + a_1c_1 - a_2c_2 + i(a_1c_2 + a_2c_1) = ab + ac. \end{aligned}$$

6° Taking the usual 1 from $\mathbb{R}$,

$$(a_1 + ia_2)1 = (a_1 + ia_2)(1 + i0) = a_11 - 0a_2 + i(a_10 + a_21) = a_1 + ia_2.$$

8° Note first that for $a \neq 0, a_1^2 + a_2^2 \neq 0$. The number

$$a^{-1} = \frac{a_1}{a_1^2 + a_2^2} - \frac{ia_2}{a_1^2 + a_2^2}$$

has the property

$$\begin{aligned} aa^{-1} = (a_1 + ia_2)\left(\frac{a_1}{a_1^2 + a_2^2} - \frac{ia_2}{a_1^2 + a_2^2}\right) &= \frac{a_1^2 + a_2^2}{a_1^2 + a_2^2} + i\left(\frac{-a_1a_2 + a_2a_1}{a_1^2 + a_2^2}\right) \\ &= 1. \end{aligned}$$

As already noted, any polynomial $x^n + b_1x^{n-1} + \cdots + b_{n-1}x + b_n$ with complex $b_1, \ldots, b_n$ can be factored as

(4) $x^n + b_1x^{n-1} + \cdots + b_{n-1}x + b_n = (x - c_1)\ldots(x - c_n)$ $(c_1, \ldots, c_n$ in $\mathbb{C})$.

This is the **fundamental theorem of algebra** ; see Further Reading. The case $n = 2, b_1$ and b_2 real, is easy:

$$\begin{aligned} x^2 + b_1x + b_2 &= \left(x + \frac{b_1}{2}\right)^2 + b_2 - \frac{b_1^2}{4} \\ &= \left(x + \frac{b_1}{2}\right)^2 - a^2 \end{aligned}$$

where $a = \left(-b_2 + \frac{b_1^2}{4}\right)^{1/2}$ if $-b_2 + \frac{b_1^2}{4} \geq 0; a = i\left(-\frac{b_1^2}{4} + b_2\right)^{1/2}$ if $-b_2 + \frac{b_1^2}{4} < 0$. Now

(5) $$x^2 + b_1x + b_2 = \left(x + \frac{b_1}{2} + a\right)\left(x + \frac{b_1}{2} - a\right)$$

is the desired factorization.

It follows from (4) that the equation (1) has at most n solutions, since the solutions are the distinct members of $c_1, c_2, \ldots, c_n$.

Example 1 Solve the quadratic equation

(6) $$x^2 + 3x + 4 = 0.$$

Solution. Taking $b_1 = 3, b_2 = 4, -b_2 + \frac{b_1^2}{4} = -\frac{7}{4}$ in (5),

$$x^2 + 3x + 4 = \left(x + \frac{3}{2} + \frac{i\sqrt{7}}{2}\right)\left(x + \frac{3}{2} - \frac{i\sqrt{7}}{2}\right).$$

Now the solutions of (6) are $x = -\frac{3}{2} \pm \frac{i\sqrt{7}}{2}$.

Definition 1 Let $a = a_1 + ia_2$. The **conjugate** of a is the number $\bar{a} = a_1 - ia_2$ (Figure 1).

Obviously the conjugate of $\bar{a}$ is a. Note the formula

$$a\bar{a} = |a|^2$$

which is an immediate consequence of $(a_1+ia_2)(a_1-ia_2) = a_1^2-i^2a_2^2 = a_1^2+a_2^2$. The length $|a|$ of a is usually called the **modulus** of a.

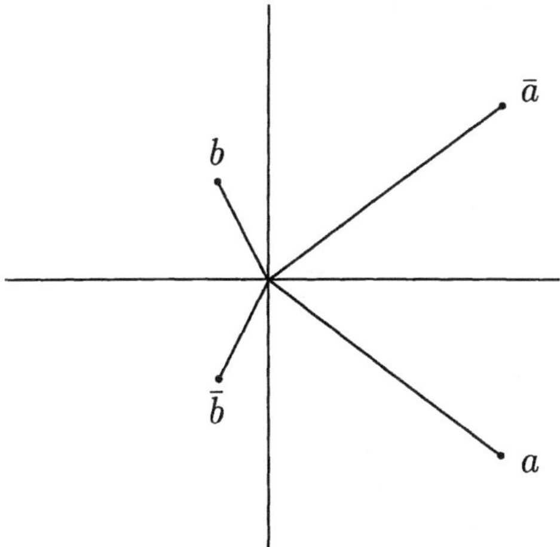

Figure 1. Conjugate of a complex number.

Proposition 2 *Let a and b be in* $\mathbb{C}$. *Then*

$$\overline{a+b} = \bar{a} + \bar{b},$$
$$\overline{ab} = \bar{a}\bar{b}$$

and

$$|ab| = |a|\ |b|.$$

Proof. We have

$$\begin{aligned}\overline{a+b} &= \overline{a_1 + b_1 + i(a_2 + b_2)} \\ &= a_1 + b_1 - i(a_2 + b_2) = a_1 - ia_2 + b_1 - ib_2 = \bar{a} + \bar{b}; \\ \overline{ab} &= \overline{a_1 b_1 - a_2 b_2 + i(a_1 b_2 + a_2 b_1)} \\ &= a_1 b_1 - a_2 b_2 - i(a_1 b_2 + a_2 b_1) = (a_1 - ia_2)(b_1 - ib_2) = \bar{a}\bar{b}.\end{aligned}$$

The last result of the proposition now follows since

$$|ab|^2 = ab\overline{ab} = ab\bar{a}\bar{b} = a\bar{a}b\bar{b} = |a|^2|b|^2.$$

One notable consequence of Proposition 2 is that the non-real solutions of an equation with real coefficients $b_1, \ldots, b_n$,

$$x^n + b_1 x^{n-1} + \cdots + b_{n-1}x + b_n = 0, \tag{7}$$

occur **in conjugate pairs**. If a is a non-real solution of (7), that is, $a = a_1 + ia_2$ with $a_2 \neq 0$, then

$$a^n + b_1 a^{n-1} + \cdots + b_{n-1} a + b_n = 0.$$

The conjugate of 0 is 0; so Proposition 2, applied repeatedly, gives

$$(8) \quad 0 = \overline{a^n + b_1 a^{n-1} + \cdots + b_{n-1} a + b_n} = \bar{a}^n + b_1 \bar{a}^{n-1} + \cdots + b_{n-1} \bar{a} + b_n.$$

Here we use the fact that b_j is real, and so $\bar{b}_j = b_j$. Now (8) asserts that $\bar{a}$ is also a solution of (7).

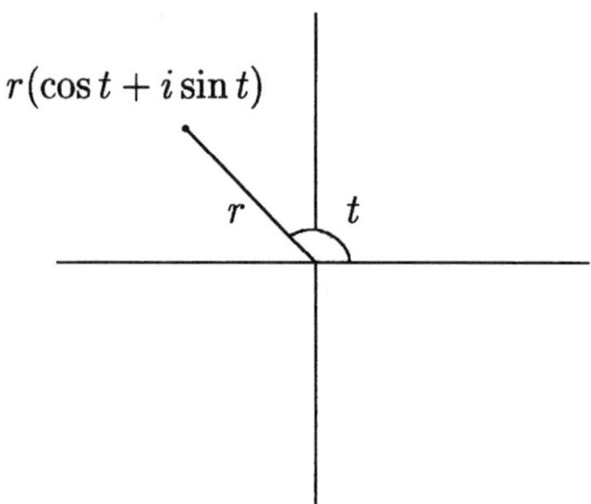

Figure 2. Polar form of a complex number.

Example 2 Find the solutions of

$$x^4 + 2x^2 + 9 = 0.$$

Solution. We have

$$\begin{aligned} x^4 + 2x^2 + 9 &= x^4 + 6x^2 + 9 - 4x^2 \\ &= (x^2 + 3)^2 - 4x^2 \\ &= (x^2 + 3 + 2x)(x^2 + 3 - 2x) \\ &= (x + 1 + i\sqrt{2})(x + 1 - i\sqrt{2})(x - 1 + i\sqrt{2})(x - 1 - i\sqrt{2}). \end{aligned}$$

The solutions are $-1 \pm i\sqrt{2}, 1 \pm i\sqrt{2}$; we have arranged them in conjugate pairs.

The geometric properties of multiplication of complex numbers are attractive and useful. Given a in $\mathbb{C}, a \neq 0$, the normalized complex number

$a/|a|$ has modulus 1 (Chapter 1). Thus

(9) $$\frac{a}{|a|} = \cos t + i \sin t$$

where t is any angle through which 1 is rotated anticlockwise to obtain a. For instance, if $a = 1 - i$, we could take $t = -\pi/4$ or $t = 7\pi/4$. For any a, there is one choice of t with $0 \leq t < 2\pi$, which we call the **argument** of a, written Arg a.

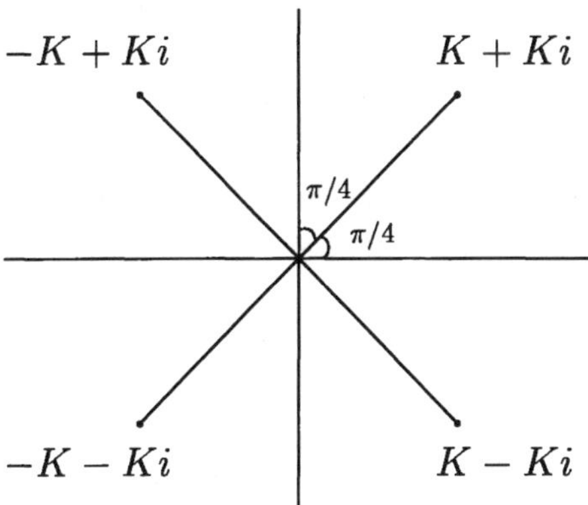

Figure 3. The numbers $K \pm Ki, -K \pm Ki$.

Example 3 Show that Arg $i = \pi/2$.

Solution. $\frac{i}{|i|} = i = \cos\frac{\pi}{2} + i \sin\frac{\pi}{2}$ since $\cos\frac{\pi}{2} = 0, \sin\frac{\pi}{2} = 1$. Since $\frac{\pi}{2}$ is in the required interval $0 \leq t < 2\pi$, the result follows.

We rewrite (9) as

(10) $$a = |a|(\cos t + i \sin t), \text{ with } t = \text{Arg } a.$$

The right side of (10) is the **polar form** of a (Figure 2).

Example 4 Let $K > 0$. Write the numbers $K+Ki, K-Ki, -K+Ki, -K-Ki$ in polar form.

Solution. It is clear that the arguments of these four numbers are $\pi/4, 7\pi/4, 3\pi/4$ and $5\pi/4$ respectively (Figure 3). All four numbers have modulus $\sqrt{K^2 + K^2} = \sqrt{2}K$. Hence the polar forms are

$$K + Ki = \sqrt{2}K\left(\cos\frac{\pi}{4} + i \sin\frac{\pi}{4}\right), \qquad K - Ki = \sqrt{2}K\left(\cos\frac{7\pi}{4} + i \sin\frac{7\pi}{4}\right),$$

$$-K + Ki = \sqrt{2}K\left(\cos\frac{3\pi}{4} + i \sin\frac{3\pi}{4}\right), \quad -K - Ki = \sqrt{2}K\left(\cos\frac{5\pi}{4} + i \sin\frac{5\pi}{4}\right).$$

The algebra of the matrices

$$\begin{bmatrix} a_1 & -a_2 \\ a_2 & a_1 \end{bmatrix}$$

(a_1, a_2 real) mimics that of $\mathbb{C}$. To be precise, define T on $\mathbb{C}$ by

$$T(a_1 + ia_2) = \begin{bmatrix} a_1 & -a_2 \\ a_2 & a_1 \end{bmatrix}.$$

It is obvious that T is one-to-one and that $T(a+b) = Ta + Tb$. Further,

$$TaTb = \begin{bmatrix} a_1 & -a_2 \\ a_2 & a_1 \end{bmatrix} \begin{bmatrix} b_1 & -b_2 \\ b_2 & b_1 \end{bmatrix} = \begin{bmatrix} a_1b_1 - a_2b_2 & -a_1b_2 - a_2b_1 \\ a_1b_2 + a_2b_1 & a_1b_1 - a_2b_2 \end{bmatrix} = T(ab).$$

Since the product of rotations through t and u is a rotation through $t+u$, we know that

(11) $$\begin{bmatrix} \cos t & -\sin t \\ \sin t & \cos t \end{bmatrix} \begin{bmatrix} \cos u & -\sin u \\ \sin u & \cos u \end{bmatrix} = \begin{bmatrix} \cos(t+u) & -\sin(t+u) \\ \sin(t+u) & \cos(t+u) \end{bmatrix}.$$

We now employ **Euler's notation** for $\cos t + i \sin t$,

$$e^{it} = \cos t + i \sin t.$$

The left side of (11) is $T(e^{it})T(e^{iu})$, which is $T(e^{it}e^{iu})$. The right side is $T(e^{i(t+u)})$. Since T is one-to-one,

(12) $$e^{it}e^{iu} = e^{i(t+u)}.$$

Equation (12), and its consequence

(13) $$(e^{it})^n = e^{int} \quad (n = 1, 2, \dots),$$

are both referred to as **De Moivre's theorem** .

If we write two complex numbers a and b in polar form,

$$a = |a|e^{it} \quad , \; b = |b|e^{iu},$$

it follows from (12) that

$$ab = |a|\;|b|e^{i(t+u)}.$$

Since $|a|\,|b| = |ab|$, this is the polar form of ab (except that $t+u$ might exceed 2π). So, to multiply a and b we 'multiply moduli and add arguments'.

It is useful to note that

$$e^{i2\pi k} = 1$$

for any integer k, since $\cos 2\pi k = \sin 2\pi k = 1$. This can be used to simplify products. For example

$$(2e^{i\pi/6})^{27} = 2^{27}e^{i27\pi/6} = 2^{27}e^{i27\pi/6}e^{-i4\pi} = 2^{27}e^{i\pi/2} = 2^{27}i.$$

Let m be a positive integer. A polynomial equation which we can now solve completely is

$$z^m - 1 = 0.$$

There are m distinct solutions, since

$$z = e^{i2\pi k/m} \quad (k = 0, 1, \ldots, m-1) \tag{14}$$

satisfies $z^m = e^{2\pi ik} = 1$. The polynomial $x^m - 1$ has factors $x - 1, x - e^{2\pi i/m}, \ldots, x - e^{2\pi i(m-1)/m}$, and in fact we must have

$$x^m - 1 = (x-1)(x - e^{2\pi i/m}) \cdots (x - e^{2\pi i(m-1)/m}) \tag{15}$$

as the coefficient of x^m is 1 on both sides of (15).

The solutions (14) of $z^m = 1$, which we refer to as **mth roots of unity**, are equally spaced around the circle

$$|z| = 1$$

(Figure 4).

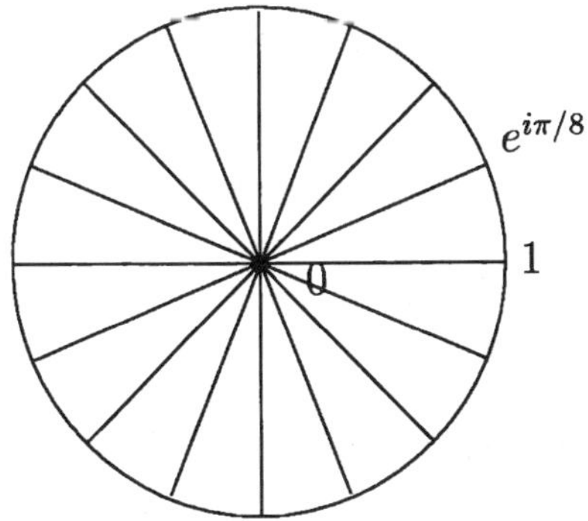

Figure 4. The ends of the spokes satisfy $z^{16} = 1$.

Note that $\overline{e^{it}} = e^{-it}$ since $\cos(-t) = \cos t, \sin(-t) = -\sin t$. The equation

$$e^{it}e^{-it} = 1$$

is both a special case of (12) and a special case of the equation $z\bar{z} = |z|^2$.

Example 5 Express $\cos 5t$ as a polynomial in $\cos t$.

Solution. We use the abbreviations

$$\text{Re}(a + ib) = a, \ \text{Im}(a + ib) = b.$$

(Re stands for **real part** , Im for **imaginary part** .) From (13),

$$\begin{aligned}\cos 5t + i \sin 5t = (\cos t + i \sin t)^5 = \cos^5 t + 5i \cos^4 t \sin t - 10 \cos^3 t \sin^2 t \\ - 10i \cos^2 t \sin^3 t + 5 \cos t \sin^4 t + i \sin^5 t.\end{aligned}$$

Equating real parts,

$$\begin{aligned}\cos 5t &= \text{Re}(\cos 5t + i \sin 5t) = \cos^5 t - 10 \cos^3 t \sin^2 t + 5 \cos t \sin^4 t \\ &= \cos^5 t - 10 \cos^3 t(1 - \cos^2 t) + 5 \cos t(1 - \cos^2 t)^2 \\ &= 16 \cos^5 t - 20 \cos^3 t + 5 \cos t.\end{aligned}$$

Example 6 Solve the equation

$$z^6 = 1 + i.$$

Solution. Write the equation in polar form,

$$z^6 = 2^{1/2} e^{i\pi/4}.$$

One solution is

$$z_0 = 2^{1/12} e^{i\pi/24}.$$

We obtain six solutions (which must be all the solutions) by multiplying z_0 by $e^{2\pi ik/6} (k = 0, 1, \ldots, 5)$, since

$$(z_0 e^{2\pi ik/6})^6 = (2^{1/12} e^{i\pi/24})^6 (e^{2\pi ik/6})^6 = 1 + i.$$

The solutions are $2^{1/12} e^{i(\pi/24 + k\pi/3)}, k = 0, 1, \ldots, 5$. A similar procedure applies to any equation $z^m = c$ with $c \neq 0$. There are m solutions equally spaced around a circle of center 0 and radius $|c|^{1/m}$.

Example 7 Show that

$$1 + e^{i\pi/7} + e^{2\pi i/7} + \cdots + e^{i13\pi/7} = 0.$$

Solution. $x^{14} - 1 = (x-1)(x - e^{i\pi/7}) \dots (x - e^{i13\pi/7})$. Equate coefficients of x^{13} to get

$$0 = -(1 + e^{i\pi/7} + \cdots + e^{i13\pi/7}).$$

2 Fields

The following definition gives greater generality to our work with no greater effort. Briefly, it says that a field F is a set that has an addition and multiplication which obey the rules $1^\circ - 8^\circ$ that we listed for $\mathbb{C}$.

Definition 2 Let F be a set whose members $a, b, \dots$ can be combined by two operations, addition and multiplication, satisfying the following rules for all a, b, c in F:

1° $a + b$ and ab are in F whenever a and b are in F.

2° $a + b = b + a$ and $ab = ba$.

3° $a + (b + c) = (a + b) + c$ and $a(bc) = (ab)c$.

4° $a(b + c) = ab + ac$.

5° There is a member 0 of the set F such that $a + 0 = a$ for all a.

6° There is a member 1 of the set $F, 1 \neq 0$, such that $a1 = a$ for all a.

7° For each a in F there is $-a$ in F such that

$$a + (-a) = 0.$$

8° For each a in $F, a \neq 0$, there is a^{-1} in F such that $aa^{-1} = 1$.

Then F is said to be a **field** .

The examples of fields we have already encountered are $F = \mathbb{R}, F = \mathbb{C}$. There are finite fields, containing p elements, for any given prime p, and these are useful in number theory and cryptography. See Further Reading. A simple way to construct a field is to take a subset of $\mathbb{C}$ that is closed under addition, multiplication, subtraction and division.

Proposition 3 *Let F be a subset of $\mathbb{C}$ containing at least one nonzero number. Suppose that $a \pm b, ab$ are in F whenever a and b are in F. Suppose that a^{-1} is in F whenever a is in $F, a \neq 0$. Then F is a field.*

Proof. Take any nonzero a in F, then $0 = a - a$ and $1 = aa^{-1}$ are both in F.

Now run through the rules $1° - 8°$. Rule $1°$ follows from our hypotheses. Rules 2, 3, 4 hold because they are true in $\mathbb{C}$. We know 0 and 1 are in F, so $5°$ and $6°$ hold. Rule $7°$ follows since $-a = 0 - a$ is in F. Rule $8°$ is simply a hypothesis in the Proposition. This proves Proposition 3.

Example 8 Show that the set $\mathbb{Q}$ of rational numbers r/s (r, s integers, $s \neq 0$) is a field.

Proof. It is easy to verify that $a \pm b$ and ab are rational numbers when a and b are in $\mathbb{Q}$. Similarly if $a = r/s \neq 0$, then $r \neq 0$ and $a^{-1} = s/r$ is in $\mathbb{Q}$. The result now follows from Proposition 3.

My students sometimes think this cannot be all that is required, and go laboriously through $1° - 8°$! Let us try another one.

Example 9 Show that the set $\mathbb{Q}(\sqrt{5})$ of real numbers $r + s\sqrt{5}$ (r, s rational) is a field.

Solution. It is easy to see that $(r_1 + s_1\sqrt{5}) \pm (r_2 + s_2\sqrt{5})$ is in $\mathbb{Q}(\sqrt{5})$ if r_1, s_1, r_2, s_2 are rational. Moreover,

$$(r_1 + s_1\sqrt{5})(r_2 + s_2\sqrt{5}) = r_1 s_2 + 5s_1 s_2 + (r_1 s_2 + r_2 s_1)\sqrt{5}$$

is in $\mathbb{Q}(\sqrt{5})$.

Finally, $(r_1 + s_1\sqrt{5})^{-1} = \frac{r_1 - s_1\sqrt{5}}{r_1^2 - 5s_1^2}$. If r_1 and s_1 are nonzero rationals we cannot have $r_1^2 = 5s_1^2$ (see Further Reading), so the denominator is nonzero and $(r_1 + s_1\sqrt{5})^{-1}$ is in $\mathbb{Q}(\sqrt{5})$. This result now follows from Proposition 3.

In **abstract algebra** the idea is to take a simple set of rules, such as $1° - 8°$, and use them to define an 'algebraic structure' such as a group, ring or field. This not only avoids duplication of effort but gives insights and unexpected applications. See Further Reading.

When a field F is mentioned below, it is acceptable to think of F as one of the pair $\mathbb{R}, \mathbb{C}$. These are the most useful fields in linear algebra.

3 Vector spaces

In the abstract algebra spirit, we define a **vector space** to be a set V that obeys the rules that we listed for $\mathbb{R}^n$ in Chapter 1. The 'scalars' will be members of a given field F. The outcome of this abstraction is an extremely useful area of mathematics. It lies at the heart of partial differential equations, integral equations and Fourier analysis.

Definition 3 Let F be a field. Let V be a set and suppose that the sum $\boldsymbol{u}+\boldsymbol{v}$ of any two elements of V is defined and is a member of V. Suppose that the scalar product $c\boldsymbol{v}$ of any c in F and $\boldsymbol{v}$ in V is defined and is in V. Suppose that for all a, b in F and $\boldsymbol{u}, \boldsymbol{v}, \boldsymbol{w}$ in V,

1° $\boldsymbol{u}+\boldsymbol{v}=\boldsymbol{v}+\boldsymbol{u}$;

2° $(\boldsymbol{u}+\boldsymbol{v})+\boldsymbol{w}=\boldsymbol{u}+(\boldsymbol{v}+\boldsymbol{w})$;

3° there is a member $\mathbf{0}$ of V such that $\boldsymbol{v}+\mathbf{0}=\boldsymbol{v}$ for all v;

4° for each $\boldsymbol{v}$ in V there is a member $-\boldsymbol{v}$ of V such that

$$\boldsymbol{v}+(-\boldsymbol{v})=\mathbf{0};$$

5° $1\boldsymbol{v}=\boldsymbol{v}$;

6° $(ab)\boldsymbol{v}=a(b\boldsymbol{v})$;

7° $a(\boldsymbol{u}+\boldsymbol{v})=a\boldsymbol{u}+a\boldsymbol{v}$;

8° $(a+b)\boldsymbol{v}=a\boldsymbol{v}+b\boldsymbol{v}$.

Then V is said to be a **vector space over** F. The members of V are **vectors**, the members of F are **scalars** .

In Chapters 1–7 we worked exclusively with $F=\mathbb{R}$ and the vector spaces: $V=\mathbb{R}^n$, or $V=$ a subspace of $\mathbb{R}^n$. Recall that in Definition 3 of Chapter 3, the essential property of a subspace V is that $\boldsymbol{u}+\boldsymbol{v}$ and $c\boldsymbol{u}$ are in V whenever $\boldsymbol{u}, \boldsymbol{v}$ are in V and c is a scalar. You can easily check that a subspace of $\mathbb{R}^n$ is a vector space over $\mathbb{R}$.

Given any field F, the set of tuples $(x_1, \dots, x_n)$ with each x_i in F is written F^n. We define addition and scalar multiplication in F^n in exactly the same way as in $\mathbb{R}^n$ (Chapter 1). Now you can check that F^n is a vector space over F. Do be cautious in expecting F^n to have notions corresponding to length and inner product! We can manage this in the case of $\mathbb{C}^n$; see Section 5.

Function spaces are a much-studied class of vector spaces in which the vectors are functions. We give three examples of function spaces after a few preparatory observations. Let X be a set. Whenever we have functions $f : X \to \mathbb{C}$ and $g : X \to \mathbb{C}$ we define $f + g$ as

$$(f+g)(x) = f(x) + g(x).$$

We define cf, for a complex number c, by

$$(cf)(x) = cf(x).$$

This is of course exactly the customary way of adding functions, and multiplying them by constants, in calculus. Observe that $f + g, cf$ also map X into $\mathbb{C}$.

Let a and b be fixed real numbers, $a < b$. Let $[a, b]$ be the set of real x with $a \leq x \leq b$. The set $[a, b]$ is a **closed interval** . We use $[a, b]$ in the role of X in Example 12 below. First, however, we make sure that you have noted the implications of using complex valued functions. If $f : [a, b] \to \mathbb{C}$, we can split up f as

$$f = f_1 + if_2$$

where f_1, f_2 map X to $\mathbb{C}$.

Example 10 Let

$$f(x) = (1+i)e^{ix^2} + (-2+i)x, a \leq x \leq b.$$

Find real functions f_1 and f_2 such that $f = f_1 + if_2$.

Solution.

$$\begin{aligned} f(x) &= (1+i)(\cos x^2 + i \sin x^2) + (-2+i)x \\ &= \cos x^2 - \sin x^2 - 2x + i(\cos x^2 + \sin x^2 + x). \end{aligned}$$

So $f_1(x) = \cos x^2 - \sin x^2 - 2x, f_2(x) = \cos x^2 + \sin x^2 + x$.

For a complex-valued function $f = f_1 + if_2, f : [a, b] \to \mathbb{C}$, suppose f_1, f_2 are differentiable. We define the derivative f' to be

$$f' = f_1' + if_2'$$

and more generally the nth derivative $f^{(n)}$ of f is $f_1^{(n)} + if_2^{(n)}$ if f_1, f_2 have derivatives of order n. Now for x_0 in $[a, b]$,

(16)
$$\frac{f(x)-f(x_0)}{x-x_0} = \frac{f_1(x)-f_1(x_0)}{x-x_0} + i\left(\frac{f_2(x)-f_2(x_0)}{x-x_0}\right) \to f'(x_0) \text{ as } x \to x_0,$$

so we can argue exactly as in a calculus course to prove

$$(f+g)' = f' + g', (fg)' = f'g + fg'$$

for f, g differentiable complex valued functions on $[a, b]$. If $g : [a, b] \to \mathbb{R}$ and $f : \mathbb{R} \to \mathbb{C}$ are differentiable, then the chain rule

$$(f(g(x)))' = f'(g(x))g'(x)$$

is proved exactly as in calculus.

Example 11 Find f' (f as in Example 8).

Solution. $f_1' = -2x\sin x^2 - 2x\cos x^2 - 2, f_2' = -2x\sin x^2 + 2x\cos x^2 + 1$, so

$$\begin{aligned} f'(x) &= -2x\sin x^2(1+i) + 2x\cos x^2(-1+i) - 2 + i \\ &= 2ix(1+i)(\cos x^2 + i\sin x^2) + (-2+i), \end{aligned}$$

since $(1+i)i = -1+i$. Briefly,

$$f'(x) = 2ix(1+i)e^{ix^2} + (-2+i).$$

Example 12 Show that $(cf(x))' = cf'(x)$ for f differentiable on $[a, b]$ and c in $\mathbb{C}$.

Solution. In view of (16), $(c)' = 0$. So $(cf(x))' = (c)'f(x) + cf'(x) = cf'(x)$.

It is extremely useful to define e^a for any complex $a = a_1 + ia_2$ by

$$e^a = e^{a_1}e^{ia_2},$$

that is,

$$e^a = e^{a_1}(\cos a_2 + i\sin a_2).$$

If a, b are any complex numbers,

$$e^a e^b = e^{a_1}e^{ia_2}e^{b_1}e^{ib_2} = e^{a_1+b_1}e^{i(a_2+b_2)} = e^{a+b},$$

which is a generalization of the rule for real a, b. Note that

$$(e^{ix})' = -\sin x + i\cos x = ie^{ix}$$

so that, using the chain rule to differentiate e^{ia_2x},

$$\begin{aligned}(e^{ax})' &= (e^{a_1x})'e^{ia_2x} + e^{a_1x}(e^{ia_2x})' \\ &= a_1e^{a_1x}e^{ia_2x} + ia_2e^{a_1x}e^{ia_2x} = ae^{ax}.\end{aligned} \tag{17}$$

Again, this generalizes the rule for real a.

We could now do Example 11 more succinctly:

$$((1+i)e^{ix^2} + (-2+i)x)' = (1+i)2ixe^{ix^2} + (-2+i)$$

by Example 12, (17) and the chain rule.

Example 13 Let $D[a,b]$ denote the set of functions $f : [a,b] \to \mathbb{C}$ that have derivatives of all orders. Show that $D[a,b]$ is a vector space over $\mathbb{C}$.

Solution. We refer to Definition 3. If f, g are in $D[a,b]$, then $(f+g)^{(k)}$ exists and is equal to $f^{(k)} + g^{(k)}$. Since k is arbitrary, $f+g$ is in $D[a,b]$. Similarly cf is in $D[a,b]$. As for the rules $1° - 8°$, with $F = \mathbb{R}$, none presents any difficulty. For example, $\mathbf{0}$ is the constant *function* 0.

With practice, it becomes easy to think of functions as being vectors! Nevertheless it will pay us to work from scratch in proving general results of the type encountered in Chapter 3, in the context of a vector space. This will be our objective in the next section.

Another function space that has lots of deep and interesting properties is

Example 14 (the **Schwartz class**) . Let S_n denote the set of functions $f : \mathbb{R}^n \to \mathbb{C}$ that have partial derivatives $\frac{\partial f}{\partial x_j}, \frac{\partial^2 f}{\partial x_i \partial x_j}, \ldots,$ of all orders, which have the following **decay property**: for any partial derivative Df, and any $m > 0$,

$$\lim_{|\boldsymbol{x}| \to \infty} |\boldsymbol{x}|^m Df(\boldsymbol{x}) = 0.$$

Show that S_n is a vector space over $\mathbb{C}$.

Examples of functions in S_n would include $P(\boldsymbol{x})e^{-a|\boldsymbol{x}|^2}$ where $a > 0$ and P is a polynomial in $x_1, \ldots, x_n$; one could also incorporate expressions such as $\frac{1}{1+x_1^2+2x_2^2}$, $\sin x_1^2$, $\cos(x_1 + x_2 + x_n)$, into $P(\boldsymbol{x})$.

Solution. As in Example 13, we need only show that $f+g$ and cf qualify for membership of S_n whenever f and g are in S_n. We have

$$D(f+g) = Df + Dg\ ,\ D(cf) = cDf$$

for any partial derivative D. So not only do $f+g$ and cf have partial derivatives of all orders, but, given any D and $m > 0$,

$$\begin{aligned}\lim_{|\boldsymbol{x}|\to\infty} |\boldsymbol{x}|^m D(f+g)(\boldsymbol{x}) &= \lim_{|\boldsymbol{x}|\to\infty} |\boldsymbol{x}|^m Df(\boldsymbol{x}) + \lim_{|\boldsymbol{x}|\to\infty} |\boldsymbol{x}|^m Dg(\boldsymbol{x}) \\ &= 0.\end{aligned}$$

Similarly $\lim_{|\boldsymbol{x}|\to\infty} |\boldsymbol{x}|^m D(cf)(\boldsymbol{x}) = 0$. This proves that $f+g$ and cf are in S_n.

The following vector space will be useful in Section 6.

Example 15 Let $E[a,b]$ be the set of bounded complex valued functions on $[a,b]$ that are continuous except for at most finitely many points x in $[a,b]$. Show that $E[a,b]$ is a vector space over $\mathbb{C}$.

Solution. Since the sum of functions continuous at x is continuous at x, and a constant multiple of a function continuous at x is continuous at x, the argument goes exactly as in Example 13.

4 General results about vector spaces

Throughout this section, F is a field, and V is a vector space over $\mathbb{R}$. The results will look familiar, because we have done them for $F = \mathbb{R}, V =$ a subspace of $\mathbb{R}^n$.

Proposition 4 *We have*

(i) If $\boldsymbol{v} + \boldsymbol{w} = \boldsymbol{v}$, *then* $\boldsymbol{w} = \mathbf{0}$.

(ii) $a\mathbf{0} = \mathbf{0}$ *for* a *in* F.

(iii) $0\boldsymbol{v} = \mathbf{0}$ *for* $\boldsymbol{v}$ *in* V.

(iv) $(-a)\boldsymbol{v} = -(a\boldsymbol{v})$ *for* a *in* F, $\boldsymbol{v}$ *in* V.

(v) If $\boldsymbol{v} \neq \mathbf{0}$, *then* $a\boldsymbol{v} = \mathbf{0}$ *implies that* $a = 0$.

You may find it surprising that these results need proof. Remember, a vector space is simply a set with certain rules, and these are all we have to go on in establishing new facts.

Proof.

(i) Add $-\boldsymbol{v}$ to both sides of

$$\boldsymbol{v} + \boldsymbol{w} = \boldsymbol{v}.$$

Then

$$-\boldsymbol{v} + (\boldsymbol{v} + \boldsymbol{w}) = -\boldsymbol{v} + \boldsymbol{v} = \mathbf{0}.$$

Regroup the left side using 2°: we get

$$(-\boldsymbol{v} + \boldsymbol{v}) + \boldsymbol{w} = \mathbf{0} + \boldsymbol{w} = \boldsymbol{w}.$$

Thus $\boldsymbol{w} = \mathbf{0}$.

I have not cited all the uses of $1° - 8°$ here. You should fill them in, and similarly below

(ii) Since $a\mathbf{0} = \boldsymbol{a}(\mathbf{0} + \mathbf{0}) = a\mathbf{0} + a\mathbf{0}$, we have $a\mathbf{0} = \mathbf{0}$, by (i).

(iii) We have $0\boldsymbol{v} = (0 + 0)\boldsymbol{v} = 0\boldsymbol{v} + 0\boldsymbol{v}$, so $0\boldsymbol{v} = \mathbf{0}$, by (i).

(iv) As $\mathbf{0} = (a + (-a))\boldsymbol{v}$ by (iii), we get

$$\mathbf{0} = a\boldsymbol{v} + (-a)\boldsymbol{v}.$$

By definition,

$$(-a)\boldsymbol{v} = -(a\boldsymbol{v}).$$

(v) If $a\boldsymbol{v} = \mathbf{0}$ and $a \neq 0$, then we use (ii) to get

$$\mathbf{0} = a^{-1}\mathbf{0} = a^{-1}(a\boldsymbol{v}) = (a^{-1}a)\boldsymbol{v} = 1\boldsymbol{v} = \boldsymbol{v}.$$

You are now doing abstract algebra!

Definition 4 Let W be a subset (not empty) of V. If $\boldsymbol{u} + \boldsymbol{v}$ and $c\boldsymbol{u}$ are in W whenever $\boldsymbol{u}, \boldsymbol{v}$ are in W and c is in F, W is a **subspace** of V.

Observe that W is a vector space too! All the rules $1° - 2°, 5° - 8°$ are satisfied, because they are known to hold in V. As for 3°, $\mathbf{0} = 0\boldsymbol{w}$ for any given $\boldsymbol{w}$ in W, so $\mathbf{0}$ is in W. Now $(-1)\boldsymbol{w} = -\boldsymbol{w}$ by Proposition 4 (iv), so $-\boldsymbol{w}$ is in W for any $\boldsymbol{w}$ in W; this means that 4° is satisfied.

Example 16 Let W be the set of functions f in $D[0,1]$ such that $f(1/2) = 0$ (Figure 5). Show that W is a subspace of $D[0,1]$.

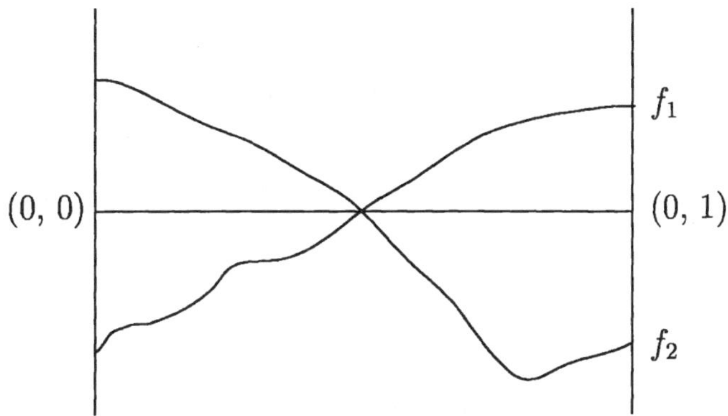

Figure 5. Functions f_1 and f_2 in the subspace W in Example 16.

Solution. Let f and g be in W. Then $(f+g)(1/2) = f(1/2) + g(1/2) = 0$. Similarly $(cf)(1/2) = cf(1/2) = 0$, for c in $\mathbb{C}$.

This kind of checking of closure under addition and scalar multiplication is usually easy, as you will find when you try it in the Exercises.

Definition 5 Given $\boldsymbol{v}_1, \ldots, \boldsymbol{v}_k$ in V, any vector $c_1\boldsymbol{v}_1 + \cdots + c_k\boldsymbol{v}_k$ with c_i in F is a **linear combination** of $\boldsymbol{v}_1, \ldots, \boldsymbol{v}_k$.

As in Chapter 3, we sometimes omit 'linear.'

Definition 6 The set of all possible linear combinations of $\boldsymbol{v}_1, \ldots, \boldsymbol{v}_k$ is the **linear span** of $\boldsymbol{v}_1, \ldots, \boldsymbol{v}_k$, written $\text{Span}\{\boldsymbol{v}_1, \ldots, \boldsymbol{v}_k\}$.

Proposition 5 $\text{Span}\{\boldsymbol{v}_1, \ldots, \boldsymbol{v}_k\}$ *is a subspace of* V.

Proof. Let $c_1\boldsymbol{v}_1 + \cdots + c_k\boldsymbol{v}_k$ and $d_1\boldsymbol{v}_1 + \cdots + d_k\boldsymbol{v}_k$ be members of $\text{Span}\{\boldsymbol{v}_1, \ldots, \boldsymbol{v}_k\}$. Then

$$c_1\boldsymbol{v}_1 + \cdots + c_k\boldsymbol{v}_k + d_1\boldsymbol{v}_1 + \cdots + d_k\boldsymbol{v}_k = (c_1+d_1)\boldsymbol{v}_1 + \cdots + (c_k+d_k)\boldsymbol{v}_k$$

is in $\text{Span}\{\boldsymbol{v}_1, \ldots, \boldsymbol{v}_k\}$; so is

$$c(c_1\boldsymbol{v}_1 + \cdots + c_k\boldsymbol{v}_k) = (cc_1)\boldsymbol{v}_1 + \cdots + (cc_k)\boldsymbol{v}_k$$

for any c in F.

Definition 7 V is **finite dimensional** if $V = \text{Span}\{\boldsymbol{v}_1, \dots, \boldsymbol{v}_k\}$ for some set $\boldsymbol{v}_1, \dots, \boldsymbol{v}_k$ in V.

Of course, F^n is finite dimensional because it is the linear span of $\boldsymbol{e}_1 = (1, 0, \dots, 0), \dots, \boldsymbol{e}_n = (0, \dots, 0, 1)$. However, the function spaces of Examples 13 and 14 are **infinite-dimensional** , that is, not finite-dimensional. To prove this we first establish some properties of linear independence.

Definition 8 The set of $\boldsymbol{v}_1, \dots, \boldsymbol{v}_k$ in V is **linearly independent** if the only solution of the equation

$$c_1\boldsymbol{v}_1 + \cdots + c_k\boldsymbol{v}_k = \boldsymbol{0} \quad (c_1, \dots, c_k \text{ in } F)$$

is the zero solution $c_1 = \cdots = c_k = 0$.

If $\boldsymbol{v}_1, \dots, \boldsymbol{v}_k$ is not a linearly independent set, it is a **linearly dependent** set.

Example 17 Show that in the vector space $\mathbb{C}$ over the field $\mathbb{C}$, 1 and i are linearly dependent.

Solution. $1i - i1 = 0$. Since 1 and $-i$ are in $\mathbb{C}$, the result follows.

Of course $1 = (1, 0)$ and $i = (0, 1)$ are linearly independent in the vector space $\mathbb{C}$ over the field $\mathbb{R}$. This suggests careful attention to the field of scalars in any example or application.

Proposition 6 *Let $\boldsymbol{v}_1, \dots, \boldsymbol{v}_n$ be linearly independent. Every member $\boldsymbol{v}$ of Span$\{\boldsymbol{v}_1, \dots, \boldsymbol{v}_n\}$ has a* unique *representation*

$$\boldsymbol{v} = c_1\boldsymbol{v}_1 + \cdots + c_n\boldsymbol{v}_n$$

with the c_i in F.

Proof. Suppose $\boldsymbol{v}$ has representations

$$\boldsymbol{v} = c_1\boldsymbol{v}_1 + \cdots + c_n\boldsymbol{v}_n = d_1\boldsymbol{v}_1 + d_2\boldsymbol{v}_2 + \cdots + d_n\boldsymbol{v}_n.$$

Then $(c_1 - d_1)\boldsymbol{v}_1 + \cdots + (c_n - d_n)\boldsymbol{v}_n = \boldsymbol{0}$. By linear independence, $c_1 - d_1 = 0, \dots, c_n - d_n = 0$. The representations are the same.

Example 18 Let $f_1, \dots, f_n$ be functions in $D[a,b]$. Suppose that the **Wronskian determinant**

$$W(x) = \det \begin{bmatrix} f_1(x) & \cdots & f_n(x) \\ f_1'(x) & \cdots & f_n'(x) \\ & \vdots & \\ f_1^{(n-1)}(x) & \cdots & f_n^{(n-1)}(x) \end{bmatrix}$$

is not the zero function. Show that $f_1, \dots, f_n$ are linearly independent.

Solution. Let $c_1, \dots, c_n$ be a solution of

$$c_1 f_1 + \cdots + c_n f_n = 0.$$

Remember, this is an equation between *functions*: it means that

$$c_1 f_1(x) + \cdots + c_n f_n(x) = 0 \tag{18}$$

for every x in $[a,b]$. It follows from (18) that

$$\begin{aligned} c_1 f_1'(x) + \cdots + c_n f_n'(x) &= 0, \\ &\cdots \\ c_1 f_1^{(n-1)}(x) + \cdots + c_n f^{(n-1)}(x) &= 0. \end{aligned} \tag{19}$$

Fix a real x in $[a,b]$ for which $W(x) \neq 0$. Then the only solution of the linear system (18), (19) is $c_1 = \cdots = c_n = 0$. So $f_1, \dots, f_n$ are linearly indcpendent.

Example 19 Show that $x, e^x, \sin x$ are linearly independent in $D[0,\pi]$.

Solution. The Wronskian is

$$\begin{aligned} W(x) &= \det \begin{bmatrix} x & e^x & \sin x \\ 1 & e^x & \cos x \\ 0 & e^x & -\sin x \end{bmatrix} \\ &= -e^x \det \begin{bmatrix} 1 & x & \sin x \\ 1 & 1 & \cos x \\ 1 & 0 & -\sin x \end{bmatrix} \end{aligned}$$

(take out factor e^x from column 2; interchange columns 1, 2)

$$= -e^x \det \begin{bmatrix} 1 & x & \sin x \\ 0 & 1-x & \cos x - \sin x \\ 0 & -x & -2\sin x \end{bmatrix} \begin{matrix} \text{II} - \text{I} \\ \text{III} - \text{I} \\ \\ \end{matrix}$$
$$= -e^x(-2(1-x)\sin x + x(\cos x - \sin x)).$$

If we substitute $x = \pi$, for instance,

$$W(\pi) = e^\pi \pi \neq 0$$

and independence follows.

Example 20 Let $b > a$. Given $f \neq 0$ in $D[a,b]$, show that $f, xf, \ldots, x^{n-1}f$ are linearly independent.

Solution. Pick x_0 such that $f(x_0) \neq 0$. As f is continuous, $f(x)$ will be nonzero in a short interval $[c,d]$ with $c \leq x_0 \leq d, c < d$.

Suppose there are $c_1, \ldots, c_n$, not all zero, such that

$$c_1 f + c_2(xf) + \cdots + c_n(x^{n-1}f) = 0.$$

Then $c_1 + c_2 x + \cdots + c_n x^{n-1} = 0$ in $[c,d]$. This is impossible because the polynomial has at most $n-1$ zeros.

Proposition 7

(i) Let $\boldsymbol{v}_1, \ldots, \boldsymbol{v}_n$ be a linearly dependent set with $\boldsymbol{v}_1 \neq \mathbf{0}$. Then

$$\boldsymbol{v}_j = c_1\boldsymbol{v}_1 + \cdots + c_{j-1}\boldsymbol{v}_{j-1} \tag{20}$$

for some j and $c_1, \ldots, c_{j-1}$ in F.

(ii) Let $\boldsymbol{v}_1, \ldots, \boldsymbol{v}_n$ be a linearly independent set. Then no equation (20) holds.

Proof. This is very simple. In part (i), choose $d_1, \dots, d_n$ not all zero, with

$$d_1\boldsymbol{v}_1 + \cdots + d_n\boldsymbol{v}_n = \mathbf{0}.$$

Let j be largest such that $d_j \neq 0$. We cannot have $j = 1$ since $d_1\boldsymbol{v}_1 \neq \mathbf{0}$. Now

$$\boldsymbol{v}_j = -d_j^{-1}(d_1\boldsymbol{v}_1 + \cdots + d_{j-1}\boldsymbol{v}_{j-1}),$$

giving (20).

In part (ii), the equation (20) rearranges to

$$c_1\boldsymbol{v}_1 + \cdots + c_{j-1}\boldsymbol{v}_{j-1} + (-1)\boldsymbol{v}_j = \mathbf{0}$$

which is not possible since $-1 \neq 0$.

Corollary 1 *If $\boldsymbol{v}_1, \dots, \boldsymbol{v}_n$ are in V and $\boldsymbol{v}_1, \dots, \boldsymbol{v}_k$ are linearly independent (where $0 \leq k < n$), we can find a linearly independent subset $\boldsymbol{v}_1, \dots, \boldsymbol{v}_k$, $\boldsymbol{v}_{i_1}, \dots, \boldsymbol{v}_{i_r}$ whose linear span is* $\mathrm{Span}\{\boldsymbol{v}_1, \dots, \boldsymbol{v}_n\}$.

Proof. If $\boldsymbol{v}_1, \dots, \boldsymbol{v}_n$ are independent, we have the result. If not, delete from $\boldsymbol{v}_1, \dots, \boldsymbol{v}_n$ any element $\boldsymbol{v}_j$ that is a combination of $\boldsymbol{v}_1, \dots, \boldsymbol{v}_{j-1}$. Of course $\boldsymbol{v}_1, \dots, \boldsymbol{v}_k$ do not get deleted. Say the set that remains is $\boldsymbol{v}_1, \dots, \boldsymbol{v}_k$, $\boldsymbol{v}_{i_1}, \dots, \boldsymbol{v}_{i_r}$. In any combination $c_1\boldsymbol{v}_1 + \cdots + c_n\boldsymbol{v}_n$, the deleted elements can be replaced by combinations of non-deleted elements. Hence $\mathrm{Span}\{\boldsymbol{v}_1, \dots, \boldsymbol{v}_k, \boldsymbol{v}_{i_1}, \dots, \boldsymbol{v}_{i_r}\} = \mathrm{Span}\{\boldsymbol{v}_1, \dots, \boldsymbol{v}_k\}$. By Proposition 7 (i), $\boldsymbol{v}_1, \dots, \boldsymbol{v}_k, \boldsymbol{v}_{i_1}, \dots, \boldsymbol{v}_{i_r}$ is linearly independent.

Definition 9 A **basis** of a vector space V is a linearly independent set $\boldsymbol{v}_1, \dots, \boldsymbol{v}_n$ such that $\mathrm{Span}\{\boldsymbol{v}_1, \dots, \boldsymbol{v}_n\} = V$.

Proposition 8 *If V is a finite-dimensional space and $V = \mathrm{Span}\{\boldsymbol{u}_1, \dots, \boldsymbol{u}_m\}$, some subset of $\boldsymbol{u}_1, \dots \boldsymbol{u}_m$ is a basis of V.*

Proof. This is an immediate consequence of the Corollary with $\boldsymbol{u}_1, \dots, \boldsymbol{u}_m$ in place of $\boldsymbol{v}_1, \dots, \boldsymbol{v}_n$.

Proposition 9 *If $\boldsymbol{v}_1, \dots, \boldsymbol{v}_n$ is a basis of V and $\boldsymbol{w}_1, \dots, \boldsymbol{w}_m$ in V are linearly independent, then $m \leq n$.*

Proof. Since $\boldsymbol{w}_m$ is a combination of $\boldsymbol{v}_1, \dots, \boldsymbol{v}_n$, the set $\boldsymbol{w}_m, \boldsymbol{v}_1, \dots, \boldsymbol{v}_n$ is dependent. By the Corollary, there is a new basis $\boldsymbol{w}_m, \boldsymbol{v}_{i_1}, \dots, \boldsymbol{v}_{i_k}$ of V with $k \leq n-1$. Repeat the procedure with $\boldsymbol{w}_{m-1}, \boldsymbol{w}_m, \boldsymbol{v}_{i_1}, \dots, \boldsymbol{v}_{i_k}$, extracting a basis $\boldsymbol{w}_{m-1}, \boldsymbol{w}_m, \boldsymbol{v}_{j_1}, \dots, \boldsymbol{v}_{j_s}$ with $s \leq n-2$. Repeating the procedure we eventually get a basis $\boldsymbol{w}_2, \dots, \boldsymbol{w}_m, \boldsymbol{v}_{r_1}, \dots, \boldsymbol{v}_{r_t}$. There must be at least one $\boldsymbol{v}_i$ in this basis because $\boldsymbol{w}_1$ is not a combination of $\boldsymbol{w}_2, \dots, \boldsymbol{w}_m$. We deleted *at least* $m-1$ $\boldsymbol{v}$'s. Hence $m-1 \leq n-1$ and $m \leq n$.

Corollary 2 *Any two bases of the finite dimensional vector space V have the same number of elements. This number is the* **dimension** *of V, written* $\dim V$.

Proof. If $\boldsymbol{w}_1, \dots, \boldsymbol{w}_m$ and $\boldsymbol{v}_1, \dots, \boldsymbol{v}_n$ are two bases, then $m \leq n$ from Proposition 9. Similarly $n \leq m$; and so $n = m$.

The dimension of F^n is of course n, since $\boldsymbol{e}_1, \dots, \boldsymbol{e}_n$ is a basis of F^n, in just the same way as for $F = \mathbb{R}$ (Chapter 3).

Proposition 10 *A subspace W of a finite dimensional vector space V is finite dimensional. In fact,* $\dim W \leq \dim V$.

This is not as obvious as it looks, because a given basis of V might not have any member of W in it.

Proof. We may suppose W has a nonzero member, say $\boldsymbol{w}_1$. Pick $\boldsymbol{w}_2$ not in $\text{Span}\{\boldsymbol{w}_1\}$, then $\boldsymbol{w}_3$ not in $\text{Span}\{\boldsymbol{w}_1, \boldsymbol{w}_2\}$ and so on. Of course $\boldsymbol{w}_1, \dots, \boldsymbol{w}_m$ are linearly independent (Proposition 7 (i)). The process stops when every member of W is in $\text{Span}\{\boldsymbol{w}_1, \dots, \boldsymbol{w}_m\}$. This must happen, with $m \leq \dim V$, by Proposition 9. At this stage, $\boldsymbol{w}_1, \dots, \boldsymbol{w}_m$ has both the properties needed for a basis of W.

Corollary 3 *Let $\boldsymbol{w}_1, \dots, \boldsymbol{w}_k$ be a linearly independent set in V where* $\dim V = n$. *There is a basis $\boldsymbol{w}_1, \dots, \boldsymbol{w}_k$, $\boldsymbol{w}_{k+1}, \dots, \boldsymbol{w}_n$ of V.*

Proof. We follow the proof of Proposition 10 with $V = W$, and the first k vectors $\boldsymbol{w}_1, \dots, \boldsymbol{w}_k$ having already been provided. This gives the desired basis $\boldsymbol{w}_1, \dots, \boldsymbol{w}_k, \boldsymbol{w}_{k+1}, \dots, \boldsymbol{w}_n$.

Of course $D[a,b]$ is infinite dimensional, since it contains an arbitrarily large number of linearly independent elements, $1, x, \dots, x^k$. Likewise S_n is infinite dimensional, because $e^{-|\boldsymbol{x}|^2}, x_1 e^{-|\boldsymbol{x}|^2}, x_1^2 e^{-|\boldsymbol{x}|^2}, \dots, x_1^k e^{-|\boldsymbol{x}|^2}$ are linearly independent (a very slight variation of Example 20).

5 Inner product spaces

We now set up the analogue of inner product, length and distance in an abstract algebra spirit.

Definition 10 Let V be a vector space over $F = \mathbb{R}$ or $F = \mathbb{C}$. Suppose that for any $\boldsymbol{u}, \boldsymbol{v}$ in V there is defined a number $\langle \boldsymbol{u}, \boldsymbol{v} \rangle$ in F, the **inner product** of $\boldsymbol{u}$ and $\boldsymbol{v}$, with the following properties for $\boldsymbol{u}, \boldsymbol{v}, \boldsymbol{w}$ in V and c in F.

1° $\langle \boldsymbol{u} + \boldsymbol{v}, \boldsymbol{w} \rangle = \langle \boldsymbol{u}, \boldsymbol{w} \rangle + \langle \boldsymbol{v}, \boldsymbol{w} \rangle$.

2° $\langle c\boldsymbol{u}, \boldsymbol{v} \rangle = c\langle \boldsymbol{u}, \boldsymbol{v} \rangle$.

3° $\langle \boldsymbol{u}, \boldsymbol{v} \rangle = \overline{\langle \boldsymbol{v}, \boldsymbol{u} \rangle}$

4° $\langle \boldsymbol{u}, \boldsymbol{u} \rangle > 0$ for $\boldsymbol{u} \neq \boldsymbol{0}$.

Then V is an **inner product space**.

I should explain that the complex conjugate in 3° can be omitted in the case $F = \mathbb{R}$, since the inner product is a real number. The idea of bringing in the conjugate is that it 'fits' with 4°. You will see what I mean in the following two examples.

Example 21 In $\mathbb{C}^n$, show that the inner product

$$\langle \boldsymbol{u}, \boldsymbol{v} \rangle = u_1\bar{v}_1 + \cdots + u_n\bar{v}_n (\boldsymbol{u} = (u_1, \ldots, u_n), \boldsymbol{v} = (v_1, \ldots, v_n))$$

obeys the rules 1° – 4°.

Solution. We have

$$1^\circ \ \langle \boldsymbol{u} + \boldsymbol{v}, \boldsymbol{w} \rangle = \sum_{j=1}^{n} (u_j + v_j)\bar{w}_j = \sum_{j=1}^{n} u_j\bar{w}_j + \sum_{j=1}^{n} v_j\bar{w}_j = \langle \boldsymbol{u}, \boldsymbol{w} \rangle + \langle \boldsymbol{v}, \boldsymbol{w} \rangle;$$

$$2^\circ \ \langle c\boldsymbol{u}, \boldsymbol{v} \rangle = \sum_{j=1}^{n} cu_j\bar{v}_j = c\sum_{j=1}^{n} u_j\bar{v}_j = c\langle \boldsymbol{u}, \boldsymbol{v} \rangle;$$

$$3^\circ \ \langle \boldsymbol{u}, \boldsymbol{v} \rangle = \sum_{j=1}^{n} u_j\bar{v}_j = \overline{\sum_{j=1}^{n} \bar{u}_j v_j} = \overline{\langle \boldsymbol{v}, \boldsymbol{u} \rangle}.$$

4° $\langle \boldsymbol{u}, \boldsymbol{u} \rangle = \sum_{j=1}^{n} u_j \bar{u}_j = \sum_{j=1}^{n} |u_j|^2.$

This is a sum of non-negative numbers. If $\boldsymbol{u} \neq \mathbf{0}$, the sum is positive. Of course this only works because of the conjugation in the definition of $\langle \boldsymbol{u}, \boldsymbol{v} \rangle$.

Example 22 Let $h(x)$ be a given positive continuous function on $[a, b]$. Show that $E[a, b]$ is an inner product space over $\mathbb{C}$ with the inner product $\langle f, g \rangle = \int_a^b f(x)\bar{g}(x)h(x)dx$. The space $E[a, b]$ is defined in Example 15. A function that is 0 except at points of discontinuity is regarded as the zero function; we use this convention to make rule 4° work. See Figure 6.

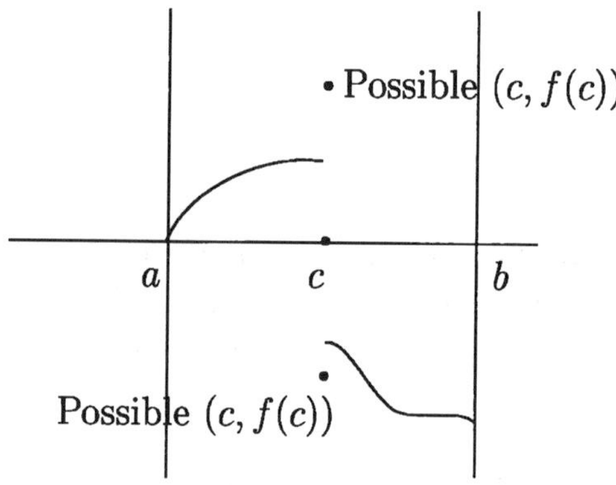

Figure 6. All functions obtained by choosing a point $(c, f(c))$ are considered as the same vector in $E[a, b]$.

Solution. We have not defined the integral of a complex valued function, so we do so. If

$$f = f_1 + i f_2$$

(f_1, f_2 real functions in $E[a, b]$) then we define

$$\int_a^b f dx = \int_a^b f(x)dx = \int_a^b f_1(x) + i \int_a^b f_2(x)dx. \tag{21}$$

It is very easy to prove that $\int_a^b \bar{f} dx = \overline{\int_a^b f dx}$, and that $\int_a^b (f + g)dx = \int_a^b f\ dx + \int_a^b g\ dx$. The proof that $\int_a^b cf\ dx = c \int_a^b f\ dx$, for f in $E[a, b]$ and

c in $\mathbb{C}$, is a bit more interesting. Let $c = c_1 + ic_2$, then

$$\int_a^b cf\ dx = \int_a^b (c_1 f_1 - c_2 f_2)x + i \int_a^b (c_1 f_1 + c_2 f_1)dx$$
$$= c_1 \int_a^b f_1 dx - c_2 \int_a^b f_2 dx + ic_1 \int_a^b f_2\ dx + ic_2 \int_a^b f_1\ dx$$

(by the definition (21), and then by standard properties of integrals)

$$= (c_1 + ic_2)\left(\int_a^b f_1\ dx + i\int_a^b f_2\ dx\right) = c\int_a^b f\ dx.$$

With that out of the way, we are ready to tackle 1° – 4°.

Let p, q, r be functions in $E[a, b]$ and c in $\mathbb{C}$. Then

1° $$\langle p+q, r\rangle = \int_a^b (p(x) + q(x))\overline{r(x)}h(x)dx$$
$$= \int_a^b p(x)\overline{r(x)}h(x)dx + \int_a^b q(x)\overline{r(x)}h(x)dx$$
$$= \langle p, r\rangle + \langle q, r\rangle.$$

2° $$\langle cp, q\rangle = \int_a^b cp(x)\overline{q(x)}h(x)dx = c\int_a^b p(x)\overline{q(x)}h(x)dx$$
$$= c\langle p, q\rangle.$$

3° $$\langle p, q\rangle = \int_a^b p(x)\overline{q(x)}dx = \overline{\int_a^b \overline{p(x)}q(x)dx} = \overline{\langle q, p\rangle}.$$

4° We have $\langle p, p\rangle = \int_a^b |p(x)|^2 h(x)dx$. Suppose that p is not 0. There is a point x_0 in $[a, b]$ where p is continuous and $p(x_0) \neq 0$. The integrand is ≥ 0, and $h(x)$ is always positive. Now $|p|^2 h$ is continuous at x_0, so it is positive on a short interval $[c, d]$ in $[a, b]$ containing x_0. So $\int_a^b |p(x)|^2 h(x)dx > 0$, as required.

Definition 11 Let V be an inner product space. The **length** of $\boldsymbol{u}$ is $\|\boldsymbol{u}\| = \langle \boldsymbol{u}, \boldsymbol{u}\rangle^{1/2}$. The **distance** from $\boldsymbol{u}$ to $\boldsymbol{v}$ is $d(\boldsymbol{u}, \boldsymbol{v}) = \|\boldsymbol{u} - \boldsymbol{v}\|$.

In Example 22 with $h(x) = 1$, the distance between f and g is $\left(\int_a^b |f(x) - g(x)|^2 dx\right)^{1/2}$. This is sometimes called the *mean square* distance. By drawing a sketch you will note that f and g can have $d(f, g)$ arbitrarily small even though $|f(x) - g(x)| > 1$ for some values of x. For more about types of distance in function spaces, see Further Reading.

Here are some results which show that geometry is alive and well in an inner product space V. We imitate the arguments in Chapter 1.

Proposition 11 *Let V be an inner product space over F and let $\boldsymbol{u}, \boldsymbol{v}$ be in V and c in F. (Remember, F is $\mathbb{R}$ or $\mathbb{C}$.) Then*

(i) $\|c\boldsymbol{v}\| = |c|\ \|\boldsymbol{v}\|$;

(ii) $|\langle \boldsymbol{u}, \boldsymbol{v}\rangle| \le \|\boldsymbol{u}\|\ \|\boldsymbol{v}\|$ *(Cauchy-Schwarz inequality);*

(iii) $\|\boldsymbol{u} + \boldsymbol{v}\| \le \|\boldsymbol{u}\| + \|\boldsymbol{v}\|$ *(triangle inequality).*

Proof. We begin by observing that

$$\langle \boldsymbol{0}, \boldsymbol{v}\rangle = 0. \tag{22}$$

To get this, $\langle \boldsymbol{0}, \boldsymbol{v}\rangle = \langle \boldsymbol{0}+\boldsymbol{0}, \boldsymbol{v}\rangle = \langle \boldsymbol{0}, \boldsymbol{v}\rangle + \langle \boldsymbol{0}, \boldsymbol{v}\rangle$, from which (22) follows. We also observe that $\langle \boldsymbol{u}, c\boldsymbol{v}\rangle = \bar{c}\langle \boldsymbol{u}, \boldsymbol{v}\rangle$. To see this,

$$\langle \boldsymbol{u}, c\boldsymbol{v}\rangle = \overline{\langle c\boldsymbol{v}, \boldsymbol{u}\rangle} = \overline{c\langle \boldsymbol{v}, \boldsymbol{u}\rangle} = \bar{c}\overline{\langle \boldsymbol{v}, \boldsymbol{u}\rangle} = \bar{c}\langle \boldsymbol{u}, \boldsymbol{v}\rangle.$$

Now we proceed with the proof.

(i) $\|c\boldsymbol{v}\| = \langle c\boldsymbol{v}, c\boldsymbol{v}\rangle^{1/2} = (c\bar{c}\langle \boldsymbol{v}, \boldsymbol{v}\rangle)^{1/2} = |c|\ \|\boldsymbol{v}\|$.

(ii) This is obvious from (22) if $\boldsymbol{v} = \boldsymbol{0}$. If $\boldsymbol{v} \neq \boldsymbol{0}$, we can use the properties of inner product to get

$$\begin{aligned} 0 &\le \left\langle \boldsymbol{u} - \frac{\langle \boldsymbol{u}, \boldsymbol{v}\rangle}{\langle \boldsymbol{v}, \boldsymbol{v}\rangle}\boldsymbol{v}, \boldsymbol{u} - \frac{\langle \boldsymbol{u}, \boldsymbol{v}\rangle}{\langle \boldsymbol{v}, \boldsymbol{v}\rangle}\boldsymbol{v}\right\rangle \\ &= \langle \boldsymbol{u}, \boldsymbol{u}\rangle - \frac{\langle \boldsymbol{u}, \boldsymbol{v}\rangle}{\langle \boldsymbol{v}, \boldsymbol{v}\rangle}\langle \boldsymbol{v}, \boldsymbol{u}\rangle - \frac{\overline{\langle \boldsymbol{u}, \boldsymbol{v}\rangle}}{\langle \boldsymbol{v}, \boldsymbol{v}\rangle}\langle \boldsymbol{u}, \boldsymbol{v}\rangle + \frac{\langle \boldsymbol{u}, \boldsymbol{v}\rangle\overline{\langle \boldsymbol{u}, \boldsymbol{v}\rangle}}{\langle \boldsymbol{v}, \boldsymbol{v}\rangle^2}\langle \boldsymbol{v}, \boldsymbol{v}\rangle \\ &= \langle \boldsymbol{u}, \boldsymbol{u}\rangle - \frac{|\langle \boldsymbol{u}, \boldsymbol{v}\rangle|^2}{\langle \boldsymbol{v}, \boldsymbol{v}\rangle}. \end{aligned}$$

Now $|\langle \boldsymbol{u}, \boldsymbol{v}\rangle|^2 \le \langle \boldsymbol{u}, \boldsymbol{u}\rangle\langle \boldsymbol{v}, \boldsymbol{v}\rangle$, and the result follows on taking the square root.

(iii) From (ii), we have

$$\begin{aligned}\|\boldsymbol{u}+\boldsymbol{v}\|^2 = \langle \boldsymbol{u}+\boldsymbol{v}, \boldsymbol{u}+\boldsymbol{v}\rangle &= \langle \boldsymbol{u},\boldsymbol{u}\rangle + \langle \boldsymbol{u},\boldsymbol{v}\rangle + \langle \boldsymbol{v},\boldsymbol{u}\rangle + \langle \boldsymbol{v},\boldsymbol{v}\rangle \\ &\le \langle \boldsymbol{u},\boldsymbol{u}\rangle + 2\langle \boldsymbol{u},\boldsymbol{u}\rangle^{1/2}\langle \boldsymbol{v},\boldsymbol{v}\rangle^{1/2} + \langle \boldsymbol{v},\boldsymbol{v}\rangle \\ &= (\langle \boldsymbol{u},\boldsymbol{u}\rangle^{1/2} + \langle \boldsymbol{v},\boldsymbol{v}\rangle^{1/2})^2 = (\|\boldsymbol{u}\| + \|\boldsymbol{v}\|)^2.\end{aligned}$$

It is interesting to pursue the idea of *orthogonality* in the abstract. Vectors $\boldsymbol{u}, \boldsymbol{v}$ in the inner product space V are **orthogonal** if $\langle \boldsymbol{u}, \boldsymbol{v}\rangle = 0$. If $\boldsymbol{v}_1, \dots, \boldsymbol{v}_n$ are nonzero vectors in V, and any pair $\boldsymbol{v}_i, \boldsymbol{v}_j (i \neq j)$ are orthogonal, then $\boldsymbol{v}_1, \dots, \boldsymbol{v}_n$ is an **orthogonal set** . If further each $\|\boldsymbol{v}_i\| = 1$, the set is **orthonormal** .

Example 23 Consider the inner product space $E[0, 2\pi]$ in Example 22 with $h(x) = 1$. Show that $e^{-inx}, e^{-i(n-1)x}, \dots, e^{-ix}, 1, e^{ix}, \dots, e^{inx}$ is an orthogonal set.

Solution. We have, for $j \neq k$,

$$\begin{aligned}\langle e^{ijx}, e^{ikx}\rangle &= \int_0^{2\pi} e^{ijx}\overline{e^{ikx}}dx \\ &= \int_0^{2\pi} e^{i(j-k)x}dx.\end{aligned}$$

We note that the fundamental theorem of calculus works for complex functions: $\int_a^b f'dx = f(b) - f(a)$ if f' is continuous. (Just read the equation as $\int_a^b f_1'dx + i\int_a^b f_2'dx = f_1(b) + if_2(b) - f_1(a) - if_2(a)$.) So

$$\begin{aligned}\int_0^{2\pi} e^{i(j-k)x}dx &= \frac{e^{i(j-k)2\pi}}{i(j-k)} - \frac{e^{i(j-k)0}}{i(j-k)} \\ &= \frac{1}{i(j-k)} - \frac{1}{i(j-k)} = 0.\end{aligned}$$

We note in passing that $\left(\int_0^{2\pi} |e^{ijx}|^2 dx\right)^{1/2} = (2\pi)^{1/2}$: all the vectors $e^{-inx}, \dots, e^{inx}$ have length $(2\pi)^{1/2}$.

Proposition 12

(i) An orthogonal set $\boldsymbol{v}_1, \dots, \boldsymbol{v}_n$ is linearly independent.

(ii) Let $\boldsymbol{u}$ be in the linear span of the orthogonal set $\boldsymbol{v}_1, \dots, \boldsymbol{v}_n$. Then

$$\boldsymbol{u} = \frac{\langle \boldsymbol{u}, \boldsymbol{v}_1 \rangle}{\|\boldsymbol{v}_1\|^2} \boldsymbol{v}_1 + \cdots + \frac{\langle \boldsymbol{u}, \boldsymbol{v}_n \rangle}{\|\boldsymbol{v}_n\|^2} \boldsymbol{v}_n,$$

$$\|\boldsymbol{u}\|^2 = \sum_{i=1}^{n} \frac{|\langle \boldsymbol{u}, \boldsymbol{v}_i \rangle|^2}{\|\boldsymbol{v}_i\|^2}. \tag{23}$$

If $\boldsymbol{v}_1, \dots, \boldsymbol{v}_n$ is an orthonormal set, then

$$\|\boldsymbol{u}\|^2 = \sum_{i=1}^{n} |\langle \boldsymbol{u}, \boldsymbol{v}_i \rangle|^2. \tag{24}$$

Proof. (i) Suppose $c_1, \dots, c_n$ are in F and

$$c_1\boldsymbol{v}_1 + \cdots + c_n\boldsymbol{v}_n = \boldsymbol{0}.$$

Then, for each i,

$$\langle c_1\boldsymbol{v}_1 + \cdots + c_n\boldsymbol{v}_n, \boldsymbol{v}_i \rangle = 0.$$

However,

$$\begin{aligned} \langle c_1\boldsymbol{v}_1 + \cdots + c_n\boldsymbol{v}_n, \boldsymbol{v}_i \rangle &= c_1\langle \boldsymbol{v}_1, \boldsymbol{v}_i \rangle + \cdots + c_n\langle \boldsymbol{v}_n, \boldsymbol{v}_i \rangle \\ &= c_i\langle \boldsymbol{v}_i, \boldsymbol{v}_i \rangle. \end{aligned}$$

Thus $0 = c_i\langle \boldsymbol{v}_i, \boldsymbol{v}_i \rangle$. Since $\boldsymbol{v}_i \neq \boldsymbol{0}$, we have $\langle \boldsymbol{v}_i, \boldsymbol{v}_i \rangle \neq 0$ and $c_i = 0$. This proves independence.

(ii) We have $\boldsymbol{u} = c_1\boldsymbol{v}_1 + \cdots + c_n\boldsymbol{v}_n$ for some c_i in F. So

$$\langle \boldsymbol{u}, \boldsymbol{v}_i \rangle = \langle c_1\boldsymbol{v}_1 + \cdots + c_n\boldsymbol{v}_n, \boldsymbol{v}_i \rangle = c_i\langle \boldsymbol{v}_i, \boldsymbol{v}_i \rangle$$

giving the value of c_i claimed in the statement of (ii). Also

$$\begin{aligned} \langle \boldsymbol{u}, \boldsymbol{u} \rangle &= \langle c_1\boldsymbol{v}_1 + \cdots + c_n\boldsymbol{v}_n, c_1\boldsymbol{v}_1 + \cdots + c_n\boldsymbol{v}_n \rangle \\ &= \sum_{i=1}^{n} \langle c_i\boldsymbol{v}_i, c_i\boldsymbol{v}_i \rangle \\ \text{(omitting zero terms)} \quad &= \sum_{i=1}^{n} |c_i|^2 \langle \boldsymbol{v}_i, \boldsymbol{v}_i \rangle = \sum_{i=1}^{n} \frac{|\langle \boldsymbol{u}, \boldsymbol{v}_i \rangle|^2}{\langle \boldsymbol{v}_i, \boldsymbol{v}_i \rangle} \end{aligned}$$

as claimed. The equation (24) is just the special case $\|\boldsymbol{v}_i\| = 1 (i = 1, \dots, n)$.

6 Fourier series

It was claimed by Fourier in the early nineteenth century that an arbitrary function $f(x)$ with period 2π can be written as an infinite series

$$f(x) = \frac{a_0}{2} + \sum_{n=1}^{\infty} (a_n \cos nx + b_n \sin nx), \tag{25}$$

where $a_n = \frac{1}{\pi}\int_0^{2\pi} f(x)\cos nx dx, b_n = \frac{1}{\pi}\int_0^{2\pi} f(x)\sin nx dx$. The numbers a_n, b_n were later called **Fourier coefficients** of f. The infinite series in (25) is the **Fourier series** of f. Some years later, Dirichlet proved that (25) holds for a 'nice function' f. For example, if f is a complex valued function on $[0, 2\pi]$, with continuous derivative, we can extend f to the real line by defining $f(x + 2k\pi) = f(x)$ for any integer k. This creates a difficulty at multiples of 2π if $f(0) \neq f(2\pi)$. We get around this by redefining f at these points: $f(2k\pi) = \frac{f(0)+f(2\pi)}{2}$, for $k = 0, \pm 1, \pm 2, \ldots$. Dirichlet showed that (25) holds for all x. See Further Reading. In this section we prove that among all trigonometric polynomials of degree at most n,

$$s(x) = c_0 + \sum_{k=1}^{n} (c_k \cos kx + d_k \sin kx), \tag{26}$$

the closest to f in mean square is the partial sum of the Fourier series,

$$s_n(x) = \frac{a_0}{2} + \sum_{k=1}^{n} (a_k \cos kx + b_k \sin kx).$$

A **trigonometric polynomial** $s(x)$ is an expression (26) with complex c_k and d_k. The **degree** of s is the largest k for which $|c_k| + |d_k|$ is positive. We can rewrite $s(x)$ in the form

$$\sum_{j=-n}^{n} A_j e^{ijx} \tag{27}$$

by using $\cos kx = \frac{e^{ikx}+e^{-ikx}}{2}$ and $\sin kx = \frac{e^{ikx}-e^{-ikx}}{2}$. On the other hand, we could start with (27), and get back to an expression (26) by using $e^{ijx} = \cos jx + i \sin jx$ (which is $\cos(-j)x - i\sin(-jx)$, if j is negative).

Proposition 13 *Let $v_1, \ldots, v_m$ be an orthogonal set in the inner product space V. Let $W = \mathrm{Span}\{v_1, \ldots, v_m\}$. Given v in V, the minimum value of $d(v, w)$ for w in W is attained for $w = w'$,*

$$w' = \frac{\langle v, v_1\rangle}{\|v_1\|^2} v_1 + \cdots + \frac{\langle v, v_m\rangle}{\|v_m\|^2} v_m;$$

w' is the **projection** *of v into W.*

Proof. We observe that $v - w'$ is orthogonal to v_i for each i, since

$$\langle v - w', v_i\rangle = \langle v, v_i\rangle - \langle w', v_i\rangle = \langle v, v_i\rangle - \frac{\langle v, v_i\rangle}{\langle v_i, v_i\rangle} \langle v_i, v_i\rangle$$

(omitting zero terms)

$$= 0.$$

Consequently, $v - w'$ is orthogonal to any point of W. In particular, given w in W,

$$\langle v - w', w' - w\rangle = 0.$$

As in the proof of Proposition 11 (iii),

$$\begin{aligned} \|v - w\|^2 &= \|v - w' + w' - w\|^2 \\ &= \langle v - w', v - w'\rangle + \langle v - w', w' - w\rangle \\ &\quad + \langle w' - w, v - w'\rangle + \langle w' - w, w' - w\rangle \\ &= \|v - w'\|^2 + \|w - w'\|^2. \end{aligned}$$

This expression obviously takes the smallest value at $w = w'$ (and nowhere else).

Corollary 4 *Let f be in $E[0, 2\pi]$. Then the minimum of*

$$\int_0^{2\pi} |f - s|^2 dx$$

taken over all trigonometric polynomials s of degree $\leq n$, is

$$\int_0^{2\pi} |f(x) - s_n(x)|^2 dx.$$

Proof. We apply Proposition 13 with $V = E[0, 2\pi]$ and $\langle f, g\rangle = \int_a^b f(x)\overline{g(x)}dx$; now $\int_0^{2\pi} |f - s|^2 dx$ becomes $d(f, s)^2$, and s is an arbitrary vector in $\text{Span}\{e^{-inx}, \ldots, e^{inx}\}$. We know that $e^{-inx}, \ldots, e^{inx}$ is an orthogonal set by Example 23. The vector $\boldsymbol{w}'$ in Proposition 13, with $m = 2n + 1$ and f in place of $\boldsymbol{v}$, is

$$\sum_{k=-n}^{n} c_k e^{ikx}$$

where $c_k = \frac{\langle f, e^{ikx}\rangle}{2\pi}$, since e^{ikx} has length $(2\pi)^{1/2}$. The expression $c_0 e^{i0x}$ is the constant

$$\frac{\langle f, 1\rangle}{2\pi} = \frac{1}{2\pi}\int_0^{2\pi} f(y)dy = \frac{1}{2}\,a_0$$

in the notation of Fourier coefficients. For $k \neq 0$,

$$c_k e^{ikx} = \frac{e^{ikx}}{2\pi}\int_0^{2\pi} f(y)e^{-iky}dy = \begin{cases} \dfrac{e^{ikx}}{2}(a_k - ib_k) & \text{if } k > 0 \\ \dfrac{e^{ikx}}{2}(a_{-k} + ib_{-k}) & \text{if } k < 0. \end{cases}$$

Let $k > 0$. Combining k and $-k$,

$$\begin{aligned} c_k e^{ikx} + c_{-k}e^{-ikx} &= \frac{e^{ikx}}{2}(a_k - ib_k) + \frac{e^{-ikx}}{2}(a_k + ib_k) \\ &= a_k \cos kx + b_k \sin kx. \end{aligned}$$

So $\boldsymbol{w}'$ is simply $s_n(x)$.

Example 24 Find the closest trigonometric polynomial, of degree at most n, to f, where $f(x) = 1\ (0 \le x < 1), f(x) = 0\ (1 \le x \le 2\pi)$. Here 'closest' refers to mean square distance.

Solution. The required trigonometric polynomial is $s_n(x) = \sum_{k=-n}^{n} c_k e^{ikx}$, where $c_0 = \frac{1}{2\pi}\int_0^{2\pi} f dx = \frac{1}{2\pi}$,

$$c_k = \frac{1}{2\pi}\int_0^{2\pi} f(x)e^{-ikx}dx = \frac{1}{2\pi}\int_0^1 e^{-ikx}dx = \frac{e^{-ik} - 1}{-2\pi ik}\ (k \neq 0).$$

Thus

$$s_n(x) = \frac{1}{2\pi} + \sum_{\substack{k=-n \\ k\neq 0}}^{n} \frac{(e^{-ik}-1)}{-2\pi ik} e^{ikx} = \frac{1}{2\pi} + \sum_{\substack{k=-n \\ k\neq 0}}^{n} \frac{e^{ik(x-1)}}{-2\pi ik} + \sum_{\substack{k=-n \\ k\neq 0}}^{n} \frac{e^{ikx}}{2\pi ik}.$$

We can simplify on combining $\frac{\cos k(x-1)}{k}$ with $\frac{\cos(-k(x-1))}{-k}$ to get 0, and $\frac{\sin k(x-1)}{k}$ with $\frac{\sin(-k(x-1))}{-k}$ to get $\frac{2\sin k(x-1)}{k}$.

Similarly for the other terms. So

$$\begin{aligned} s_n(x) &= \frac{1}{2\pi} + \frac{1}{\pi}\sum_{k=1}^{n} \frac{\sin kx - \sin k(x-1)}{k} \\ &= \frac{1}{2\pi} + \frac{1}{\pi}\sum_{k=1}^{n} \left(\sin kx \frac{(1-\cos k)}{k} + \cos kx \frac{\sin k}{k}\right). \end{aligned}$$

7 Eigenvalues and eigenvectors of matrices over $\mathbb{C}$

The results of Chapters 4, 5 and 6 are valid if we replace 'real numbers' by 'complex numbers' throughout, and replace $\mathbb{R}^n$ by $\mathbb{C}^n$. If you reread the chapters in this way, vectors are in $\mathbb{C}^n$, scalars are in $\mathbb{C}$, matrices have entries in $\mathbb{C}$; determinants are complex numbers. Linear mappings will be of the form $T : \mathbb{C}^m \to \mathbb{C}^n$. Kernel and image of T are subspaces of $\mathbb{C}^m$ and $\mathbb{C}^n$ respectively. Characteristic polynomials have complex coefficients, eigenvalues are in $\mathbb{C}$, and eigenvectors are in $\mathbb{C}^n$. We can always factor the characteristic polynomial as

(28) $$c^n + b_1c^{n-1} + \cdots + b_{n-1}c + b_n = (c-c_1)\ldots(c-c_n),$$

where the c_i are complex numbers.

Example 25 Let $T : \mathbb{C}^2 \to \mathbb{C}^2$ have matrix

$$A = \begin{bmatrix} \cos a & -\sin a \\ \sin a & \cos a \end{bmatrix}$$

where $0 < a < \pi$ or $\pi < a < 2\pi$. Diagonalize A **over** $\mathbb{C}$ (that is, diagonalize using complex matrices).

Solution. Note that we can no longer interpret T as a rotation (Example 1, Chapter 6). The characteristic polynomial of A is

$$(c - \cos a)^2 + \sin^2 a = (c - \cos a + i \sin a)(c - \cos a - i \sin a).$$

The eigenvalues are $e^{\pm ia}$.

Eigenspace for $c = e^{ia}$. The coefficient matrix of $(e^{ia}I - A)\boldsymbol{x} = \mathbf{0}$ is

$$\begin{bmatrix} i\sin a & \sin a \\ -\sin a & i \sin a \end{bmatrix} \sim \begin{bmatrix} 1 & -i \\ 0 & 0 \end{bmatrix} \begin{matrix} \mathrm{I} \times \frac{1}{i \sin a} \\ \mathrm{II} + (\sin a)\, \mathrm{I} \end{matrix} .$$

A basis of the eigenspace is $\boldsymbol{v}_1 = (i, 1)$.

Eigenspace for $c = e^{-ia}$. Repeat the above calculation with i replaced by $-i$. A basis of the eigenspace is $(-i, 1)$.

Exactly as in the proof of Proposition 3 of Chapter 6, this leads to

$$A = PDP^{-1}, \tag{29}$$

where

$$P = \begin{bmatrix} i & -i \\ 1 & 1 \end{bmatrix}, \quad D = \begin{bmatrix} e^{ia} & \\ & e^{-ia} \end{bmatrix}.$$

A real matrix A is **diagonalizable over $\mathbb{C}$**, but **nondiagonalizable over $\mathbb{R}$**, if $A = PDP^{-1}$ where D is diagonal and has a non-real entry. This is the case in (29).

Example 26 Let

$$A = \begin{bmatrix} 1 & -3 & 3 \\ 3 & -5 & 3 \\ 6 & -6 & 4 \end{bmatrix}.$$

Diagonalize A over $\mathbb{C}$.

Solution. We can repeat the solution of Example 4 of Chapter 6 verbatim, because A could already be diagonalized over $\mathbb{R}$.

Example 27 Diagonalize $\begin{bmatrix} -i & 2 \\ 1 & i \end{bmatrix}$ over $\mathbb{C}$.

Solution. The characteristic polynomial is

$$(c+i)(c-i)-2=c^2-1=(c-1)(c+1).$$

Eigenspace for $c=1$. The coefficient matrix of $(I-A)\boldsymbol{x}=\mathbf{0}$ is

$$\begin{bmatrix} 1+i & -2 \\ -1 & 1-i \end{bmatrix} \sim \begin{bmatrix} 1 & -1+i \\ 0 & 0 \end{bmatrix} \begin{matrix} \text{I} \times \frac{1}{1+i} \\ \text{II}+\text{I} \end{matrix}$$

since

$$(30) \qquad -2/(1+i)=-1+i.$$

A basis of the eigenspace is $\boldsymbol{v}_1=(1-i,1)$.

Eigenspace for $c=-1$. The coefficient matrix of $(-I-A)\boldsymbol{x}=\mathbf{0}$ is

$$\begin{bmatrix} -1+i & -2 \\ -1 & -1-i \end{bmatrix} \sim \begin{bmatrix} 1 & 1+i \\ 0 & 0 \end{bmatrix} \begin{matrix} \text{I} \times \frac{1}{-1+i} \\ \text{II}+\text{I} \end{matrix}$$

using (30) again. A basis of the eigenspace is $\boldsymbol{v}_2=(-1-i,1)$. Now

$$A=PDP^{-1},$$

where

$$P=\begin{bmatrix} 1-i & -1-i \\ 1 & 1 \end{bmatrix}, \quad D=\begin{bmatrix} 1 & \\ & -1 \end{bmatrix}.$$

Example 28 Given the transition matrix

$$A=\begin{bmatrix} 0.4 & 0.2 & 0 \\ 0.1 & 0.7 & 0.7 \\ 0.5 & 0.1 & 0.3 \end{bmatrix},$$

find the steady state vector and show that it is independent of the initial probability vector.

Solution. There are no complex probabilities! However, a transition matrix can have complex *eigenvalues*, as we see below. The characteristic polynomial is

$$f(c)=\det\begin{bmatrix} c-0.4 & -0.2 & 0 \\ -0.1 & c-0.7 & -0.7 \\ -0.5 & -0.1 & c-0.3 \end{bmatrix}=\det\begin{bmatrix} c-0.4 & -0.2 & 0 \\ -0.1 & c-0.7 & -0.7 \\ c-1 & c-1 & c-1 \end{bmatrix} \begin{matrix} \text{III}+\text{I} \\ \text{III}+\text{II} \end{matrix}.$$

Factor out $c-1$ from row 3 and subtract column 1 from columns 2, 3:

$$\begin{aligned} f(c) &= (c-1)\det\begin{bmatrix} c-0.4 & 0.2-c & 0.4-c \\ -0.1 & c-0.6 & -0.6 \\ 1 & 0 & 0 \end{bmatrix} \\ &= (c-1)(c^2-0.4c+0.12). \end{aligned}$$

The quadratic factor has zeros $0.2 \pm i\sqrt{0.08}$ of modulus less than 1. The proof of Proposition 8 of Chapter 6 works equally well here, since

$$\lim_{n\to\infty} |c^n| = \lim_{n\to\infty} |c|^n = 0$$

for a complex number c with $|c| < 1$. The eigenvalues of A are distinct, so we can apply Proposition 10 of Chapter 6. Hence there is a steady state vector. Proposition 11 of Chapter 6 shows that the steady state vector is the probability vector $\boldsymbol{q}$ for which $A\boldsymbol{q} = \boldsymbol{q}$, which we now find.

The coefficient matrix of $(I-A)\boldsymbol{x} = \boldsymbol{0}$ is

$$\begin{bmatrix} 0.6 & -0.2 & 0 \\ -0.1 & 0.3 & -0.7 \\ -0.5 & -0.1 & 0.7 \end{bmatrix} \sim \begin{bmatrix} 1 & -\frac{1}{3} & 0 \\ 0 & \frac{4}{15} & -\frac{7}{10} \\ 0 & 0 & 0 \end{bmatrix} \begin{matrix} \text{III}+\text{I} \\ \text{III}+\text{II} \\ \text{I}\times\frac{5}{3} \\ \text{II}+\text{I}\times\frac{1}{10} \end{matrix}$$

$$\sim \begin{bmatrix} 1 & 0 & -\frac{7}{8} \\ 0 & 1 & -\frac{21}{8} \\ 0 & 0 & 0 \end{bmatrix} \begin{matrix} \text{II}\times\frac{15}{4} \\ \text{I}+\text{II}\times\frac{1}{3}. \end{matrix}$$

It follows that $\boldsymbol{q} = (7/36, 21/36, 8/36)$.

Example 29 Without using Section 4 of Chapter 7, show that the eigenvalues of a real symmetric matrix are real.

Solution. Suppose that the complex number c_1 is an eigenvalue of A. Then

$$A\boldsymbol{x} = c_1\boldsymbol{x}\ ,\ \boldsymbol{x} \in \mathbb{C}^n,\ \boldsymbol{x} \neq \boldsymbol{0}.$$

Writing $\bar{\boldsymbol{x}}$ for the vector $(\bar{\boldsymbol{x}}_1, \dots, \bar{\boldsymbol{x}}_n)$,

$$\bar{\boldsymbol{x}}^t A\boldsymbol{x} = \bar{\boldsymbol{x}}^t c_1\boldsymbol{x} = c_1|\boldsymbol{x}|^2. \tag{31}$$

Proposition 1 (ii) of Chapter 7 (which works for $\boldsymbol{x}, \boldsymbol{y}$ in $\mathbb{C}^n$) permits us to rewrite (31) as

$$\boldsymbol{x}^t A\bar{\boldsymbol{x}} = c_1|\boldsymbol{x}|^2.$$

Take the conjugate to get

$$\bar{\boldsymbol{x}}^t A\boldsymbol{x} = \bar{c}_1|\boldsymbol{x}|^2.$$

Comparing with (31), we see that $c_1 = \bar{c}_1$ and c_1 is real.

8 Linear differential equations

Let V be the vector space over $\mathbb{C}$ of complex-valued functions on $\mathbb{R}$ with derivatives of all orders. We write $D^n y$ for $y^{(n)}(t)$. So, with $a_0 \neq 0$, and a given function y in V,

$$(a_0 D^n + a_1 D^{n-1} + \cdots + a_{n-1}D + a_n)y = \sum_{j=0}^{n} a_j y^{(n-j)}.$$

The mapping $L = a_0 D^n + \cdots + a_n, L : V \to V$, is linear. Its **auxiliary polynomial** is $f(c) = a_0 c^n + a_1 c^{n-1} + \cdots + a_n$, which we factor as

$$f(c) = a_0(c - c_1)^{h(1)} \cdots (c - c_k)^{h(k)} \quad (c_j \text{ distinct numbers in } \mathbb{C}).$$

Now $L = a_0(D - c_1 I)^{h(1)} \ldots (D - c_k I)^{h(k)}$, where $I : V \to V$ is the identity, and the product of linear mappings is their composition.

We wish to solve the linear differential equation

$$(32) \qquad Ly = 0.$$

If we use language similar to Chapter 4, this amounts to finding Ker L.

Proposition 14 *The n functions $t^s e^{c_j t}$ $(0 \leq s < h(j), j = 1, \ldots, k)$ are linearly independent solutions of (32).*

Proof. Observe that

$$(33) \qquad \begin{aligned} (D - cI)t^s e^{ct} &= st^{s-1}e^{ct} + ct^s e^{ct} - ct^s e^{ct} \\ &= st^{s-1}e^{ct}. \end{aligned}$$

Evidently, then, repeated application of $D - cI$ gives

$$(D - cI)^h t^s e^{ct} = 0 \quad \text{if } h > s;$$

in particular

$$(D - c_j I)^{h(j)} t^s e^{ct} = 0 \quad \text{if } 0 \le s < h(j).$$

Now if we group the factors of L suitably we get

$$L(t^s e^{ct}) = M(D - c_j I)^{h(j)} t^s e^{ct} = 0, \quad \text{for } 0 \le s < h(j).$$

Here M is a product of $n - h(j)$ factors $D - c_m I$.

To prove the linear independence, suppose that

$$(34) \quad A_1 e^{c_1 t} + \cdots + A_{h-1} t^{h-1} e^{c_1 t} + \cdots + Z_1 e^{c_k t} + \cdots + Z_{h(k)} t^{h(k)-1} e^{c_k t} = 0$$

where $h = h(1)$. As the arrangement of $c_1, \ldots, c_k$ is arbitrary, we need only show the A_i are 0.

Now

$$(D - cI) t^j e^{dt} = j t^{j-1} e^{dt} + (d - c) t^j e^{dt}.$$

Hence

$$(35) \qquad (D - cI)((b_0 t^m + \cdots + b_m) e^{dt}) = (b_0 (d - c) t^m + Q(t)) e^{dt}$$

where Q has degree less than m. Applying $D - cI$ with $c \neq d$ has not changed the *degree* of the polynomial factor attached to e^{dt}.

If some $A_\ell \neq 0$, pick the largest ℓ. Applying $(D - c_2 I)^{h(2)} \cdots (D - c_k I)^{h(k)}$ to (34),

$$(36) \qquad (D - c_2 I)^{h(2)} \cdots (D - c_k I)^{h(k)} (A_1 + A_2 t + \cdots + A_\ell t^\ell) e^{c_1 t} = 0$$

(the other terms disappear when this operator is applied). However, using (35) repeatedly with $d = c_1$ and with $c = c_2$ ($h(2)$ times), $c = c_3$ ($h(3)$ times) and so on, the left-hand side of (36) is a nonzero polynomial times $e^{c_1 t}$. This is absurd. Thus $A_1 = \ldots = A_h = 0$. This proves Proposition 14.

Given $b_0, b_1, \ldots, b_{n-1}$ there is *just one* solution of $Ly = 0$ satisfying the **initial conditions** $D^j y = b_j$ at $t = 0$ (see Further Reading). Now $Ly = 0$ cannot have $n + 1$ linearly independent solutions $y_1, \ldots, y_{n+1}$. To see this, the linear system corresponding to $b_0 = 0, \ldots, b_{n-1} = 0$ would be

$$\sum_{j=1}^{n+1} a_j (D^\ell y_j)(0) = 0 \quad (\ell = 0, 1, \ldots, n - 1).$$

This has *infinitely many* solutions $a_1, \ldots, a_{n+1}$. Proposition 14 now yields a corollary, to the effect that we have already identified *all* the solutions of $Ly = 0$.

Corollary 5 *The general solution of $Ly = 0$ is the linear span of the n solutions in Proposition 14.*

We now look at *systems* of equations of the shape

$$Dy_i = \sum_{j=1}^{n} a_{ij}y_j \quad (y_i \text{ in } V). \tag{37}$$

We agree to use the following notation for the derivative of a vector: $D\boldsymbol{y} = (Dy_1, \ldots, Dy_n)$ for $\boldsymbol{y} = (y_1, \ldots, y_n)$, each y_i in V. The system to be solved is

$$D\boldsymbol{y} = A\boldsymbol{y}. \tag{38}$$

If it happens that A is diagonalizable over $\mathbb{C}$,

$$A = PBP^{-1},$$

with P invertible and

$$B = \begin{bmatrix} b_1 & & \\ & \ddots & \\ & & b_n \end{bmatrix},$$

the situation is simple, since

$$D\boldsymbol{y} = PBP^{-1}\boldsymbol{y}, \quad P^{-1}(D\boldsymbol{y}) = B(P^{-1}\boldsymbol{y}).$$

Obviously $P^{-1}(D\boldsymbol{y}) = D(P^{-1}\boldsymbol{y})$. Writing $\boldsymbol{z} = P^{-1}\boldsymbol{y}$,

$$D\boldsymbol{z} = B\boldsymbol{z} \quad , \quad \text{or } Dz_j = b_j z_j \text{ for each } j,$$

which has solution

$$z_j = k_j e^{b_j t}, \quad k_j \text{ in } \mathbb{C}. \tag{39}$$

Now $\boldsymbol{y}$ is given by $\boldsymbol{y} = P\boldsymbol{z}$, $\boldsymbol{z}$ as in (39).

Example 30 Find the general real solution of

$$\begin{aligned} Dy_1 &= 3y_1 - 4y_2 \\ Dy_2 &= y_1 + 3y_2. \end{aligned}$$

Solution. Here $A = \begin{bmatrix} 3 & -4 \\ 1 & 3 \end{bmatrix}$ with characteristic polynomial $(c-3)^2 + 4$. We proceed to find the *complex* solutions $\boldsymbol{y}$ first.

Eigenspace for $c = 3 + 2i$. The coefficient matrix for $((3+2i)I - A)\boldsymbol{x} = \boldsymbol{0}$ is

$$\begin{bmatrix} 2i & 4 \\ -1 & 2i \end{bmatrix} \sim \begin{bmatrix} 1 & -2i \\ 0 & 0 \end{bmatrix} \begin{matrix} \scriptstyle \mathrm{I} \times \frac{1}{2i} \\ \scriptstyle \mathrm{II} + \mathrm{I}. \end{matrix}$$

A basis of the eigenspace is $(2i, 1)$.

Repeat this calculation with $-i$ in place of i to get a basis $(-2i, 1)$ of the eigenspace for $c = 3 - 2i$.

Now $A = PBP^{-1}$,

$$P = \begin{bmatrix} 2i & -2i \\ 1 & 1 \end{bmatrix}, \quad B = \begin{bmatrix} 3+2i & \\ & 3-2i \end{bmatrix}.$$

The general complex solution is

$$\begin{aligned} \boldsymbol{y} &= \begin{bmatrix} 2i & -2i \\ 1 & 1 \end{bmatrix} \begin{bmatrix} ke^{3+2i)t} \\ \ell e^{(3-2i)t} \end{bmatrix} \qquad (k, \ell \text{ in } \mathbb{C}) \\ &= \begin{bmatrix} 2ike^{(3+2i)t} - 2i\ell e^{(3-2i)t} \\ ke^{(3+2i)t} + \ell e^{(3-2i)t} \end{bmatrix}. \end{aligned}$$

To get a real solution we certainly need $y_2 e^{-3t}$ real. That is, we need k_1, k_2, ℓ_1, ℓ_2 to satisfy

$$\operatorname{Im}\{(k_1 + ik_2)(\cos 2t + i \sin 2t) + (\ell_1 + i\ell_2)(\cos 2t - i \sin 2t)\} = 0.$$

Explicitly,

$$(k_1 - \ell_1)\sin 2t + (k_2 + \ell_2)\cos 2t = 0.$$

So $\ell_2 = -k_2, \ell_1 = k_1$, and so $\ell = \bar{k}$. Now, since $z + \bar{z} = 2\operatorname{Re} z$,

$$\boldsymbol{y} = 2\operatorname{Re} \begin{bmatrix} 2ike^{(3+2i)t} \\ ke^{(3+2i)t} \end{bmatrix} = \begin{bmatrix} 4e^{3t}(-k_1 \sin 2t - k_2 \cos 2t) \\ 2e^{3t}(k_1 \cos 2t - k_2 \sin 2t) \end{bmatrix}$$

with arbitrary real k_1, k_2.

The method needs modification if A is not diagonalizable over $\mathbb{C}$.

Example 31 Find the general complex solution of

$$
\begin{aligned}
Dy_1 &= -y_2 \\
Dy_2 &= y_1 - y_2 - y_3 \\
Dy_3 &= y_1 - 2y_3.
\end{aligned}
\tag{40}
$$

Solution. Here $A = \begin{bmatrix} 0 & -1 & 0 \\ 1 & -1 & -1 \\ 1 & 0 & -2 \end{bmatrix}$ has characteristic polynomial

$$\det \begin{bmatrix} c & 1 & 0 \\ -1 & c+1 & 1 \\ -1 & 0 & c+2 \end{bmatrix} = c(c+1)(c+2) + (c+1)$$

(expanding by row 1)

$$= (c+1)(c^2+2c+1) = (c+1)^3.$$

So -1 is the only eigenvalue. The coefficient matrix of $(-I - A)\boldsymbol{x} = \mathbf{0}$ is

$$\begin{bmatrix} -1 & 1 & 0 \\ -1 & 0 & 1 \\ -1 & 0 & 1 \end{bmatrix} \sim \begin{bmatrix} 1 & -1 & 0 \\ 0 & -1 & 1 \\ 0 & 0 & 0 \end{bmatrix} \begin{matrix} \text{III} - \text{II} \\ \text{I} \times -1 \\ \text{II} + \text{I} \end{matrix} \sim \begin{bmatrix} 1 & 0 & -1 \\ 0 & 1 & -1 \\ 0 & 0 & 0 \end{bmatrix} \begin{matrix} \text{I} - \text{II} \\ \text{II} \times -1 \end{matrix}.$$

A basis for the eigenspace is $(1,1,1)$; A is not diagonalizable. The Jordan form cannot be

$$J_1 = \begin{bmatrix} -1 & 1 & 0 \\ 0 & -1 & 0 \\ 0 & 0 & -1 \end{bmatrix},$$

since if $AP = PJ_1$ the first and third columns of P are linearly independent eigenvectors of A. So the Jordan form is

$$J_2 = \begin{bmatrix} -1 & 1 & 0 \\ 0 & -1 & 1 \\ 0 & 0 & -1 \end{bmatrix}.$$

Let $P = [\boldsymbol{v}_1\ \boldsymbol{v}_2\ \boldsymbol{v}_3]$, $AP = PJ_2$. Then $A\boldsymbol{v}_1 = -\boldsymbol{v}_1, A\boldsymbol{v}_2 = \boldsymbol{v}_1 - \boldsymbol{v}_2$, $A\boldsymbol{v}_3 = \boldsymbol{v}_2 - \boldsymbol{v}_3$. Take $\boldsymbol{v}_1 = (1,1,1)$. To avoid repetition, we work with the augmented matrix for $(A+I)\boldsymbol{x} = \boldsymbol{b}$. Here $b_2 = b_3$ for a consistent system. The augmented matrix is

$$[A+I|\boldsymbol{b}] = \begin{bmatrix} 1 & -1 & 0 & b_1 \\ 1 & 0 & -1 & b_2 \\ 1 & 0 & -1 & b_2 \end{bmatrix} \sim \begin{bmatrix} 1 & -1 & 0 & b_1 \\ 0 & 1 & -1 & b_2 - b_1 \\ 0 & 0 & 0 & 0 \end{bmatrix} \begin{matrix} \text{III} - \text{II} \\ \text{II} - \text{I} \\ \\ \end{matrix}$$

$$\sim \begin{bmatrix} 1 & 0 & -1 & b_2 \\ 0 & 1 & -1 & b_2 - b_1 \\ 0 & 0 & 0 & 0 \end{bmatrix} \begin{matrix} \text{I} + \text{II} \\ \\ \\ \end{matrix}.$$

In order to calculate a suitable $\boldsymbol{v}_2$, take $\boldsymbol{b} = (1,1,1)$. Choose the last coordinate 0 for $\boldsymbol{v}_2$, then $\boldsymbol{v}_2 = (1,0,0)$. Now take $\boldsymbol{b} = \boldsymbol{v}_2$ so that we can find a suitable $\boldsymbol{v}_3$. Choose the last coordinate 0 for $\boldsymbol{v}_3$, then $\boldsymbol{v}_3 = (0,-1,0)$, and

$$A = PJ_2P^{-1},$$

$$P = \begin{bmatrix} 1 & 1 & 0 \\ 1 & 0 & -1 \\ 1 & 0 & 0 \end{bmatrix}, J_2 = \begin{bmatrix} -1 & 1 & 0 \\ 0 & -1 & 1 \\ 0 & 0 & -1 \end{bmatrix}.$$

The system (40) becomes $D\boldsymbol{z} = J_2\boldsymbol{z}$, where $\boldsymbol{z} = P^{-1}\boldsymbol{y}$. That is,

$$Dz_1 = -z_1 + z_2, \quad Dz_2 = -z_2 + z_3, \quad Dz_3 = -z_3.$$

We have $z_3 = k_1e^{-t}$. Then

$$Dz_2 + z_2 = k_1e^{-t}.$$

We find a particular solution $z_2 = k_1te^{-t}$, using (33). Clearly the general solution is obtained by adding to k_1te^{-t} an arbitrary solution of $(D+I)z = 0$. So $z_2 = k_1te^{-t} + k_2e^{-t}$. Similarly,

$$Dz_1 + z_1 = k_1te^{-t} + k_2e^{-t}$$

is equivalent to

$$z_1 = \frac{k_1t^2}{2}e^{-t} + k_2te^{-t} + k_3e^{-t}.$$

Finally, with k_1, k_2, k_3 arbitrary real numbers,

$$\begin{aligned} \boldsymbol{y} = P\boldsymbol{z} &= (z_1 + z_2, z_1 - z_3, z_1) \\ &= \Big(\frac{k_1 t^2 e^{-t}}{2} + (k_1 + k_2)te^{-t} + (k_2 + k_3)e^{-t}, \frac{k_1 t^2}{2}\, e^{-t} + k_2 te^{-t} \\ &\qquad + (-k_1 + k_3)e^{-t}, \frac{k_1 t^2}{2}\, e^{-t} + k_2 te^{-t} + k_3 e^{-t}\Big). \end{aligned}$$

Linear differential equations become much more interesting when a_1, a_1, $a_2, \ldots$ are functions of t in the linear mapping L. See Further Reading.

Exercises for Chapter 8

Reminder: attempt Exercises a.b after reading Section a.

1.1 Compute $z\bar{z}$ and $|z|$ for the following complex numbers z:

(i) $3+4i$ (ii) $1-2i$ (iii) $5+6i$ (iv) $\sqrt{3}+\sqrt{5}i$ (v) $12-5i$.

1.2 Show that $|1/z| = 1/|z|$ for $z \neq 0$.

1.3 Compute w/z for the following pairs of complex numbers:

(i) $w = 1+i, z = 2-5i$ (ii) $w = 3i, z = 4+7i$
(iii) $w = 9+2i, z = 1-i$ (iv) $w = 10, z = 1+2i$
(v) $w = 5+12i, z = 5-12i$.

1.4 Find the general solution of the linear system

$$\begin{aligned}(1+i)x_1 + ix_2 &= 3\\ (2+i)x_1 + 2ix_2 &= 1-i.\end{aligned}$$

1.5 Show that $z^2 = a_1 + ia_2$ has solutions $z = x+iy$,

$$x = \pm\left(\frac{1}{2}\left(a_1 + \sqrt{a_1^2+a_2^2}\right)\right)^{1/2}, \quad y = a_2/2x$$

provided $a_2 \neq 0$.

1.6 Solve the quadratic equation $c^2 - (4+4i)c - 3 + 4i = 0$.

Hint: Complete the square and use Exercise 1.5.

1.7 Find the fourth roots of $-6+6i$.

1.8 Find the fifth roots of $32i$.

2.1 Let F be a field and suppose $ax = ay$ where a, x, y are in $F, a \neq 0$. Show that $x = y$, giving a careful accounting of your use of the rules $1^\circ - 8^\circ$ of a field.

2.2 Which of the following subsets of $\mathbb{C}$ are fields?

(i) the set $\mathbb{Z}$ of integers $0, \pm 1, \pm 2, \ldots$;

(ii) the set of $a + b\sqrt{7}$ (a, b in $\mathbb{Q}$);

(iii) the set of $a + b2^{1/3}$ (a, b in $\mathbb{Q}$);

(iv) the set of $a + bi$ (a, b in $\mathbb{Z}$);

(v) the set of $a + bi$ (a, b in $\mathbb{Q}$).

3.1 Let V_n be the set of polynomials $a_0x^n + a_1x^{n-1} + \cdots + a_{n-1}x + a_n$ with coefficients a_i that are arbitrary members of a field F. Addition of polynomials, and scalar multiplication by a member of F, are defined in the obvious way. Show that V_n is a vector space and $\dim V_n = n + 1$.

3.2 Let V be the set of infinite sequences $(x_1, x_2, x_3, \ldots)$ with x_i arbitrary real numbers. The sum is defined by

$$(x_1, x_2, \ldots) + (y_1, y_2, \ldots) = (x_1 + y_1, x_2 + y_2, \ldots);$$

and we define $c(x_1, x_2, \ldots) = (cx_1, cx_2, \ldots)$. Show that V is a vector space over $\mathbb{R}$. Show that V is infinite dimensional.

3.3 Let $V(m, n)$ be the set of $m \times n$ real matrices, with addition and scalar multiplication as in Chapter 4. Show that $V(m, n)$ is a vector space over $\mathbb{R}$. What is $\dim V$?

3.4 Let V be the set of real continuous functions f defined on $\mathbb{R}$ such that $\int_{-\infty}^{\infty} f(x)^2 dx$ is finite. Show that V is a vector space over $\mathbb{R}$. Show that V is infinite dimensional.

3.5 Let $\boldsymbol{v}, \boldsymbol{w}$ and $\boldsymbol{u}$ be members of the vector space V with $\boldsymbol{v} + \boldsymbol{w} = \boldsymbol{v} + \boldsymbol{u}$. Show that $\boldsymbol{w} = \boldsymbol{u}$, giving a careful accounting of your use of the rules $1° - 8°$ of a vector space.

3.6 Show that the Schwartz class S_n is **closed under multiplication**, that is, fg is in S_n for f, g in S_n.

4.1 Which of the following are subspaces of $D[0, 1]$?

(i) the set of f in $D[0, 1]$ with $f^{(3)}(1/2) + f^{(5)}(1) = 0$.

(ii) the set of f in $D[0, 1]$ with $f(1) = 1$.

(iii) the set of f in $D[0, 1]$ with $f(0) < f(1)$.

(iv) the set of f in $D[0,1]$ with $\int_0^1 f(t)dt = 0$.

4.2 Let V be a vector space and let V_1, V_2 be subspaces of V. Show that $V_1 \cap V_2$ is a subspace of V. Show that $V_1 + V_2$, the set of $\boldsymbol{v}_1 + \boldsymbol{v}_2$ ($\boldsymbol{v}_1$ in V_1, $\boldsymbol{v}_2$ in V_2) is a subspace of V.

4.3 In Exercise 4.2 suppose that $\boldsymbol{w}_1, \dots, \boldsymbol{w}_m$ is a basis of $V_1 \cap V_2$; that $\boldsymbol{w}_1, \dots, \boldsymbol{w}_m, \boldsymbol{v}_1, \dots, \boldsymbol{v}_k$ is a basis of V_1; and that $\boldsymbol{w}_1, \dots, \boldsymbol{w}_m, \boldsymbol{u}_1, \dots, \boldsymbol{u}_\ell$ is a basis of V_2. Show that $\boldsymbol{w}_1, \dots, \boldsymbol{w}_m, \boldsymbol{v}_1, \dots, \boldsymbol{v}_k, \boldsymbol{u}_1, \dots, \boldsymbol{u}_\ell$ is a basis of $V_1 + V_2$.

4.4 Let $\boldsymbol{v}_1, \dots, \boldsymbol{v}_m$ be linearly independent in $\mathbb{C}^n$. Show that $\bar{\boldsymbol{v}}_1, \dots, \bar{\boldsymbol{v}}_m$ are linearly independent.

4.5 Let W be the set of symmetric $n \times n$ real matrices. Show that W is a subspace of $V(n,n)$ in Exercise 3.3.

4.6 Let W be the set of polynomials P in the space V_2 (Exercise 3.1) for which $P(0) = -P(1)$. Show that W is a subspace of V_2. Find a basis of W.

Hint: It is best to think of this as a problem in $\mathbb{R}^3$.

4.7 Show that $1, \frac{1}{1+t}, \frac{1}{1+t^2}$ is a linearly independent set in $D[0,1]$.

5.1 Let $b_1, \dots, b_n$ be given positive numbers and write

$$\langle \boldsymbol{x}, \boldsymbol{y} \rangle = \sum_{j=1}^{n} b_j x_j \bar{y}_j \quad \text{for } \boldsymbol{x}, \boldsymbol{y} \text{ in } \mathbb{C}^n.$$

Show that $\mathbb{C}^n$ is an inner product space with this definition of inner product. Show that this result fails if some b_i is *not* positive.

5.2 Let V be the inner product space $D[0,1]$ with $\langle f, g \rangle = \int_0^1 f(t)\overline{g(t)}dt$. Verify Cauchy's inequality $|\langle f, g \rangle| \le \|f\| \, \|g\|$ and the triangle inequality $\|f+g\| \le \|f\| + \|g\|$ when $f(t) = 1+t$ and $g(t) = \sin t$ on $[0,1]$.

5.3 Let $\boldsymbol{w}_1, \dots, \boldsymbol{w}_n$ be linearly independent in the inner product space V. Let $\boldsymbol{v}_1 = \boldsymbol{w}_1$ and define $\boldsymbol{v}_2, \dots, \boldsymbol{v}_n$ in succession by

$$\boldsymbol{v}_k = \boldsymbol{w}_k - \sum_{j-1}^{k-1} \frac{\langle \boldsymbol{w}_k, \boldsymbol{v}_j \rangle}{\|\boldsymbol{v}_j\|^2} \boldsymbol{v}_j \quad (k = 2, \dots, n).$$

Show that $\boldsymbol{v}_1, \ldots, \boldsymbol{v}_n$ is an orthogonal set and $\text{Span}\{\boldsymbol{v}_1, \ldots, \boldsymbol{v}_n\} = \text{Span}\{\boldsymbol{w}_1, \ldots, \boldsymbol{w}_n\}$.

5.4 Find an orthogonal basis of $\text{Span}\{1, t, t^2\}$, if $V = D[-1, 1]$ with $\langle f, g\rangle = \int_{-1}^{1} f(t)\overline{g(t)}dt$.

5.5 Let $A^* = (\bar{A})^t$ for a complex matrix A, where $A = [a_{ij}], \bar{A} = [\bar{a}_{ij}]$. A is **unitary** if A is $n \times n$ with $AA^* = I$. Show that a unitary matrix A has $|\det A| = 1$ and that the columns of A are an orthonormal basis of $\mathbb{C}^n$, with the inner product of Example 21.

6.1 Find the closest trigonometric polynomial in mean square, of degree at most 2, to $f(t) = t$ $(0 \leq t \leq 2\pi)$.

6.2 Find the closest trigonometric polynomial in mean square, of degree at most 1, to $f(t) = |\sin t|$ $(0 \leq t \leq 2\pi)$.

7.1 Find the eigenvalues of

$$\begin{bmatrix} 2+4i & 4-7i \\ i & 2 \end{bmatrix}$$

and find bases of the eigenspaces in $\mathbb{C}^2$.

Hint: Use Exercise 1.6.

7.2 Define A^* as in Exercise 5.5. Let A be a complex matrix satisfying $A^* = A$ (a **Hermitian** matrix) . Show that the eigenvalues of A are real.

Hint: Since a real symmetric matrix is a special case of a Hermitian matrix, adapt the method of Example 29.

7.3 Let A be a real matrix and c an eigenvalue of A in $\mathbb{C}$, with eigenvector $\boldsymbol{x}$ in $\mathbb{C}^n$. Show that $\bar{\boldsymbol{x}}$ is in the eigenspace for $\bar{c}$.

7.4 Show that in Exercise 7.3 the eigenspaces for c and $\bar{c}$ have the same dimension.

Hint: Use Exercise 4.4.

7.5 Let A be a complex matrix one of whose eigenvalues c is a kth root of unity, $c \neq 1$. Show that $\lim_{n\to\infty} A^n$ does not exist.

Hint: Consider $A^n\boldsymbol{x}$ when $\boldsymbol{x}$ is an eigenvector for c.

8.1 For each of the following linear differential equations, find a solution that satisfies the given initial conditions.

(i) $(D^2 - 12D + 36I)y = 0,\ y(0) = 0, Dy(0) = 1.$

(ii) $(D^3 - 6D^2 + 12D - 8I)y = 0,\ y(0) = 1, Dy(0) = 2, D^2y(0) = 4.$

(iii) $(D^3 - 2D^2 - 2D + 4I)y = 0,\ y(0) = -1, Dy(0) = \frac{1}{2}, D^2y(0) = 0.$

(iv) $(D^3 - 3D^2 + D - 3I)y = 0,\ y(0) = 3, Dy(0) = 0, D^2y(0) = 1.$

(v) $(D^2 - 4D + 5)y = 0,\ y(0) = 1, Dy(0) = -2.$

8.2 Let $L = (D - c_1 I)\dots(D - c_n I)$ where $c_1, \dots, c_n$ are distinct. Show that the determinant of the linear system $y(0) = b_1, \dots, D^{n-1}y(0) = b_n$ that determines the solution of $Ly = 0$ is a Vandermonde determinant.

8.3 Find the general solution of the system

$$\begin{aligned} Dy_1 &= 2y_1 - 3y_2 + 3y_3, \\ Dy_2 &= 3y_1 - 4y_2 + 3y_3, \\ Dy_3 &= 6y_1 - 6y_2 + 5y_3. \end{aligned}$$

8.4 Find the solution of the system

$$\begin{aligned} Dy_1 &= -12y_2 \\ Dy_2 &= -y_1 + 4y_2 \end{aligned}$$

satisfying $y_1(0) = 1, y_2(0) = -1.$

8.5 Find the general solution of the system

$$\begin{aligned} Dy_1 &= -4y_1 + y_2 - y_3 \\ Dy_2 &= -7y_1 + 4y_2 - y_3 \\ Dy_3 &= -6y_1 + 6y_2 - 3y_3. \end{aligned}$$

8.6 Find the solution of the system

$$\begin{aligned} Dy_1 &= \frac{7}{2}\, y_1 - \frac{1}{2}\, y_2 - y_3 \\ Dy_2 &= \frac{3}{2}\, y_1 + \frac{3}{2}\, y_2 - y_3 \\ Dy_3 &= y_1 + y_3 \end{aligned}$$

satisfying $y_1(0) = -1, y_2(0) = 0, y_3(0) = 3.$

Further Reading

The books listed below give more details of various topics that were mentioned in the text.

Chapter 4

Linear mappings as approximations to vector functions: [7]. The theory of volume in $\mathbb{R}^n$: [7].

Chapter 6

Use of computer programs to do linear algebra computations: [5]. Approximate numerical determination of eigenvalues: [5]. Jordan form of a square matrix: [1].

Chapter 7

Use of quadratic forms to determine behavior of a function at a stationary point: [7].

Chapter 8

The fundamental theorem of algebra: [2]. Irrationality of square roots of integers: [4]. Finite fields and their applications: [4]. The abstract approach in algebra: [4]. The geometry of function spaces: [6]. The convergence of a Fourier series at individual points: [7]. Linear differential equations in general: [3].

Bibliography

[1] S. Axler, *Linear algebra done right.* Second edition, Springer 1997.

[2] J. Bak and D. Newman, *Complex analysis.* Springer 1982.

[3] M. Braun, *Differential equations and their applications.* Fourth edition, Springer 1993.

[4] L. N. Childs, *A concrete approach to higher algebra.* Second edition, Springer 1995.

[5] C. G. Cullen, *An introduction to numerical linear algebra.* PWS, Boston 1994.

[6] I. J. Maddox, *Elements of functional analysis.* Second edition, Cambridge 1998.

[7] M. H. Protter and C. B. Morrey, *A first course in real analysis.* Second edition, Springer 1991.

Answers to Selected Exercises

Please use these answers only as a last resort. In some cases they only partially answer the question.

Chapter 1

1.1 $\left(-\frac{3}{5}, \frac{4}{5}\right), (1, 0)$

2.1 $\frac{4}{\sqrt{10}}$

2.3 30 seconds, 120 meters

2.5 $\left(\frac{27}{5}, \frac{16}{5}\right)$

4.3 $\frac{43}{24}$

4.7 $\frac{\sqrt{165}}{3}$

5.1 $\frac{1}{\sqrt{5}}$

5.3 $\left(\frac{5}{3}, \frac{1}{3}, \frac{14}{3}, \frac{2}{3}, 1\right)$

Chapter 2

1.1
$$\begin{aligned} -2x_1 + 3x_2 + 6x_3 &= -1 \\ 3x_2 &= 2 \\ 4x_1 + 5x_2 - x_3 &= 1 \end{aligned}$$

2.1 $\begin{bmatrix} 1 & -2 & 3 \\ 1 & 2 & 4 \end{bmatrix}$

2.3 $\begin{bmatrix} 1 & -1 & 0 & 0 & 4 \\ 0 & 1 & -1 & 0 & 1 \\ 2 & -1 & -1 & -1 & 0 \end{bmatrix}$

3.1 Is in r. e. f.

3.3 Is not in r. e. f.

3.5 Is in r. e. f.

3.7 Yes

4.1 Unique solution $(2, -1, 0)$

4.3 $\left(\frac{6}{5}x_3 + \frac{3}{5}x_4 - \frac{14}{5}, -\frac{2}{5}x_3 - \frac{17}{10}x_4 + \frac{11}{10}, x_3, x_4\right)$

4.5 $(38x_3 + 28x_4 - 17, -11x_3 - 7x_4 + 5, x_3, x_4)$

4.7 Inconsistent

4.9 $b = -20$

4.11 Yes

4.13 $x_1^2 + x_2^2 + 29x_1 - 21x_2 - 42 = 0$

4.15 Infinitely many spheres. One is $x_1^2 + x_2^2 + x_3^2 - 7x_1 - 9x_2 - x_3 = 0$

6.1 $\boldsymbol{I} = (5, -1, 6, 1)$

Chapter 3

2.1 (i) $(a_1 + 6a_3)\boldsymbol{w}_1 + (a_2 - 9a_3)\boldsymbol{w}_2$

(ii) $\left(a_1 + \frac{2}{3}a_2\right)\boldsymbol{w}_1 + \left(a_3 - \frac{1}{9}a_2\right)\boldsymbol{w}_2$

(iii) $\left(a_2 + \frac{3}{2}a_1\right)\boldsymbol{w}_2 + \left(\frac{a_1}{6} + a_3\right)\boldsymbol{w}_3$

2.3 (i) false, (ii) true, (iii) false

2.5 Independent

2.7 Dependent

3.1 Nul $A = \mathbf{0}$

3.3 $(-3, 4, 1, 0, 0), (0, -1, 0, -1, 1)$

3.7 In each case, V_1 is the plane $x_1 = 0$. (i) $W = V$, (ii) W is the plane $x_2 = 0$, (iii) W is the line $t\boldsymbol{e}_3$, t real, (iv) W is the line $t\boldsymbol{e}_1, t$ real.

5.1 $\boldsymbol{a}_1, \boldsymbol{a}_2, \boldsymbol{a}_3, \boldsymbol{a}_4$

5.3 $\boldsymbol{a}_1, \boldsymbol{a}_2, \boldsymbol{a}_4$

Chapter 4

1.3 Parallellogram with vertices (0, 0), (1, 2), (1, 1), (2, 3)

1.5 Matrix of TS is $\begin{bmatrix} 1 & 2 & 0 \\ 0 & 1 & -4 \\ 1 & 1 & 1 \end{bmatrix}$.

3.1 $\begin{bmatrix} b_{11} & 0 & b_{13} \\ 0 & b_{22} & 0 \\ b_{13} & 0 & b_{11} \end{bmatrix}$. Not all of the form $x_1 A + x_2 I$.

4.1 $b_0 = -\frac{1}{2}$. For $b \neq -\frac{1}{2}$, inverse is $\frac{1}{1+2b} \begin{bmatrix} 2b & -b & 1-2b \\ 1-2b & 2b & 2b-3 \\ -2 & 1 & 4 \end{bmatrix}$

4.3 $$\begin{bmatrix} \dfrac{1}{a_{11}} & 0 & 0 \\ \dfrac{-a_{21}}{a_{11}a_{22}} & \dfrac{1}{a_{22}} & 0 \\ \dfrac{a_{21}a_{32} - a_{31}a_{22}}{a_{11}a_{22}a_{33}} & \dfrac{-a_{32}}{a_{22}a_{33}} & \dfrac{1}{a_{33}} \end{bmatrix}$$

4.5 $$\begin{bmatrix} -\frac{1}{6} & -\frac{1}{6} & 0 & \frac{1}{2} \\ -\frac{7}{6} & -\frac{1}{6} & 1 & \frac{1}{2} \\ \frac{5}{6} & -\frac{7}{6} & 0 & \frac{1}{2} \\ -\frac{1}{6} & \frac{5}{6} & 0 & -\frac{1}{2} \end{bmatrix}$$

6.1 $\frac{1}{4(b_1^2+b_2^2)}\begin{bmatrix} 2b_1^2+5b_1b_2+2b_2^2 & 2b_1^2-3b_1b_2-2b_2^2 \\ 2b_1^2+3b_1b_2-2b_2^2 & 2b_1^2-5b_1b_2+2b_2^2 \end{bmatrix}$

Chapter 5

1.1 (i) Even; (ii) Even; (iii) Odd; (iv) Even

1.5 $A = B = I$

2.2 22

3.2 13

4.3 $\begin{bmatrix} ka_{21} & -ka_{11} \\ -a_{21} & a_{11} \end{bmatrix}$

4.6 $\boldsymbol{x} = (-1, 6, 7)$

5.4 17

Chapter 6

1.6 $a_{11}a_{22} + a_{11}a_{33} + a_{22}a_{33} - a_{23}a_{32} - a_{12}a_{21} - a_{13}a_{31}$

2.1 $P = \begin{bmatrix} 1 & 3 \\ 0 & 1 \end{bmatrix}$

2.3 Not diagonalizable

2.5 Not diagonalizable

2.9 Dimension 2, 3, 4 respectively

3.1 $P = \begin{bmatrix} -1 & \frac{1}{4} & 0 \\ -1 & \frac{3}{4} & 0 \\ 1 & 0 & 1 \end{bmatrix}$

6.5 $\frac{1}{162}(50, 55, 57)$

6.7 $\left(\frac{2}{5}, \frac{3}{5}, 0, 0\right)$

Chapter 7

1.1 $2x_1^2 - 18x_1x_2 + 7x_2^2$

1.3 $2x_1^2 + 4x_2^2 + 6x_3^2 - 5x_4^2 + 6x_1x_2 + 2x_1x_4 - 4x_2x_3 + 2x_2x_4$

1.5 $\begin{bmatrix} 0 & \frac{1}{2} & 1 \\ \frac{1}{2} & 0 & \frac{5}{2} \\ 1 & \frac{5}{2} & 0 \end{bmatrix}$

2.1 $(1, 2, 3), (-13, 2, 3)$

2.3 $(1, 0, 1, -1, 0), (-2, 3, 4, 2, 3), (-2, 3, -10, -12, 17)$

2.5 $\frac{1}{39}(17, -6, 7, -15, -18)$

4.1 $P = \begin{bmatrix} \frac{1}{\sqrt{2}} & \frac{1}{\sqrt{2}} \\ \frac{1}{\sqrt{2}} & -\frac{1}{\sqrt{2}} \end{bmatrix}$

4.3 $P = \begin{bmatrix} 1 & 0 & 0 \\ 0 & \frac{5}{\sqrt{26}} & \frac{1}{\sqrt{26}} \\ 0 & -\frac{1}{\sqrt{26}} & \frac{5}{\sqrt{26}} \end{bmatrix}$

5.1 $P = \begin{bmatrix} \frac{2}{\sqrt{5}} & 0 & \frac{1}{\sqrt{5}} \\ 0 & 1 & 0 \\ \frac{1}{\sqrt{5}} & 0 & -\frac{2}{\sqrt{5}} \end{bmatrix}$

6.1 Indefinite

7.1 Ellipse

Chapter 8

1.1 (i) 5; (ii) $\sqrt{5}$; (iii) $\sqrt{61}$; (iv) $2\sqrt{2}$; (v) 13

1.3 (i) $-\frac{3}{29} + \frac{7i}{29}$; (ii) $\frac{21}{65} + \frac{12i}{65}$; (iii) $\frac{7}{2} + \frac{11i}{2}$; (iv) $2 - 4i$; (v) $-\frac{119}{169} + \frac{120i}{169}$

1.7 $3^{1/4}2^{3/8}e^{ki\pi/16}, k = 3, 11, 19, 27$

2.2 (i) Not a field; (ii) Is a field; (iii) Not a field; (iv) Not a field; (v) Is a field

3.3 Dimension is mn

4.1 (i) Is a subspace; (ii) Not a subspace; (iii) Not a subspace; (iv) Is a subspace

4.6 $1-2x, 1-2x^2$

5.2 $\langle f,g\rangle = 0.37\ldots, \|f\|\;\|g\| = 1.3\ldots$

5.4 $1, t, t^2 - \frac{1}{3}$

6.1 $\pi - 2\sin t - \sin 2t$

7.1 Eigenvalues $i, 4+3i$

8.1 (i) $\frac{t}{6}e^{6t}$; (ii) e^{2t}; (iii) $e^{2t} - \left(1+\frac{3}{4\sqrt{2}}\right)e^{\sqrt{2}t} + \left(\frac{3}{4\sqrt{2}}-1\right)e^{-\sqrt{2}t}$; (iv) $\frac{2}{5}e^{3t} + \frac{13}{5}\cos t - \frac{6}{5}\sin t$; (v) $e^{2t}(\cos t - 4\sin t)$

8.3 $((k_1+k_2)e^{-t} + k_3e^{5t}, k_1e^{-t} + k_3e^{5t}, -k_2e^{-t} + 2k_3e^{5t}))$

8.5 $(k_1e^{-3t} + k_2te^{-3t}, k_1e^{-3t} + k_2te^{-3t} + k_3e^{3t}, -k_2e^{-3t} + k_3e^{3t})$

Index